Experiments Manual for Use with

Digital Electronics

Digital Electronics

Principles and Applications

EIGHTH EDITION

Roger L. Tokheim

Connect
Learn
Succeed™

Experiments Manual for Use with Digital Electronics: Principles and Applications,
Eighth Edition
Roger Tokheim

Published by McGraw-Hill, a business unit of The McGraw-Hill Companies, Inc., 1221
Avenue of the Americas, New York, NY 10020. Copyright © 2014 by The McGraw-Hill
Companies, Inc. All rights reserved. Printed in the United States of America. Previous
editions © 2008, 2003, and 1999.

1 2 3 4 5 6 7 8 9 0 QDB/QDB 1 0 9 8 7 6 5 4 3

ISBN: 978-0-07-752080-9
MHID: 0-07-752080-7

Contents

Editor's Foreword

The McGraw-Hill Education Trade and Technology list has been designed to provide entry-level competencies in a wide range of occupations in the electrical and electronics fields. It consists of coordinated instructional materials designed especially for career-oriented students. A textbook, an experiments manual, and online resources support each major subject area covered in the series. All of these focus on theory, practice, applications, and experiences necessary for those preparing to enter technical careers.

There are two fundamental considerations in the preparation of a text like *Experiments Manual for Use with Digital Electronics:* the needs for the learner and the needs of the employer. This text meets those needs in expert fashion. The authors and editors have drawn upon their broad teaching and technical experiences to accurately interpret and meet the needs of the student. The needs of business and industry have been identified through personal interviews, industry publications, government occupational trend reports, and reports by industry associations.

The processes used to produce and refine the series have been ongoing. Technological change is rapid, and the content has been revised to focus on current trends. Refinements in pedagogy have been defined and implemented based on classroom testing and feedback from students and instructors using the series. Every effort has been made to offer the best possible learning materials. These include animated PowerPoint presentations, circuit files for simulation, a test generator with correlated test banks, dedicated websites for both students and instructors, and basic instrumentation labs. All of these are well coordinated and have been prepared by the author.

The widespread acceptance of *Experiments Manual for Use with Digital Electronics* and the positive feedback from users confirm the basic soundness in content and design of all the components as well as their effectiveness as teaching and learning tools. Instructors will find the texts and manuals in each of the subject areas logically structured, well paced, and developed around a framework of modern objectives. Students will find the materials to be readable, lucidly illustrated, and interesting. They will also find a generous amount of self-study materials, review items, and examples to help them determine their own progress.

Charles A. Schuler, Project Editor

Basic Skills in Electricity and Electronics

Charles A. Schuler, Project Editor

New Editions in This Series
Electricity: Principles and Applications, Eighth Edition, Richard J. Fowler
Electronics: Principles and Applications, Eighth Edition, Charles A. Schuler

Preface

The *Experiments Manual for Use with Digital Electronics,* eighth edition, is designed to provide practical, hands-on experience with digital components, digital techniques, and digital circuits. The *Experiments Manual* is closely correlated with its companion textbook, *Digital Electronics: Principles and Applications,* eighth edition. Entry-level knowledge and skills for a wide range of occupations is the goal of this experiments manual and textbook combination.

Learning Features

- Hardware lab experiments
- Circuit simulations using Multisim software
- Programmable digital devices including expanded microcontroller experiments using the BASIC Stamp 2 modules
- Hardware design problems
- Hardware troubleshooting problems
- Software troubleshooting problems
- More complete Multisim circuit simulation files
- Chapter Tests

Resources

- Parts/display boards/trainers available from Dynalogic Concepts (1-800-246-4907)

- BASIC Stamp 2 materials available from Parallax, Inc. (**www.parallax.com**)
- A CD-ROM with Multisim circuit files/activities and a primer for those who are unfamiliar with the software; Multisim circuit simulation software available from National Instruments (**www.ni.com/multisim/**)
- An Online Learning Center (**www.mhhe.com/tokheim8e**) containing PowerPoint presentations on breadboarding, soldering, circuit interrupters, AFCI, and GFCI and four Hewlett-Packard simulated labs on instrumentation

About the Author

Over several decades, Roger L. Tokheim has published many textbooks and lab manuals in the areas of digital electronics and microprocessors. His books have been translated into nine languages. He taught technical subjects including electronics for more than 35 years in public schools.

Safety

Electric and electronic circuits can be dangerous. Safe practices are necessary to prevent electrical shock, fires, explosions, mechanical damage, and injuries resulting from the improper use of tools.

Perhaps the greatest hazard is electrical shock. A current through the human body in excess of 10 milliamperes can paralyze the victim and make it impossible to let go of a "live" conductor or component. Ten milliamperes is a rather small amount of current flow: It is only *ten one-thousandths* of an ampere. An ordinary flashlight can provide more than 100 times that amount of current!

Flashlight cells and batteries are safe to handle because the resistance of human skin is normally high enough to keep the current flow very small. For example, touching an ordinary 1.5-V cell produces a current flow in the microampere range (a microampere is one-millionth of an ampere). This amount of current is too small to be noticed.

High voltage, on the other hand, can force enough current through the skin to produce a shock. If the current approaches 100 milliamperes or more, the shock can be fatal. Thus, the danger of shock increases with voltage. Those who work with high voltage must be properly trained and equipped.

When human skin is moist or cut, its resistance to the flow of electricity can drop drastically. When this happens, even moderate voltages may cause a serious shock. Experienced technicians know this, and they also know that so-called low-voltage equipment may have a high-voltage section or two. In other words, they do not practice two methods of working with circuits: one for high voltage and one for low voltage. They follow safe procedures at all times. They do not assume protective devices are working. They do not assume a circuit is off even though the switch is in the OFF position. They know the switch could be defective.

Even a low-voltage, high-current-capacity system like an automotive electrical system can be quite dangerous. Short-circuiting such a system with a ring or metal watchband can cause very severe burns—especially when the ring or band welds to the points being shorted.

As your knowledge and experience grow, you will learn many specific safe procedures for dealing with electricity and electronics. In the meantime:

1. Always follow procedures.
2. Use service manuals as often as possible. They often contain specific safety information. Read, and comply with, all appropriate material safety data sheets.
3. Investigate before you act.
4. When in doubt, *do not act*. Ask your instructor or supervisor.

General Safety Rules for Electricity and Electronics

Safe practices will protect you and your fellow workers. Study the following rules. Discuss them with others, and ask your instructor about any you do not understand.

1. Do not work when you are tired or taking medicine that makes you drowsy.
2. Do not work in poor light.
3. Do not work in damp areas or with wet shoes or clothing.
4. Use approved tools, equipment, and protective devices.
5. Avoid wearing rings, bracelets, and similar metal items when working around exposed electric circuits.
6. Never assume that a circuit is off. Double-check it with an instrument that you are sure is operational.
7. Some situations require a "buddy system" to guarantee that power will not be turned on while a technician is still working on a circuit.
8. Never tamper with or try to override safety devices such as an interlock (a type of switch that automatically removes power when a door is opened or a panel removed).
9. Keep tools and test equipment clean and in good working condition. Replace insulated probes and leads at the first sign of deterioration.

10. Some devices, such as capacitors, can store a *lethal* charge. They may store this charge for long periods of time. You must be certain these devices are discharged before working around them.

11. Do not remove grounds and do not use adaptors that defeat the equipment ground.

12. Use only an approved fire extinguisher for electrical and electronic equipment. Water can conduct electricity and may severely damage equipment. Carbon dioxide (CO_2) or halogenated-type extinguishers are usually preferred. Foam-type extinguishers may also be desired in *some* cases. Commercial fire extinguishers are rated for the type of fires for which they are effective. Use only those rated for the proper working conditions.

13. Follow directions when using solvents and other chemicals. They may be toxic, flammable, or may damage certain materials such as plastics. Always read and follow the appropriate material safety data sheets.

14. A few materials used in electronic equipment are toxic. Examples include tantalum capacitors and beryllium oxide transistor cases. These devices should not be crushed or abraded, and you should wash your hands thoroughly after handling them.

Other materials (such as heat shrink tubing) may produce irritating fumes if over-heated. Always read and follow the appropriate material safety data sheets.

15. Certain circuit components affect the safe performance of equipment and systems. Use only exact or approved replacement parts.

16. Use protective clothing and safety glasses when handling high-vacuum devices such as picture tubes and cathode-ray tubes.

17. Don't work on equipment before you know proper procedures and are aware of any potential safety hazards.

18. Many accidents have been caused by people rushing and cutting corners. Take the time required to protect yourself and others. Running, horseplay, and practical jokes are strictly forbidden in shops and laboratories.

19. Never look directly into light-emitting diodes or fiber-optic cables. Some light sources, although invisible, can cause serious eye damage.

Circuits and equipment must be treated with respect. Learn how they work and the proper way of working on them. Always practice safety: your health and life depend on it.

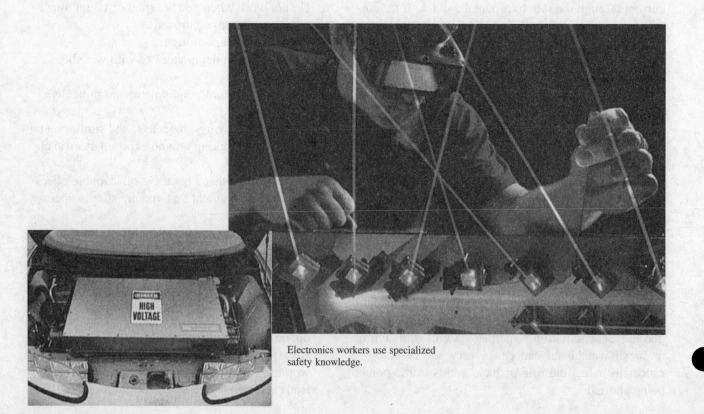

Electronics workers use specialized safety knowledge.

CHAPTER 1

Digital Electronics

TEST: DIGITAL ELECTRONICS

Answer the questions in the spaces provided.

1. A waveform that has just two distinct voltages, such as 0 V and 3.5 V, is called a(n) _____ signal.
   a. Analog
   b. Digital

2. The +5-V level of a TTL digital signal is also called a logical
   a. 0 or LOW   c. 1 or LOW
   b. 0 or HIGH   d. 1 or HIGH

3. A mechanical slide switch can be used to generate a digital signal if the output is debounced using a(n)
   a. Astable multivibrator
   b. Counter
   c. Latch

4. A mechanical normally open push-button switch can be used to generate a single digital pulse if the output is conditioned using a _____ multivibrator.
   a. Bistable
   b. Free-running
   c. Monostable

5. A continuous series of TTL-level pulses can be generated using several discrete components and a 555 _____ IC.
   a. Counter
   b. Multiplexer
   c. Timer

6. A free-running clock that produces a series of TTL-level pulses can also be called a(n)
   a. Astable multivibrator
   b. Bistable multivibrator

7. The simple-to-use instrument often employed to detect static logic levels is the
   a. Logic probe
   b. Logic pulser

8. Most logic probes are designed to detect HIGH and LOW logic levels in both _____ IC logic families.
   a. CMOS and TTL
   b. HTL and RTL

1. _____

2. _____

3. _____

4. _____

5. _____

6. _____

7. _____

8. _____

9. Assume a 5-V power supply. In a TTL logic circuit, a voltage of 3 V would be interpreted as a(n) _____ logic level.
   a. HIGH
   b. LOW
   c. Undefined

9. _____

10. Assume a 5-V power supply. In a CMOS logic circuit, a voltage of 1.5 V would be interpreted as a(n) _____ logic level.
    a. HIGH
    b. LOW
    c. Undefined

10. _____

11. Most real-world information (time, weight, light intensity, etc.) is _____ in nature.
    a. Analog
    b. Digital

11. _____

12. When complicated calculations must be done in the processing stage, the electronic system will probably be _____ in nature.
    a. Analog
    b. Digital

12. _____

13. One reason digital circuits are becoming more popular is
    a. Availability of low-cost digital ICs
    b. Total compatibility with natural world measurements

13. _____

14. Unwanted electrical interference in an electronic circuit is commonly called _____.
    a. Noise
    b. Saturation signals

14. _____

15. To switch to an alternative state such as a switch or flip-flop generating a digital output of HIGH, LOW, HIGH, LOW is referred to as _____.
    a. Polarization   b. Toggling

15. _____

16. The operator of a lab instrument called a _____ can adjust knobs to vary the shape, voltage, and frequency of an output waveform.
    a. Frequency counter
    b. Function generator

16. _____

17. Refer to Fig. 1-1. The scope screen displays two cycles each with a time duration of _____ (5 ms, 10 ms).

17. _____

18. Refer to Fig. 1-1. The frequency of the input signal to the scope is calculated at _____ (200 Hz, 500 Hz).

18. _____

19. Refer to Fig. 1-1. The scope screen shows a digital waveform having an amplitude of _____ (4, 20) volts.

19. _____

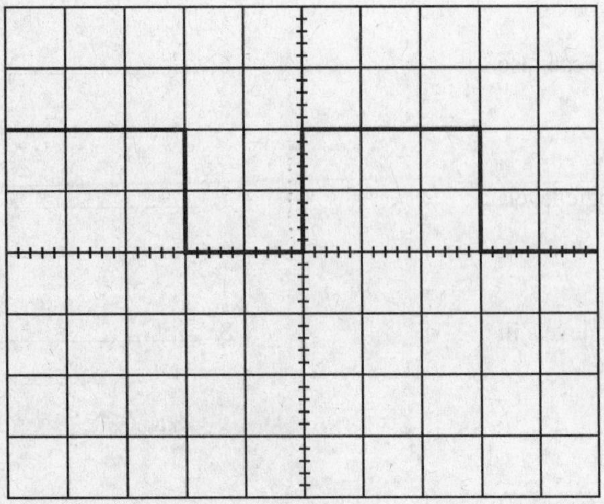

Oscilloscope settings:
   Power ON
   DC input
   Triggering mode = Auto
   Vertical deflection = 2 V/division
   Horizontal sweep time = 1 ms/division

Fig. 1-1 Oscilloscope problem.

2

1-1 LAB EXPERIMENT: CLOCK CIRCUIT

OBJECTIVE

To wire and test a free-running clock circuit.

MATERIALS

Qty.

1   555 timer IC
1   LED indicator-light
    assembly
1   1-kΩ, ¼-W resistor
1   100-kΩ, ¼-W resistor

Qty.

1   470-kΩ, ¼-W resistor
1   5-V dc regulated power
    supply
1   1-μF electrolytic capacitor
1   10-μF electrolytic capacitor

SYSTEM DIAGRAM

You will wire and operate a free-running clock circuit. This circuit will generate a TTL-level digital signal. The 555 timer IC is used to generate the continuous string of square-wave pulses. The frequency is low (1 to 2 pulses per second), and therefore the pulses may be directly observed on a simple LED output indicator. A schematic diagram for the astable multivibrator (free-running clock) circuit is shown in Fig. 1-2.

A very simple LED output indicator light assembly is shown connected to the free-running clock circuit in Fig. 1-2. A HIGH logic level is indicated when the LED lights. A LOW logic level is indicated when the LED does not light. Although very simple, the LED output indicator in Fig. 1-2 does have the disadvantage of loading the output of the IC more than recommended.

A more complicated LED output indicator-light assembly that may be used on your digital lab trainer is sketched in Fig. 1-3. This circuit contains a general-purpose NPN driver transistor. When the input voltage is HIGH, the transistor turns on (conducts), causing the LED to light. When the input voltage is LOW (near ground), the transistor is turned off, causing the LED to turn off. This commonly used circuit does not exceed the drive capabilities of the ICs energizing the output indicators.

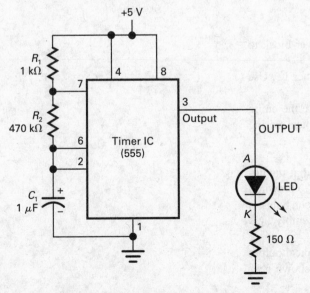

Fig. 1-2 Schematic diagram of a free-running clock circuit.

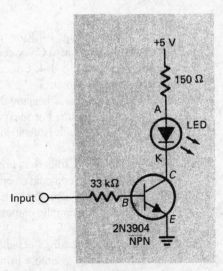

Fig. 1-3 Alternative circuit for LED indicator-light assembly.

Many digital lab trainers have the LED indicator-light assemblies prewired. If not, your instructor will tell you which LED indicator-light assembly to use in your experiments.

PROCEDURE

1. Insert the 555 IC into a mounting board. Use care because the eight pins may not match the holes in the mounting board.
2. Refer to Fig. 1-4. This is a simplified view of solderless breadboards similar to those on a digital trainer manufactured by Dynalogic Concepts.
   a. *Power block.* The four holes on the left side of the power block supply GND (like the negative of a battery). The eight holes on the right side of the power block supply +5 V. The main power switch on the trainer is used to energize the power block.
   b. *Power distribution strip.* All the holes in the top row of the power distribution strip are connected and distribute +5 V in this example. Likewise, all the holes in the bottom row are connected and distribute GND voltage in this example.
   c. *IC mounting board.* On the main IC mounting board, only the four holes in each vertical group are connected.

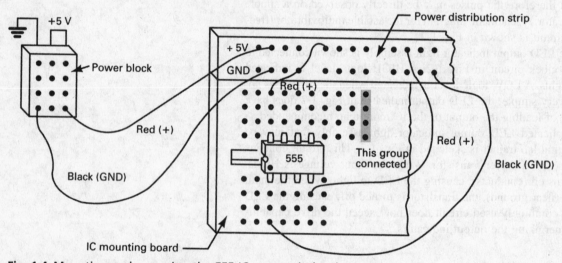

Fig. 1-4 Mounting and powering the 555 IC on a typical trainer.

3. Power OFF. Refer to Fig. 1-4. Connect power from the power block to the power distribution strip. Color-code wires as shown.
4. Power OFF. Refer to Fig. 1-4. Connect power to the 555 timer IC. Use color-coded wires as shown.
5. Power OFF. Refer to the schematic diagram in Fig. 1-2. Wire the entire free-running clock circuit. For inexperienced students, a typical wiring layout for the clock circuit is detailed in Fig. 1-5.
6. Refer to Fig. 1-5.
   a. *Output connector.* A solderless breadboard has been added at the upper right in Fig. 1-5 as a convenient method of connecting to prewired LED indicator-light assemblies. Each vertical group of four holes is connected. In this example, output LED indicator-light assembly L_1 is being used.
   b. *Output LED indicator-light assembly.* A schematic of a typical output LED indicator-light assembly using a driver transistor is shown near the top in Fig. 1-5.

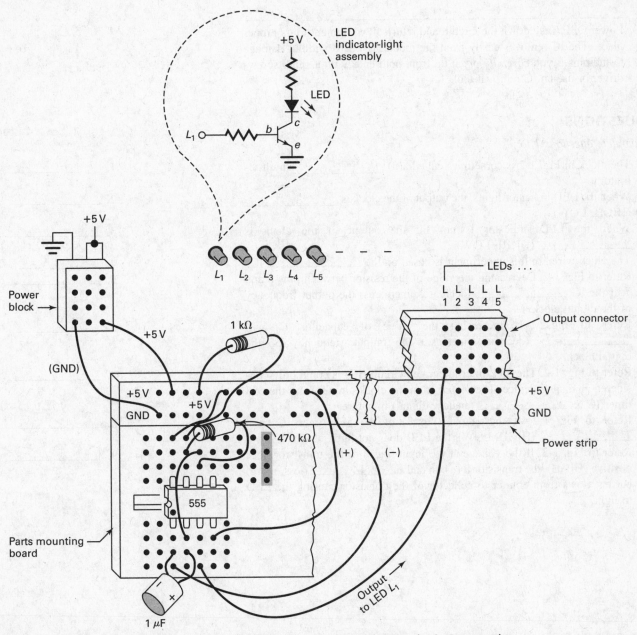

Fig. 1-5 Wiring clock circuit on digital trainer. (Trainer is DT-1000 by Dynalogic Concepts.)

7. Power ON. The output LED should flash on and off at a low frequency. A light means a HIGH or logical 1. No light means a LOW or logical 0 digital signal.
8. Have your instructor check the proper operation of your free-running clock.
9. Power OFF. Remove the 470-kΩ resistor (R_2), and replace it with the 100-kΩ resistor.
10. Power ON. What happened to the frequency of the digital signal when the value of R_2 was reduced?

_____

11. Power OFF. Remove the 1-μF (C_1) capacitor, and replace it with a 10-μF electrolytic capacitor.

12. Power ON. What happened to the frequency of the digital signal when the value of C_1 was increased?

_____

13. Power OFF. Take down the circuit, and return all equipment to its proper place. The IC removes easily from the mounting board without damage to the pins if you *carefully* pry it up from both ends with a small screwdriver or use an IC removal tool.

QUESTIONS

Complete questions 1 to 9.

1. The clock in Fig. 1-2 is sometimes called a(n) _____ multivibrator.

2. When the LED indicator lights, the output of the clock is _____ (HIGH, LOW).

3. When the LED indicator is not lit, the output of the clock is _____ (HIGH, LOW).

4. The clock wired in this experiment is based on the _____ IC.

5. Refer to Fig. 1-2. Decreasing the value of the resistor between pins 6 and 7 of the IC _____ (decreases, increases) the output frequency of the digital clock.

6. Refer to Fig. 1-2. Increasing the value of capacitor C_1 will _____ (decrease, increase) the output frequency of the digital clock.

7. Refer to Fig. 1-2. The 555 timer integrated circuit (IC) is commonly considered an analog device. The *output* from the clock circuit using the 555 timer IC is _____ (analog, digital) in nature.

8. Refer to Fig. 1-2. If the output (pin 3) of the 555 timer IC goes _____ (HIGH, LOW), the LED does not light.

9. Refer to Fig. 1-3. If the voltage at the input (base) of the transistor goes positive (HIGH), the transistor turns on and _____ (less, more) current flows from emitter to collector of the transistor causing the LED to light.

1. _____

2. _____

3. _____

4. _____

5. _____

6. _____

7. _____

8. _____

9. _____

6

1-2 LAB EXPERIMENT: ONE-SHOT MULTIVIBRATOR AND DEBOUNCED SWITCH

OBJECTIVES

1. To wire and test a one-shot multivibrator circuit.
2. To add a debounced input switch to the one-shot multivibrator.
3. *OPTIONAL:* To measure the time duration of the output pulse from the one-shot multivibrator with an oscilloscope.

MATERIALS

Qty.

1   74121 one-shot multivibrator IC
1   555 timer IC
1   LED indicator-light assembly
1   330-Ω, ¼-W resistor
1   1-kΩ, ¼-W resistor
1   33-kΩ, ¼-W resistor
2   100-kΩ, ¼-W resistor
1   5-V dc regulated power suppy

Qty.

1   0.01-μF capacitor
1   0.033-μF capacitor
1   0.1-μF capacitor
1   10-μF electrolytic capacitor
1   N.O. push-button switch (not debounced)
1   debounced switch assembly
OPTIONAL: oscilloscope

SYSTEM DIAGRAM

You will wire a monostable multivibrator circuit based on the 74121 IC. The circuit in Fig. 1-6 shows the wiring of the 74121 one-shot MV. The external components R_3 and C_1 determine the pulse width (time duration) of the positive pulse. This circuit was designed to emit a positive pulse of about 2 to 3 ms. A positive pulse of 2 to 3 ms is long enough to produce a visible flash on the attached LED indicator-light assembly. The one-shot is triggered by a positive voltage appearing at *input B* of the 74121 IC caused by the closing of SW_1. The normal Q output of the 74121 emits a short positive pulse. Remember that the pulse width is determined by the design of the multivibrator circuit and not on how long the input switch (SW_1) was pressed. To increase the pulse width of the one-shot in Fig. 1-6, the values of R_3 and/or C_1 would be increased.

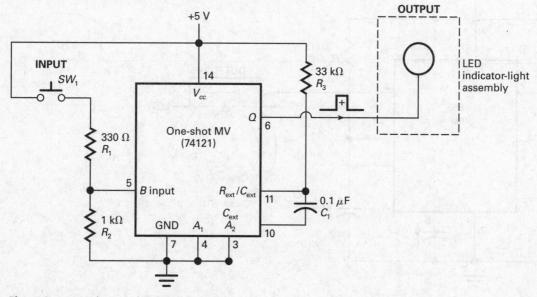

Fig. 1-6 A one-shot multivibrator circuit.

The one-shot MV circuit in Fig. 1-6 should emit only one pulse with the press and release of SW_1. As a practical matter, the 74121 IC may be triggered more than once with a single press and release of input switch SW_1 because of *switch bounce*. A revised schematic for the one-shot multivibrator is drawn in Fig. 1-7. A debounced switch provides the positive trigger voltage in the revised circuit.

The normal output Q (pin 6 on the 74121 IC) in Fig. 1-7 emits a positive pulse when the input is triggered. The complementary output $\overline{Q}$ (pin 1) is also identified in Fig. 1-7 and could be used if a negative pulse were required.

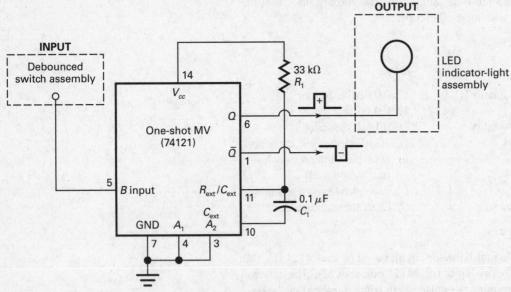

Fig. 1-7 A debounced switch input to a one-shot multivibrator circuit.

In the lab your digital trainer may have a debounced switch available as the input to the one-shot MV circuit in Fig. 1-7. If you do not have a debounced switch available, Fig. 1-8 provides a debouncing circuit based on the 555 timer IC. The output from the debouncing circuit emits a positive voltage beginning when input switch SW_1 is first pressed. The output remains

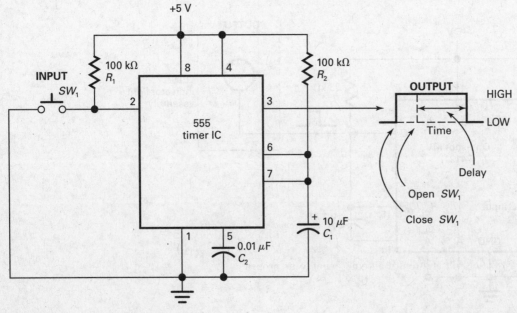

Fig. 1-8 A switch debouncing circuit.

HIGH for a time after SW_1 is released (opened). This delay time is adjustable by changing the value of capacitor C_1. Decreasing the value of C_1 decreases the delay time. The delay time designed into the circuit in Fig. 1-8 is about 1 second so it can be easily observed on the output LED.

PROCEDURE

1. Insert the 74121 IC into a mounting board.
2. Power OFF. Connect power to the IC. Color-code the wires: red for +5 V and black for GND.
3. Power OFF. Refer to Fig. 1-6. Wire the entire circuit. Pin numbers for the 74121 IC are shown on the outside of the symbol. See your instructor if you have any questions about the input switch or output LED indicator-light assemblies.
4. Power ON. Operate the one-shot multivibrator by pressing and releasing the input switch SW_1. Carefully observe the output LED. You should get a single short flash on the LED as you first press the input switch. Because of switch bounce you may also observe other flashes on the output LED. These "extra" output pulses are *not normal* and indicate false triggering of the 74121 IC.
5. Did you observe false triggering when operating the one-shot MV circuit in Fig. 1-6?

   _____

6. Power OFF. Refer to Fig. 1-7. Rewire the one-shot multivibrator circuit using a *debounced switch assembly*. The debounced switch input will eliminate false triggering, and the circuit will emit only a single positive pulse at output Q as the input switch is closed and opened. Check with your instructor on which debounced switch assembly to use (from digital trainer or the debounced switch circuit furnished in Fig. 1-8).
7. Power ON. Operate the one-shot multivibrator circuit, carefully observing the output LED. The one-shot MV circuit should *emit only one positive pulse* as the input switch closes and opens.
8. Show your instructor the normal operation of the circuit from Fig. 1-7.
9. *OPTIONAL:* Connect an oscilloscope to the normal output Q (pin 6) of the one-shot multivibrator shown in Fig. 1-7, and observe the time duration of the positive pulse emitted by the 74121 IC. It will be in the range of 2 to 3 ms. Suggested initial scope settings might include DC mode, auto triggering, 1 V per division, and 1 ms per division.
10. Connect the oscilloscope to the complementary output $\overline{Q}$ (pin 1 of the 74121 IC), and observe the output. Is the output a positive or negative pulse?

    _____

11. Power OFF. Refer to Fig. 1-7. Try replacing C_1 with a 0.033 μF capacitor to change the time duration of the positive pulse from the 74121 one-shot.
12. Power ON. Measure the time duration of the positive pulse (from pin 6 of the 74121 IC) emitted after changing the value of C_1. What is the time duration of the pulse from the revised one-shot MV?

    _____

13. Show your instructor your measurements on the circuit in Fig. 1-7 when $C_1 = 0.1$ μF and when $C_1 = 0.033$ μF. Be prepared to answer questions on the one-shot MV circuit and your measurements.
14. Power OFF. Take down the circuit and return all equipment to its proper place. Use an IC removal tool to extract the IC from the mounting board.

QUESTIONS

Complete questions 1 to 8.

1. The 74121 IC is a one-shot multivibrator and is classified as a(n) _____ (astable, monostable) MV.

2. Refer to Fig. 1-6. The false triggering that was observed during the operation of this MV circuit was caused by switch _____ (bounce, hysteresis).

3. Refer to Fig. 1-7. Which debounced switch did you use when operating this MV circuit?

4. Refer to Fig. 1-7. Why does the 74121 have two outputs?

5. Refer to Fig. 1-8. This circuit uses the popular 555 timer IC wired as a _____ (one-shot MV, switch debouncer).

6. Refer to Fig. 1-7. What is the approximate pulse width for this one-shot MV circuit?

7. Refer to Fig. 1-7. If you decrease the value of capacitor C_1 from 0.1 μF to 0.033 μF, the pulse width _____ (decreases, increases).

8. Refer to Fig. 1-8. If you decrease the value of capacitor C_1 from 10 μF to 1 μF, the time delay after the input switch opens _____ (decreases, increases).

1. _____

2. _____

3. _____

4. _____

5. _____

6. _____

7. _____

8. _____

CHAPTER 2

Numbers We Use in Digital Electronics

TEST: NUMBERS WE USE IN DIGITAL ELECTRONICS

Read each sentence and determine whether it is true or false.

1. In digital electronics we use the symbols 0 and 1; this is referred to as the hexadecimal number system.

2. Decimal 7 equals 0110 in binary.

3. The binary number 1001 equals 8 in decimal.

4. The 1 in the binary number 10000 has a weight of decimal 16.

5. The binary number 110110 equals 54 in decimal.

6. The binary number 11001000 equals 100 in decimal.

7. The decimal number 44 equals 101100 in binary.

8. The decimal number 253 equals 11111111 in binary.

9. The decimal number 14 equals 1101 in binary.

10. The binary number 1111 equals 15 in decimal.

11. A decoder is an electronic device that translates from a decimal input to binary.

12. An encoder is used between a calculator keyboard and the central processing unit.

13. Decimal 13 equals D in hexadecimal.

14. Decimal 10 equals A in hexadecimal.

15. Binary 111111 equals 3F in hexadecimal.

16. Binary 11101011 equals 7B in hexadecimal.

17. Hexadecimal 4B equals 75 in decimal.

18. Hexadecimal A6 equals 166 in decimal.

19. Decimal 108 equals 5C in hexadecimal.

20. Decimal 45 equals D2 in hexadecimal.

21. Subscripts are sometimes added to numbers to show the base of the number.

22. $1011_2 = 11_{16}$.

23. The octal number 157 equals 1101111 in binary.

24. The octal number 710 equals 558 in decimal.

25. The decimal number 198 equals 306 in octal.

26. To encode means to convert from a readable code (such as decimal) to an encrypted code (such as binary).

27. A nibble is an 8-bit data group.

28. A single binary digit (0 or 1) is called a bit.

29. An 8-bit data group used to represent a number, letter, or op code is called a byte.

30. The most common word length for modern personal computers is 4 bits.

1. _____
2. _____
3. _____
4. _____
5. _____
6. _____
7. _____
8. _____
9. _____
10. _____
11. _____
12. _____
13. _____
14. _____
15. _____
16. _____
17. _____
18. _____
19. _____
20. _____
21. _____
22. _____
23. _____
24. _____
25. _____
26. _____
27. _____
28. _____
29. _____
30. _____

2-1 LAB EXPERIMENT: USING AN ENCODER

OBJECTIVES

1. To wire a 74147 encoder integrated circuit (IC).
2. To convert decimal numbers to binary numbers using the 74147 electronic encoder.
3. To observe the priority feature of the 74147 encoder (IC).
4. To define active LOW inputs or outputs.
5. To demonstrate that TTL inputs (left unconnected) float HIGH.

NOTE: This is the hardware version of Using an Encoder.
The software version is Lab Experiment 2-4.

MATERIALS

Qty.		Qty.	
1	7404 hex inverter IC (use four sections)	1	74147 or 74LS147 encoder IC
1	keypad (0 to 9, N.O. contacts)	4	LED indicator-light assemblies
1	5-V dc regulated power supply	9	100-kΩ, resistors (optional)

SYSTEM DIAGRAM

A decimal-to-binary encoder system is drawn in Fig. 2-1. The switches (1 to 9) are the decimal inputs while the LED indicator-light assemblies form the binary output. As an example, pressing the 3 on the keypad would cause an output of OFF OFF ON ON or 0011 in binary. Likewise, pressing 9 on the keypad would cause an output of ON OFF OFF ON or 1001 in binary.

The keypad shown in Fig. 2-1 is constructed of normally open push-button switches. Each input switch is wired like the one detailed in Fig. 2-2. With the push-button switch released or open, the pull-up resistor connects input 3 (pin 13) to the 74147 IC directly to +5 V (HIGH). Continuing in Fig. 2-2, pressing input switch 3 connects input 3 (pin 13) of the IC to ground (LOW). The 74147 IC has *active LOW inputs* as shown on the schematic by the small "invert bubble" at the inputs. Therefore, only a ground voltage (about 0 V) will activate an input. A HIGH input to the 74147 IC means the input is deactivated.

As a conventional practice, the labels given an input or output on an IC are inside the logic symbol. The pin numbers used in actually wiring the ICs are shown on the outside of the symbol. As an example in Fig. 2-1, input 3 to the 74147 IC has the pin number 13 outside of the symbol while the number 3 (input 3) is listed on the inside.

The 74LS147 encoder is a low-power variation of the 74147 IC. Either may be used to perform this experiment. The embedded LS in the part number 74LS147 means low-power Schottky, which is a modern variation of the original 74147 encoder. The root numbers of the 74147 and 74LS147 are said to be identical (that is, 74147). The 74147 and 74LS147 ICs both perform the same function and have the same pin numbers.

Notice that the outputs of the 74147 IC also have bubbles. This means that they will generate a *complementary output*. A complementary output is the opposite of what you would expect. As an example, if input key 3 were pressed, the output from the 74147 IC would be HHLL (1100). If you invert or complement each bit, the HHLL (1100) becomes LLHH or the true binary output of 0011 (for decimal 3). The 7404 IC contains devices called *inverters* which complement each output to generate the true binary output. The true binary output is then displayed on the LED indicator-light assemblies. In simple terms, pressing any number (1 through 9) on the keypad shown in Fig. 2-1 will generate its binary equivalent on the LED displays.

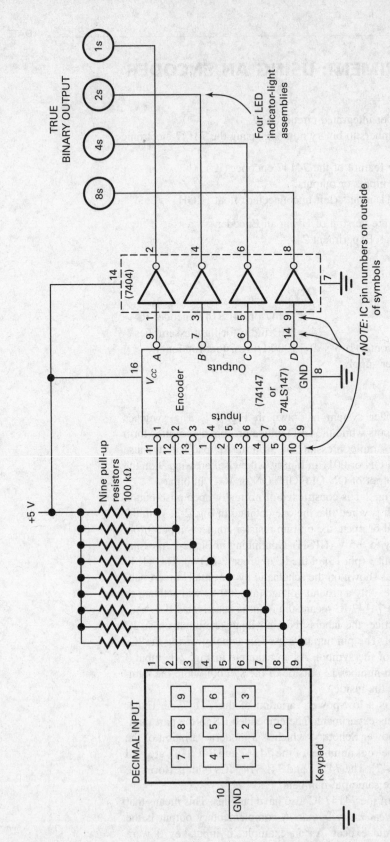

Fig. 2-1 A decimal-to-binary encoder system.

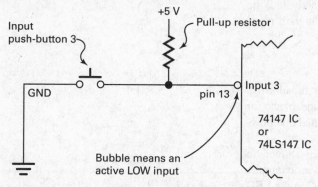

Fig. 2-2 Schematic of N.O. push-button switch as input to the 74147 encoder.

Consider the simplified version of the decimal-to-binary encoder circuit redrawn in Fig. 2-3. Comparing the simplified version to the original encoder shown in Fig. 2-1, you will notice that the nine pull-up resistors have been deleted from the simplified circuit in Fig. 2-3. You may wire and operate the simplified version shown in Fig. 2-3 in the lab. The simplified encoder will operate without the pull-up resistors because when all inputs to the 74147 are deactivated (no input switches closed), the inputs will "float HIGH." Floating inputs (not connected to +5 V or GND) will act as if they are HIGH. We say that the inputs to an IC such as this will float HIGH.

The 74147 IC is part of a family of digital ICs manufactured using the *transistor-transistor logic* (TTL) technology. The inputs on TTL ICs float HIGH when left disconnected. This is not true of all families of ICs. As an example, CMOS (*complementary metal-oxide semiconductor*) ICs may not have inputs unconnected or floating.

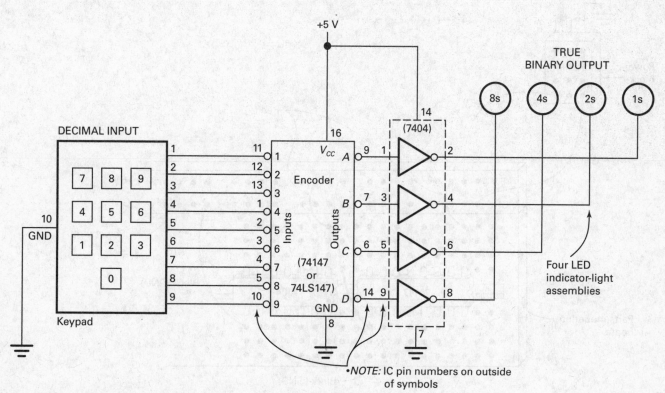

Fig. 2-3 Simplified version of decimal-to-binary encoder system.

The 74147 IC is referred to by the manufacturer as a *priority encoder*. Priority encoding means that the highest active input (1 through 9) will be decoded. As an example, in Fig. 2-3, if both inputs 1 and 6 were activated, the output on the LEDs would read binary 0110 ($0110_2 = 6_{10}$). The highest number activated will be encoded by the 74147 priority encoder IC.

PROCEDURE

1. Power OFF. Insert the 74147 and 7404 ICs into a solderless breadboard. Power the ICs using colored-coded wires (red for +5 V and black for GND). See Fig. 2-4 for an example of how ICs may be mounted and powered using a trainer by Dynalogic Concepts.

2. Power OFF. Wire the decimal-to-binary encoder circuit shown in Fig. 2-3. IC pin numbers are shown outside the symbol. As an example, Fig. 2-5 shows the four output wires leading from the 7404 IC to the LED indicator-light assemblies as they might be wired on a commercial trainer. Partial wiring of the input keypad to the 74147 IC is detailed in Fig. 2-5 including the GND connection and a few of the switch connections.
   HINT: Color coding of wires for power, inputs, outputs is useful.
   TECHNICAL NOTE: The encoder circuit without pull-up resistors in Fig. 2-3 is wired for this experiment for simplicity. The encoder circuit in Fig. 2-1 has pull-up resistors and is an example of a better design.

3. Power ON. LED lights should be all OFF. This means the binary number 0000 is appearing at the output of the system.

4. Power ON. Press the decimal digit 1 on the keypad, and record the results in the left side of Table 2-1. Record an ON or OFF for each lamp as you press each key on the keypad.

5. Power ON. Experiment with the priority feature of the 74147 encoder IC. Press at least two input switches at the same time. Observe the results.

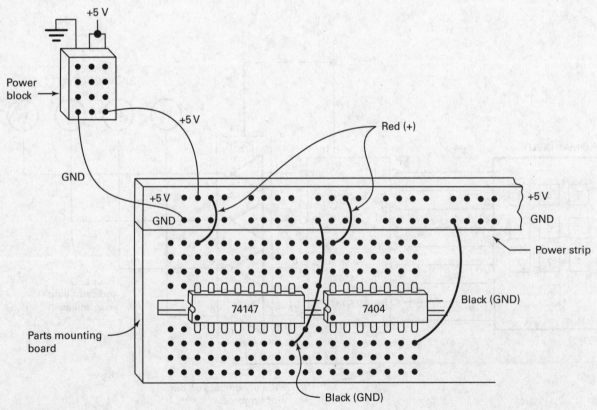

Fig. 2-4 Mounting and powering the ICs.

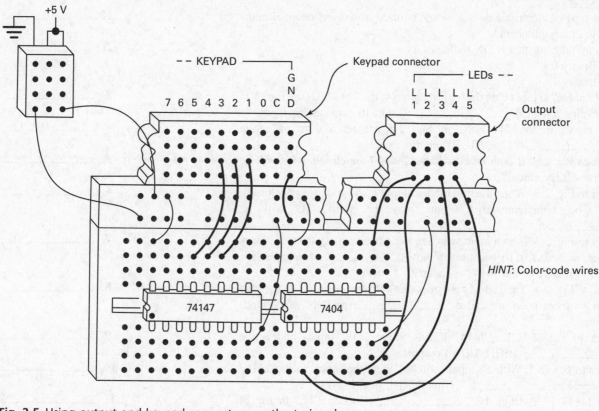

Fig. 2-5 Using output and keypad connectors on the trainer by Dynalogic Concepts, Inc.

6. Have your instructor check the proper operation of the circuit in Fig. 2-3.
7. Fill in the right side of Table 2-1 with 0s and 1s to match the lamp indications in the left side of the table.
8. Power OFF. Take down the circuit and return all equipment to its proper place. The ICs remove easily from the mounting board without damage if you *carefully* pry them up from both ends with a *small* screwdriver or use an IC removal tool.

TABLE 2-1 74147 Encoder

INPUT	OUTPUT							
Decimal digit	Lamp indicators (on or off)				Binary number (1 or 0)			
	8s	4s	2s	1s	8s	4s	2s	1s
0								
1								
2								
3								
4								
5								
6								
7								
8								
9								

QUESTIONS

Complete questions 1 to 14.

1. What type of electronic device would be used to convert decimal numbers to binary numbers?

2. A lamp indicator that is ON indicates a
   a. Binary 0
   b. Binary 1

3. The "bubble" at input 6 on the logic diagram of the 74147 IC (see Fig. 2-1) means that a logical _____ (0, 1) will activate this input.

4. The inputs to the 74147 IC in Fig. 2-1 are referred to as active _____ (LOW, HIGH) inputs.

5. Refer to Fig. 2-3. If both input switches 2 and 7 are closed, what will be the true binary output?

6. Refer to Fig. 2-3. When the true binary output reads 1001 (key 9 closed), the 4-bit complementary output from the 74147 IC will be _____ .

7. Refer to Fig. 2-3. If both input switches 1 and 6 are closed, the true binary output will read 0110 because of the _____ (multiplexing, priority) feature of the 74147 encoder IC.

8. Refer to Fig. 2-1. The 100-kΩ resistors that hold the input HIGH with the input switches open are called _____ (pull-up, wirewound) resistors.

9. Refer to Fig. 2-3. The 74147 IC has active LOW inputs and active _____ (HIGH, LOW) outputs.

10. Refer to Fig. 2-3. With all input switches open, the inputs to the 74147 IC appear to _____ (float HIGH, float LOW).

11. The 74147 IC is from the _____ (CMOS, TTL) family of digital ICs.

12. The 74LS147 IC is a _____ (CMOS, low-power) version of the 74147 TTL priority encoder.

13. Refer to Fig. 2-3. The inverters housed in the 7404 IC are said to _____ (complement, increment) the output of the encoder to true binary.

14. Refer to Fig. 2-3. If the output at outputs *D*, *C*, *B*, and *A* of the 74147 encoder IC is binary 1101, then the 7404 inverters will convert this to true binary _____ representing decimal 2.

1. _____

2. _____

3. _____

4. _____

5. _____

6. _____

7. _____

8. _____

9. _____

10. _____

11. _____

12. _____

13. _____

14. _____

18

2-2 LAB EXPERIMENT: USING A DECODER

OBJECTIVES

1. To wire a 7442 decoder IC.
2. To convert binary numbers to decimal numbers using the 7442 decoder IC.

MATERIALS

Qty.

2   7404 hex inverter ICs (use 10 sections)
4   logic switches
1   5-V dc regulated power supply
1   7442 decoder IC

Qty.

10   LED indicator-light assemblies

SYSTEM DIAGRAM

Figure 2-6 illustrates the electronic system that you will construct in this experiment. This system will electronically convert binary numbers into a decimal output.

The inputs to the 7442 decoder IC are "active HIGH" inputs. If the input switches in Fig. 2-6 were set to HLLH this would mean binary 1001. Notice the bubbles at the outputs of the 7442 decoder IC symbol in Fig. 2-6. These bubbles mean that the outputs from the 7442 IC are "active LOW." As an example, if the binary input to the 7442 IC were 1001 then output 9 would go LOW while all other outputs (0 through 8) would remain HIGH. The

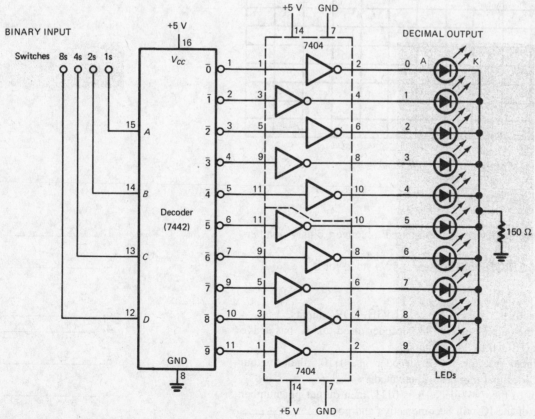

Fig. 2-6 Binary-to-decimal decoder system.

inverters in the 7404 ICs shown in Fig. 2-6 reverse the logic, so if output 9 goes LOW the bottom inverter would output a HIGH, causing LED 9 to light.

PROCEDURE

1. Insert the 7442 and the two 7404 ICs into the mounting board.
2. Power OFF. Connect power to the three ICs: use red wire for +5 V and black wire for GND. See your instructor for help in using your trainer.
3. Refer to Fig. 2-6. Wire the entire circuit (input switches, 7442 and two 7404 ICs, and LED indicator lights). Pin numbers are shown *outside* the 7442 and 7404 IC diagrams in Fig. 2-6. See your instructor if you have difficulty with the input logic switches.
4. Power ON. Place all input switches in the LOW voltage position (GND voltage). This means that all the inputs are logical 0. The zero LED output light should be ON.
5. Power ON. Set up each input switch combination shown in the left side of Table 2-2. Record the results in the right side of Table 2-2. Record 0s and 1s in the proper columns.
6. Have your instructor check the proper operation of the circuit in Fig. 2-6.
7. Power OFF. Take down the circuit, and return all equipment to its proper place. *Carefully* remove the ICs from the mounting board.

TABLE 2-2   7442 Decoder

INPUTS				DECIMAL OUTPUT									
0 = GND	1 = +5 V			Lit = 1						Not lit = 0			
8s	4s	2s	1s	9	8	7	6	5	4	3	2	1	0
0	0	0	0										
0	0	0	1										
0	0	1	0										
0	0	1	1										
0	1	0	0										
0	1	0	1										
0	1	1	0										
0	1	1	1										
1	0	0	0										
1	0	0	1										

QUESTIONS

Complete questions 1 to 6.

1. What type of electronic device would be used to convert binary numbers to decimal numbers?
2. In this experiment a HIGH voltage (near +5 V) stood for a
   a. Logical 0
   b. Logical 1
3. The 7442 IC has active _____ (HIGH, LOW) inputs.
4. The "bubbles" at the outputs of the 7442 decoder mean this IC has active _____ (HIGH, LOW) outputs.
5. If the binary input to the 7442 decoder is 0110, then output _____ (decimal number) is activated.
6. If the binary input to the 7442 decoder is 0111, then output pin number _____ of the IC will become active and go _____ (HIGH, LOW).

1. _____
2. _____
3. _____
4. _____
5. _____
6. _____

2-3 LAB EXPERIMENT: USING A CMOS BINARY COUNTER

OBJECTIVES

1. To wire a CMOS 74HC393 binary counter IC.
2. To interpret the 8-bit binary display in both hexadecimal and octal.

MATERIALS

Qty.

1 74HC393 CMOS binary counter IC
1 logic switch
1 clock (free-running)
1 clock (single-pulse)

Qty.

8 LED indicator-light assemblies
1 5-V dc regulated power supply

SYSTEM DIAGRAM

Figure 2-7 details the wiring of an 8-bit binary counter. This IC will count from binary 00000000 to 11111111. The 74HC393 IC will count each clock pulse entering pin 1 (clock or *CLK* input). The binary output is represented by eight LED indicator-light assemblies. It is suggested that the output LEDs be arranged with the LSB (1s LED) on the right and the MSB (128s LED) on the left. This arrangement makes the binary number indicated on the LEDs much easier to read.

PROCEDURE

CAUTION CMOS ICs can be damaged from static electricity. Store CMOS ICs with their pins in conductive foam or covered with aluminum foil. Try not to touch pins when handling CMOS ICs.

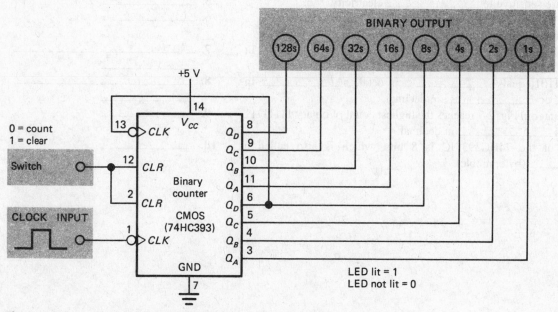

Fig. 2-7 Wiring diagram for an 8-bit binary counter system.

1. Insert the 74HC393 CMOS IC into the mounting board.
2. Power OFF. Connect power to the IC: use red wire for +5 V and black wire for GND.
3. Refer to Fig. 2-7. Power OFF. Wire the entire circuit (switch, free-running clock, IC, and LED indicator-light assemblies). Pin numbers are shown *outside* the 74HC393 IC symbol. See your instructor if you have questions about the input switch, free-running clock, or LED output indicators.
4. Power ON. Move *CLR* switch to HIGH and then back to LOW to clear output to 00000000. Operate the free-running clock at about 1 to 2 Hz, and observe the binary counting action on the LED output indicators.
5. Power OFF. Disconnect the free-running clock. Connect the single-pulse clock to the *CLK* (pin 1) input of the 74HC393 counter IC.
6. Power ON. Clear the counter. Repeatedly pulse the *CLK* input until the output reads 00001111. What does this equal in hexadecimal? What does this equal in octal?

———————————————————————————————

7. Power ON. Pulse the counter until it reads 10101011. What is the hexadecimal and octal representation of this binary output? What does this equal in decimal?

———————————————————————————————

8. Show your instructor your calculations, and demonstrate the operation of the CMOS binary counter circuit.
9. Power OFF. Take down the circuit, and return all equipment to its proper place. Carefully remove and properly store the CMOS IC in conductive foam.

QUESTIONS

Complete statements 1 to 10.

1. The 74HC393 is a _____ (CMOS, TTL) integrated circuit.
2. Refer to Fig. 2-7. Placing the *CLR* switch in the _____ (HIGH, LOW) position clears all outputs to 0.
3. Refer to Fig. 2-7. The square-wave pulses from the _____ input cause the 74HC393 IC to count upward.
4. Refer to Fig. 2-7. The highest possible count recorded at the output of the counter was _____. This equals _____ in hexadecimal, _____ in octal, or _____ in decimal.
5. CMOS ICs are sensitive to _____ electricity.
6. CMOS ICs should be stored with their pins in _____ (conductive foam, styrofoam).
7. Binary 11110011 equals _____ in octal, _____ in hexadecimal, or _____ in decimal.
8. Binary 01111101 equals _____ in octal, _____ in hexadecimal, or _____ in decimal.
9. The 8-bit counter in Fig. 2-7 reaches the highest count of binary 11111111 which equals _____ in decimal.
10. The output of the 74HC393 IC is 8 bits, which is also called a _____ (byte, nibble).

1. _____
2. _____
3. _____
4. _____ _____
   _____ _____
5. _____
6. _____
7. _____ _____
   _____
8. _____ _____
   _____
9. _____
10. _____

2-4 MULTISIM EXPERIMENT: ENCODING DECIMAL TO BINARY

OBJECTIVE

To construct and test a simple decimal-to-binary encoder system using electronic circuit simulation software.

NOTE: The hardware version of this lab is Experiment 2-1.

MATERIALS

Qty.

1   electronic circuit simulation software (such as Multisim) and a computer system.

SYSTEM DIAGRAM

The task of the *encoder system* detailed in Figs. 2-8 and 2-9 is to translate a decimal input to a binary output. The 74147 encoder IC contains the digital logic circuit that will make this conversion. For instance, pressing the 7 on the keypad would generate a binary output of 0111 on the output indicators.

The hardware version of the encoder system is shown in Fig. 2-8. In this circuit, all inputs to the 74147 IC are held HIGH by the nine pull-up resistors. The small bubbles at the inputs to the 74147 encoder IC mean the inputs (1 through 9) to the encoder require a LOW for activation. An *active LOW input* means a logical 0 (LOW) activates or turns on the function of the pin. As an example, if the 8 is pressed on the keypad, then input 8 (pin 5 on the IC) is grounded through the switch and is activated. In this example, the output would be binary 1000 at the displays.

Look at the outputs from the 74147 IC in both versions of the encoder system (Figs. 2-8 and 2-9). Invert bubbles at the output of an IC symbol means they are *active LOW outputs*. The software version of the encoder system in Fig. 2-9 suggests that the immediate output of the 74147 encoder IC will be an *inverted binary* (sometimes called the 1s complement). With the use of four inverters, the inverted binary is changed into true binary. This is displayed at the upper right in Fig. 2-9.

PROCEDURE

1. Construct the decimal-to-binary encoder system shown in Fig. 2-9 using electronic circuit simulation software.
2. Operate the encoder system. Recall that the 74147 IC has active LOW inputs.
3. Observe the true binary outputs as well as the inverted binary outputs.
4. Show your simulated circuit to your instructor. Be prepared to demonstrate the encoder system, and answer questions about its operation.

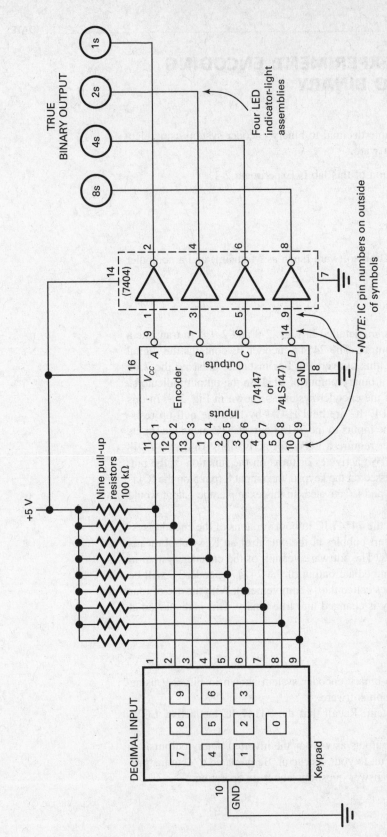

Fig. 2-8 Hardware version of decimal-to-binary encoder system.

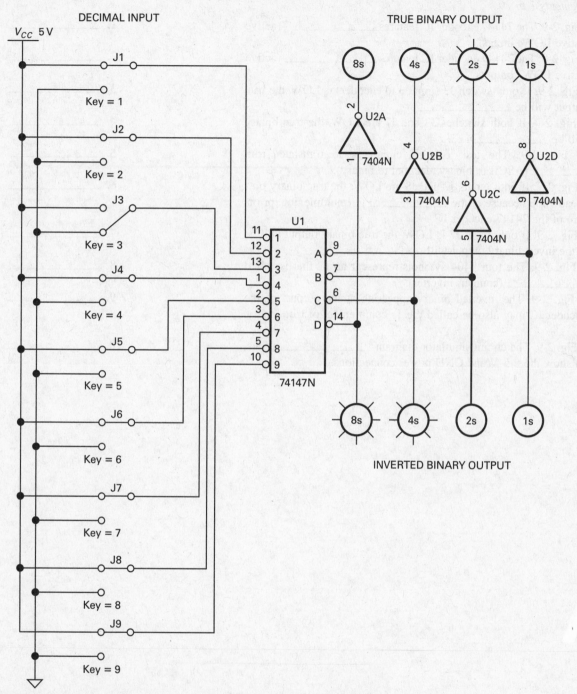

Fig. 2-9 Circuit simulator version of decimal-to-binary encoder system.
Note: Power connections to ICs are not shown in diagram but are understood to be connected.

QUESTIONS

Complete statements 1 to 10.

1. Refer to Fig. 2-9. The 74147 encoder IC features _____ (active HIGH, active LOW) inputs.
2. Refer to Fig. 2-9. The 74147 encoder IC features _____ (active HIGH, active LOW) outputs.
3. Refer to Fig. 2-9. If only switch J3 (input 3 to encoder) is LOW, the true binary output will be _____ .
4. Refer to Fig. 2-9. If both switches J3 and J7 are LOW, the true binary output will be _____ .
5. Refer to Fig. 2-9. The task of this circuit is to translate from _____ (ASCII to binary, decimal to binary).
6. Refer to Fig. 2-9. If both inputs J2 and J8 are LOW, the true binary output will read 1000 because of the _____ (multiplexing, priority) feature of the 74147 encoder IC.
7. Refer to Fig. 2-9. If only input J7 is LOW, the true binary output will be 0111 but the invert binary output will read _____ .
8. Refer to Fig. 2-9. The four 7404 symbols represent an IC that is called a(n) _____ (counter, inverter).
9. Refer to Fig. 2-9. The inverted binary appearing at the outputs of the 74147 encoder IC may also be called the 1s complement of true binary. (T or F)
10. Refer to Fig. 2-9. The circuit simulator diagram _____ (does, does not) show the +5-V and GND power connections.

1. _____
2. _____
3. _____
4. _____
5. _____
6. _____
7. _____
8. _____
9. _____
10. _____

CHAPTER 3

Logic Gates

TEST: LOGIC GATES

Answer the questions in the spaces provided.

1. Fig. 3-1(*a*) is the logic symbol for a(n) _____.
2. Fig. 3-1(*b*) is the logic symbol for a(n) _____.
3. Fig. 3-1(*c*) is the logic symbol for a(n) _____.
4. Fig. 3-1(*d*) is the logic symbol for a(n) _____.
5. Fig. 3-1(*e*) is the logic symbol for a(n) _____.
6. Fig. 3-1(*f*) is the logic symbol for a(n) _____.
7. Fig. 3-1(*g*) is the logic symbol for a(n) _____.
8. Fig. 3-1(*h*) is the logic symbol for a(n) _____.
9. Fig. 3-1(*i*) is the logic symbol for a(n) _____.
10. The Boolean expression $A + B = Y$ matches the logic symbol in Fig. 3-1(_____).
11. The Boolean expression $A \cdot B = Y$ matches the logic symbol in Fig. 3-1(_____).
12. The Boolean expression $A \oplus B = Y$ matches the logic symbol in Fig. 3-1(_____).
13. The Boolean expression $\overline{A + B} = Y$ matches the logic symbol in Fig. 3-1(_____).
14. The Boolean expression $\overline{A \cdot B} = Y$ matches the logic symbol in Fig. 3-1(_____).
15. The Boolean expression $\overline{A \oplus B} = Y$ matches the logic symbol in Fig. 3-1(_____).

1. _____
2. _____
3. _____
4. _____
5. _____
6. _____
7. _____
8. _____
9. _____
10. _____
11. _____
12. _____
13. _____
14. _____
15. _____

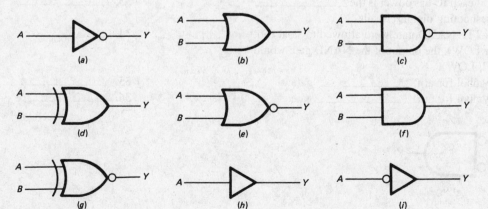

Fig. 3-1 Logic symbols.

16. List the four bits in the output column (top to bottom) of the truth table in Fig. 3-2 for the AND function.

16. _____

17. List the four bits in the output column (top to bottom) of the truth table in Fig. 3-2 for the OR function.

17. _____

18. List the four bits in the output column (top to bottom) of the truth table in Fig. 3-2 for the NAND function.

18. _____

19. List the four bits in the output column (top to bottom) of the truth table in Fig. 3-2 for the NOR function.

19. _____

20. List the four bits in the output column (top to bottom) of the truth table in Fig. 3-2 for the XOR function.

20. _____

INPUTS		OUTPUT
B	A	Y
0	0	?
0	1	?
1	0	?
1	1	?

Fig. 3-2 Truth table.

21. Write the Boolean expression for a three-input OR gate.

21. _____

22. Place an inverter at the output of an AND gate. This combination of gates will produce the _____ logic function.

22. _____

23. Place an inverter at each input of a two-input AND gate. This combination of gates will produce the _____ logic function.

23. _____

24. Place an inverter at each input of a two-input OR gate. This combination of gates will produce the _____ logic function.

24. _____

25. TTL digital ICs typically use a _____-V dc power supply.

25. _____

26. If a digital dual in-line package IC had the number 74HC08 printed on the top, it would be a(n) _____ (CMOS, IIL, TTL) unit.

26. _____

27. If a digital DIP IC had the part number 74LS08 printed on the top, it would be a _____ (CMOS, DTL, TTL) device.

27. _____

28. How should CMOS ICs be stored to guard against static electricity?

28. _____

29. Unused inputs to a _____ (CMOS, TTL) digital device should never be left "floating" or the IC may be damaged.

29. _____

30. If a digital SOIC has a part number SN74LVC00 printed on the top, it would be a low-voltage _____ (CMOS, TTL) device.

30. _____

31. Digital ICs from the 74LVCXX subfamily typically use a _____ (3-V, 9-V) dc power supply.

31. _____

32. Using your sense of smell, touch, and sight is the _____ (first, second) step in troubleshooting digital circuits.

32. _____

33. Using a logic probe to check if each IC has power is the _____ (first, second) step in troubleshooting digital circuits.

33. _____

34. If all inputs to a 7400 series TTL NAND gate were allowed to float (not connected to either HIGH or LOW), the output of the NAND gate would go _____ (HIGH, LOW).

34. _____

35. Fig. 3-3(a) is an alternate symbol for a(n) _____ gate.

35. _____

36. Fig. 3-3(b) is an alternate symbol for a(n) _____ gate.

36. _____

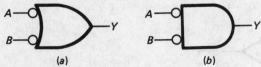

Fig. 3-3 Logic symbols.

37. The unique output for the _____ (AND, NAND) gate is a HIGH only when all inputs are HIGH.

38. The unique output for the OR gate is a _____ (HIGH, LOW) only when all inputs are LOW.

39. The unique output for the _____ (NAND, NOR) gate is a HIGH only when all inputs are LOW.

40. The output Y_1 on the truth table in Fig. 3-4 describes the three-input _____ logic function.

41. The output Y_2 on the truth table in Fig. 3-4 describes the three-input _____ logic function.

42. The output Y_3 on the truth table in Fig. 3-4 describes the three-input _____ logic function.

43. The output Y_4 on the truth table in Fig. 3-4 describes the three-input _____ logic function.

44. The output Y_5 on the truth table in Fig. 3-4 describes the three-input _____ logic function.

45. The output Y_6 on the truth table in Fig. 3-4 describes the three-input _____ logic function.

37. _____

38. _____

39. _____

40. _____

41. _____

42. _____

43. _____

44. _____

45. _____

INPUTS			OUTPUTS					
A	B	C	Y_1	Y_2	Y_3	Y_4	Y_5	Y_6
0	0	0	0	1	0	1	1	0
0	0	1	1	0	0	0	1	1
0	1	0	1	0	0	0	1	1
0	1	1	0	0	0	1	1	1
1	0	0	1	0	0	0	1	1
1	0	1	0	0	0	1	1	1
1	1	0	0	0	0	1	1	1
1	1	1	1	0	1	0	0	1

Fig. 3-4 Truth table.

46. Refer to Fig. 3-5. Surface-mount technology is used to fasten this _____ (CCMIC, SOIC) package NAND gate device to a printed circuit board.

47. Refer to Fig. 3-5. Pin 1 on the 74LVC00 logic gate IC is located at _____ (A, B, or C).

46. _____

47. _____

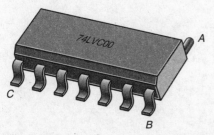

Fig. 3-5 Tiny surface-mount 74LVC00 IC.

48. The PBASIC assignment statement used with the BASIC Stamp 2 Microcomputer Module that represents the two-input NOR logic function is _____.

   a. $Y = A \& B$
   b. $Y = A \wedge B$
   c. $Y = \sim(A \mid B)$
   d. $Y = \sim(A \wedge B)$

48. _____

49. The PBASIC assignment statement used with the BASIC Stamp 2 Microcomputer Module that represents the three-input XOR logic function is _____.

   a. $Y = A \wedge B \wedge C$
   b. $Y = A \& B \& C$
   c. $Y = A \mid B \mid C$
   d. $Y = \sim(A \& B \& C)$

49. _____

3-1 LAB EXPERIMENT: AND GATES

OBJECTIVES

1. To wire and operate a two-input AND gate.
2. To wire and operate a three-input AND gate using a 7408 IC.
3. To design and implement a five-input AND function using a CMOS 74HC08 IC.

MATERIALS

Qty.   **Qty.**

1   7408 two-input TTL AND gate IC   1   LED indicator-light assembly
1   74HC08 CMOS two-input AND   5   logic switches
    gate IC   1   5-V dc regulated power supply

SYSTEM DIAGRAMS

Figure 3-6 is the electronic system you will construct to perform the two-input AND function.

INPUTS

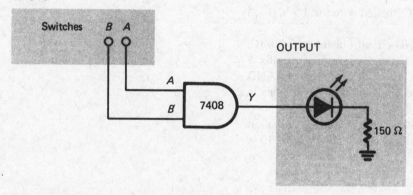

Fig. 3-6 Wiring a two-input AND gate.

The logic symbol diagram in Fig. 3-7 is the second circuit you will construct. This circuit will perform the three-input AND function.

INPUTS

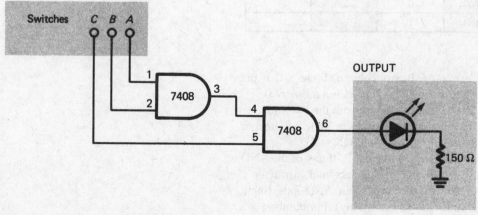

Fig. 3-7 Wiring a three-input AND gate.

The logic symbol drawn in Fig. 3-8 suggests that you will design and implement a circuit that will perform the five-input AND function. You will use a CMOS 74HC08 quad two-input AND gate IC in the design. In the design process, the pattern of connecting gates shown in Fig. 3-7 will be useful. A pin diagram for the 74HC08 IC is given in Appendix A.

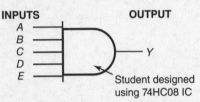

Fig. 3-8 Symbol for five-input AND gate. (Students must design using the 74HC08 IC.)

Some cautions are required when using CMOS ICs. CMOS ICs are sensitive to static electricity and must be handled carefully. CMOS ICs should be stored in conductive foam or a static-free bag when not in use.

PROCEDURE

1. Insert the 7408 IC into the mounting board.
2. Power OFF. Connect power to the 7408 IC: the red wire for +5 V (V_{CC}) and the black wire for GND.
3. Refer to Fig. 3-6. Wire the two-input AND circuit (switches, 7408 IC, and LED indicator-light assembly). Refer to pin diagrams in Appendix A of this manual. Notice that the 7408 IC actually contains four AND gates in the one package. You will test each of the four AND gates in this experiment.
4. Power ON. Move input switches A and B to each combination in the left part of the truth table in Table 3-1. Record the results (an ON or OFF) in the Light column, Table 3-1.

TABLE 3-1 2-Input AND Gate

| INPUTS | | | | OUTPUT | |
| A | | B | | | |
Voltage	Binary	Voltage	Binary	Light	Binary
LOW	0	LOW	0		
LOW	0	HIGH	1		
HIGH	1	LOW	0		
HIGH	1	HIGH	1		

5. Record the binary digits 0 or 1 in the Binary column, Table 3-1. If the light is ON, record a binary 1. If the light is OFF, record a binary 0.
6. Power OFF. Rewire the second to fourth AND gates in the 7408 IC.
7. Power ON. Test each AND gate and record the results in Table 3-2. Record the results as a binary 0 or 1 (light OFF = 0, light ON = 1).
8. Power OFF. Look at the results in Tables 3-1 and 3-2. If any of the AND gates in the 7408 IC are not working properly, contact your instructor.
9. Power OFF. Refer to Fig. 3-7. Wire the three-input AND gate (input switches, 7408 IC, and LED indicator-light assembly). Pin numbers are shown in Fig. 3-7.

TABLE 3-2   Truth Table for 7408 IC

INPUTS		OUTPUTS		
A	B	Second AND gate	Third AND gate	Fourth AND gate
0	0			
0	1			
1	0			
1	1			

10. Power ON. Operate the input switches *A*, *B*, and *C* according to the truth table in Table 3-3. Record a 0 for a LOW output voltage (GND). Record a 1 for a HIGH output voltage (near +5 V). Observe and record the results in the Output column of Table 3-3.

11. Refer to Fig. 3-8. Design a logic circuit using the CMOS 74HC08 quad two-input AND gate IC that will perform the five-input AND function. The pin diagram for the 74HC08 IC is drawn in Appendix A.

TABLE 3-3   Three-Input AND Gate

INPUTS			OUTPUT
A	B	C	
0	0	0	
0	0	1	
0	1	0	
0	1	1	
1	0	0	
1	0	1	
1	1	0	
1	1	1	

CAUTION CMOS ICs can be damaged by static electricity. Store CMOS ICs with their pins in conductive foam or covered with aluminum foil.

12. Power OFF. Insert the 74HC08 IC into the mounting board.
13. Power OFF. Connect power wires to the IC: Use red wire for +5 V and black wire for GND.
14. Power OFF. Wire the entire circuit (input switches, IC, and LED indicator-light assembly).
15. Power ON. Operate and observe the action of the five-input AND circuit. Show your instructor your design and operating circuit. Be prepared to answer questions about the circuit.
16. Power OFF. Take down the circuit, and return all equipment to its proper place.

QUESTIONS

Complete questions 1 to 10.

1. Draw a single logic symbol for a three-input AND gate. Label the inputs
   A, B, and C; label the output Y.

2. Draw a logic symbol diagram of a four-input AND gate using 3 two-input
   AND gates.

3. In this experiment a LOW voltage at the input switch stood for a
   a. Logical 0
   b. Logical 1

4. In this experiment a HIGH voltage (near +5 V) stood for a
   a. Logical 0
   b. Logical 1

5. Draw a truth table for a four-input AND gate. Label the inputs A, B, C,
   and D; label the output Y.

6. When powering the IC in this experiment, the V_{CC} pin is connected to
   the _____ of the power supply.

7. The AND gate's unique output is a _____ (0, 1) which only
   occurs when all inputs are _____ (HIGH, LOW).

8. The 7408 is a TTL IC, while the 74HC08 uses _____ (APL,
   CMOS) technology in its design and manufacture.

9. CMOS ICs are sensitive to static electricity and should be handled carefully and stored in static-proof containers or in _____ (conductive foam, water).

10. Draw a logic symbol diagram using 2 two-input NAND gates to perform
    the two-input AND function. (*HINT*: Refer to textbook, Fig. 3-38.)

3. _____

4. _____

6. _____

7. _____
   _____

8. _____

9. _____

3-2 LAB EXPERIMENT: OR GATES

OBJECTIVES

1. To wire and operate a two-input OR gate.
2. To wire and operate a five-input OR gate using a 7432 IC.
3. To design and implement a logic circuit that will perform the two-input OR function using NAND gates.

MATERIALS

Qty.   **Qty.**

1   7400 quad two-input NAND gate   5   logic switches
1   7432 two-input OR gate IC   1   5-V dc regulated power supply
1   LED indicator-light assembly

SYSTEM DIAGRAMS

Figures 3-9 and 3-10 are the electronic OR gates you will wire and test in this experiment.

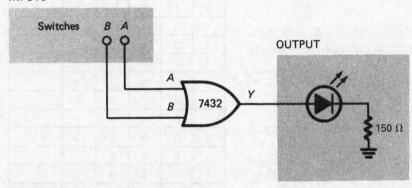

Fig. 3-9 Wiring a two-input OR gate.

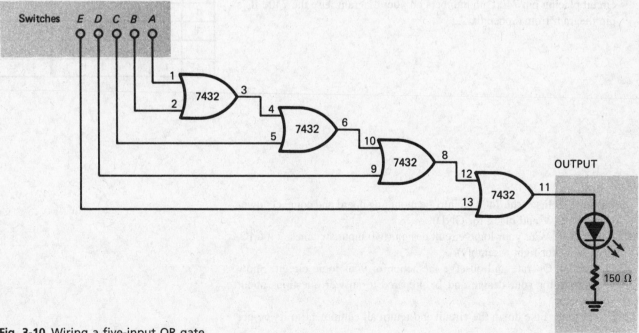

Fig. 3-10 Wiring a five-input OR gate.

PROCEDURE

1. Insert the 7432 IC into the mounting board.
2. Power OFF. Connect power to the 7432 IC: the red wire for $+5$ V (V_{CC}) and the black wire for GND.
3. Refer to Fig. 3-9. Wire the two-input OR gate (input switches, 7432 IC, and LED indicator-light assembly). See pin diagrams in Appendix A.
4. Power ON. Move input switches A and B to each combination shown in the left side of the truth table in Table 3-4. Observe and record the results in the OR gate 1 column. Record a logical 0 if the light is OFF. Record a logical 1 if the light is ON.

TABLE 3-4 Truth Table for 7432 IC

INPUTS		OUTPUTS			
A	B	OR gate 1	OR gate 2	OR gate 3	OR gate 4
0	0				
0	1				
1	0				
1	1				

5. Power OFF. Now wire and test each of the other three OR gates packaged inside the 7432 IC. Record their outputs in the right-hand columns of Table 3-4. Record 0s and 1s.
6. Power OFF. Refer to Fig. 3-10. Wire the five-input OR gate using all 4 two-input OR gates in the 7432 IC. Pin numbers are shown in Fig. 3-10.
7. Power ON. Operate and observe the output of the five-input OR gate as you place the input switches (A, B, C, D, and E) in the 32 different combinations in the Inputs column, Table 3-5. Record the outputs in the Output column, Table 3-5.
8. Using a single 7400 quad two-input NAND gate IC, *design a logic circuit* that will perform the two-input OR function ($A + B = Y$). *HINT*: Refer to the textbook for using NANDs as "universal" gates. Draw the logic circuit placing pin 7400 pin numbers on your diagram. Use the 7400 IC's pin diagram from Appendix A.

9. Power OFF. Insert the 7400 IC into the mounting board and connect power (red for $+5$ V and black for GND).
10. Power OFF. Wire your logic circuit design (two input switches, 7400 IC, LED indicator-light assembly).
11. Power ON. Operate and observe the action of your logic circuit. Show your instructor your design and be prepared to answer questions about your circuit.
12. Power OFF. Take down the circuit and return all equipment to its proper place.

TABLE 3-5 Five-Input OR Gate

INPUTS					OUTPUT
A	B	C	D	E	
0	0	0	0	0	
0	0	0	0	1	
0	0	0	1	0	
0	0	0	1	1	
0	0	1	0	0	
0	0	1	0	1	
0	0	1	1	0	
0	0	1	1	1	
0	1	0	0	0	
0	1	0	0	1	
0	1	0	1	0	
0	1	0	1	1	
0	1	1	0	0	
0	1	1	0	1	
0	1	1	1	0	
0	1	1	1	1	
1	0	0	0	0	
1	0	0	0	1	
1	0	0	1	0	
1	0	0	1	1	
1	0	1	0	0	
1	0	1	0	1	
1	0	1	1	0	
1	0	1	1	1	
1	1	0	0	0	
1	1	0	0	1	
1	1	0	1	0	
1	1	0	1	1	
1	1	1	0	0	
1	1	1	0	1	
1	1	1	1	0	
1	1	1	1	1	

QUESTIONS

Complete questions 1 to 12.

1. Draw a single logic symbol for the five-input OR gate you wired in this experiment. Label the inputs *A*, *B*, *C*, *D*, and *E*; label the output *Y*.

2. Draw a logic diagram of a three-input OR gate using 2 two-input OR gates. Label the inputs *A*, *B*, and *C* and the output *Y*.

3. When powering the IC in this experiment, the GND pin is connected to the _____ of the power supply.

   3. _____

4. A logical 0 on the truth table in this experiment means that input or output is
   a. Near GND voltage
   b. Near +5 V

   4. _____

5. A logical 1 on the truth table in this experiment means that input or output is
   a. Near GND voltage
   b. Near +5 V

   5. _____

6. A truth table with two inputs has how many switch combinations?

   6. _____

7. A truth table with three inputs has how many switch combinations?

   7. _____

8. A truth table with five inputs has how many switch combinations?

   8. _____

9. The OR gate's unique output is a _____ (0, 1), which only occurs when all inputs are _____ (HIGH, LOW).

   9. _____ _____

10. The 7432 is described by the manufacturer as a quadruple two-input OR gate from the _____ (CMOS, TTL) family of digital ICs.

    10. _____

11. The logic circuit shown in Fig. 3-11(*a*) performs the three-input _____ (AND, OR) logic function.

    11. _____

12. The logic circuit drawn in Fig. 3-11(*b*) performs the two-input _____ (NAND, OR) logic function.

    12. _____

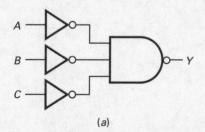

(a)

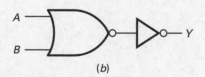

(b)

Fig. 3-11 Logic circuit.

3-3 LAB EXPERIMENT: INVERTERS

OBJECTIVES

1. To wire and operate an inverter.
2. To wire and test all six inverters in a 7404 IC.

MATERIALS

Qty.

1 7404 hex inverter IC
6 LED indicator-light assemblies

Qty.

1 logic switch
1 5-V dc regulated power supply

SYSTEM DIAGRAMS

Figures 3-12 and 3-13.

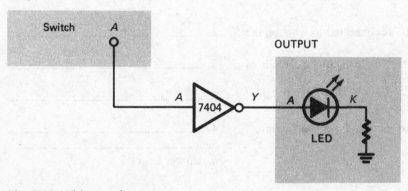

Fig. 3-12 Wiring an inverter.

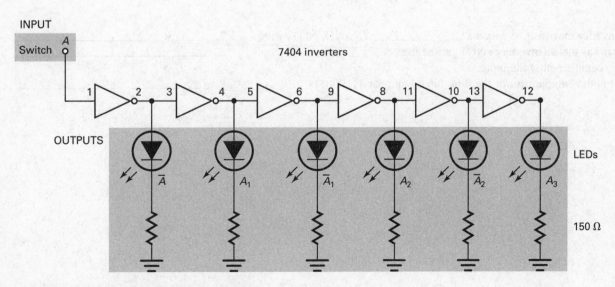

Fig. 3-13 Testing all inverters in the 7404 IC.

PROCEDURE

1. Insert a 7404 IC into the mounting board.
2. Power OFF. Connect power to the 7404 IC: the red wire for +5 V (V_{CC}) and the black wire for GND.
3. Refer to Fig. 3-12. Wire the inverter (input switch, 7404 IC, and LED indicator-light assembly). Refer to pin diagrams in Appendix A.
4. Power ON. Move the input switch to the positions shown in the Input column, Table 3-6. Observe and record the results in the Output column, Table 3-6.
5. Power OFF. Wire the circuit diagramed in Fig. 3-13 to test all six inverters in the 7404 IC package. Pin numbers are shown in Fig. 3-13.
6. Power ON. Operate the input switch, and observe the results on the six output LED indicators. Record the results in Table 3-7.
7. Power OFF. Look over the results in Table 3-7, and decide if the six inverters in the 7404 IC are working properly. Contact your instructor if you find a faulty inverter.
8. Take down the circuit, and return all equipment to its proper place.

TABLE 3-6 Truth Table—Inverters

INPUT	OUTPUT
A	$\overline{A}$
0	
1	

TABLE 3-7 Test Results—7404 IC

INPUT	OUTPUTS					
A	$\overline{A}$	A_1	$\overline{A}_1$	A_2	$\overline{A}_2$	A_3
0						
1						

QUESTIONS

Complete questions 1 to 9.

1. The power connection (V_{CC}) on the 7404 IC is connected to what on the power supply?
2. An input of near +5 V on a 7404 inverter will produce an output of _____ volts from the inverter.
3. The output of an inverter will be a logical _____, with a logical 1 input.
4. If we invert a signal twice, what output do we get from the second inverter?
5. The Boolean expression $\overline{A}$ means what?
6. Draw two logic symbols for an inverter.

7. An inverter may also be called a _____ (NAND, NOT) gate.
8. We can say that an inverter or NOT gate negates or _____ (compares, complements) the input.
9. What is the complement of the 8-bit binary number 11100011?

1. _____
2. _____
3. _____
4. _____
5. _____
6. answer below

7. _____
8. _____
9. _____

3-4 LAB EXPERIMENT: NAND AND NOR GATES

OBJECTIVES

1. To wire and operate a two-input NAND gate.
2. To wire and operate a two-input NOR gate.
3. To construct a NAND gate using an AND gate and an inverter.
4. To design a gating circuit that will perform the NOR function using a 7432 IC and a 7404 IC.

MATERIALS

Qty.

1 7400 two-input NAND gate IC
1 7404 inverter IC
1 7432 two-input OR gate IC
2 LED indicator-light assemblies

Qty.

1 7402 two-input NOR gate IC
1 7408 two-input AND gate IC
3 logic switches
1 5-V dc regulated power supply

SYSTEM DIAGRAMS

Figures 3-14, 3-15, and 3-16.

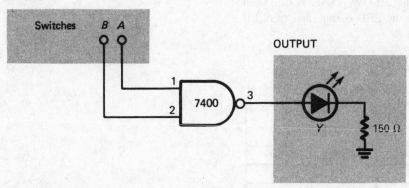

Fig. 3-14 Wiring a two-input NAND gate.

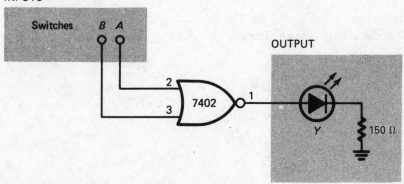

Fig. 3-15 Wiring a two-input NOR gate.

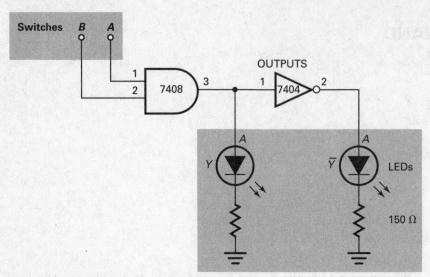

Fig. 3-16 Wiring a two-input NAND gate.

PROCEDURE

1. Insert a 7400 IC into the mounting board.
2. Power OFF. Connect power to the 7400 IC: the red wire for +5 V (V_{CC}) and the black wire for GND. Refer to pin numbers in Appendix A.
3. Refer to Fig. 3-14. Wire the two-input NAND gate (input switches, 7400 IC, and LED indicator-light assembly). Pin numbers are shown.
4. Power ON. Move the input switches A and B to the combinations shown in the left side of Table 3-8. Remember that a LOW voltage is GND and a HIGH voltagxe is near +5 V. Observe the LED output, and record if the LED is ON or OFF.

TABLE 3-8   Truth Table for 7400 IC

INPUTS				OUTPUT	
A		B		Y	
Voltage	Binary	Voltage	Binary	Light	Binary
LOW	0	LOW	0		
LOW	0	HIGH	1		
HIGH	1	LOW	0		
HIGH	1	HIGH	1		

5. Power OFF. Under the Output section fill in the Binary column, Table 3-8, with 0s and 1s.
6. Remove the 7400 IC from the mounting board.
7. Insert the 7402 IC into the mounting board, and connect power to the IC. Refer to pin diagrams in Appendix A.
8. Refer to Fig. 3-15. Wire the two-input NOR gate (input switches, 7402 IC, and LED indicator-light assembly). Pin numbers are shown.

42

9. Power ON. Move the input switches *A* and *B* to all the four combinations in the left side of the truth table in Table 3-9. Observe and record the output results in the right column, Table 3-9.

10. Power OFF. Remove the 7402 IC from the mounting board.

11. Insert the 7408 and 7404 ICs into the mounting board. Connect power to both ICs. Refer to pin numbers in Appendix A.

12. Wire the circuit diagramed in Fig. 3-16. Notice the use of two LED output indicators. Pin numbers are shown in Fig. 3-16.

13. Power ON. Move the input switches *A* and *B* to the four combinations in the left side of the truth table in Table 3-10. Observe the outputs and record in columns *Y* and $\overline{Y}$, Table 3-10.

14. Power OFF. Column *Y*, Table 3-10, should be the output of an AND gate, and column $\overline{Y}$ should be a NAND function.

15. Draw a logic symbol diagram of a three-input NOR gate. Use 2 two-input OR gates and an inverter. Label the inputs *A*, *B*, and *C*, and the output *Y*.

16. Power OFF. Construct the three-input NOR gate you just designed. Wire the 7432 and 7404 ICs, three input switches, and output LED indicator-light assembly. Refer to pin diagrams in Appendix A.

17. Power ON. Move the input switches *A*, *B*, and *C* to the eight combinations in the left side of the truth table in Table 3-11. Observe the output and record it in column *Y*, Table 3-11.

18. Power OFF. Take down the circuit, and return all equipment to its proper place.

TABLE 3-9 Truth Table for 7402 IC

INPUTS		OUTPUT
A	*B*	*Y*
0	0	
0	1	
1	0	
1	1	

TABLE 3-10 Truth Table for AND and NAND Gates

INPUTS		OUTPUTS	
A	*B*	*Y*	$\overline{Y}$
0	0		
0	1		
1	0		
1	1		

TABLE 3-11 Truth Table for three-input NOR Gate

INPUTS			OUTPUT
A	*B*	*C*	*Y*
0	0	0	
0	0	1	
0	1	0	
0	1	1	
1	0	0	
1	0	1	
1	1	0	
1	1	1	

QUESTIONS

Complete questions 1 to 8.

1. Write the Boolean expression for each of the following circuits you constructed in this experiment:
   a. Two-input NAND gate
   b. Two-input NOR gate
   c. Three-input NOR gate

2. When the indicator LED is ON in this experiment, it means the gate has an output of
   a. Binary 0
   b. Binary 1

1. a. _____
   b. _____
   c. _____

2. _____

3. The NAND gate's unique output is a _____ (0, 1), which only occurs when all inputs are _____ (HIGH, LOW).

4. The NOR gate's unique output is a _____ (0, 1), which only occurs when all inputs are _____ (HIGH, LOW).

5. The NAND function can be created by inverting the output of a(n) _____ gate.

6. The NOR function can be created by inverting the output of a(n) _____ gate.

7. The logic circuit shown in Fig. 3-17(a) performs the two-input _____ (AND, NAND, NOR) logic function.

8. The logic circuit drawn in Fig. 3-17(b) performs the two-input _____ (NAND, NOR, XOR) logic function.

3. _____ _____

4. _____ _____

5. _____

6. _____

7. _____

8. _____

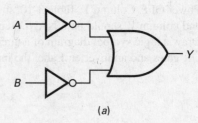

(a)

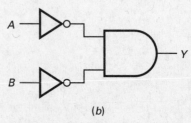

(b)

Fig. 3-17 Logic circuit for questions 7 and 8.

3-5 LAB EXPERIMENT: XOR AND XNOR GATES

OBJECTIVES

1. To wire and operate an exclusive OR (XOR) gate.
2. To construct and operate an exclusive NOR (XNOR) gate using a 7486 IC and a 7404 IC.
3. To observe that the XOR gate generates a HIGH when an odd number of inputs are HIGH.
4. To observe that the XNOR gate generates a LOW when an odd number of inputs are HIGH.

MATERIALS

Qty.

1 7404 hex inverter IC
3 logic switches
1 5-V dc regulated power supply

Qty.

1 7486 two-input exclusive OR gate IC
1 LED indicator-light assembly

SYSTEM DIAGRAMS

Figures 3-18, 3-19, 3-20, and 3-21 are on the next page.

PROCEDURE

1. Insert a 7486 IC into the mounting board.
2. Power OFF. Connect power to the 7486 IC: the red wire for +5 V (V_{CC}) and the black wire for GND.
3. Wire the two-input exclusive OR gate shown in Fig. 3-18. Wire input switches A and B, the 7486 IC, and the output LED indicator-light assembly. Pin numbers are shown in Fig. 3-18.
4. Power ON. Move the input switches to the positions shown in the left side of the truth table in Table 3-12. Observe and record the output results in the XOR column, Table 3-12.

TABLE 3-12   Truth Table for XOR and XNOR Gates

INPUTS		OUTPUTS	
		Step 4	Step 6
A	B	XOR	XNOR
0	0		
0	1		
1	0		
1	1		

5. Power OFF. Construct the two-input exclusive NOR circuit diagrammed in Fig. 3-19. Wire the input switches A and B, the 7486 and 7404 ICs, and the LED output indicator-light assembly.
6. Power ON. Place the input switches A and B in the positions shown in the left side of the truth table in Table 3-12. Observe and record the output of the gate in the XNOR column, Table 3-12.
7. Power OFF. Construct the three-input exclusive OR gate diagrammed in Fig. 3-20.

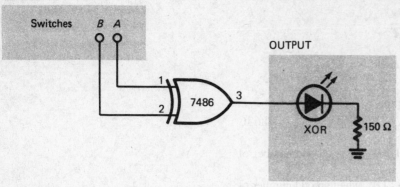

Fig. 3-18 Wiring a two-input exclusive OR gate.

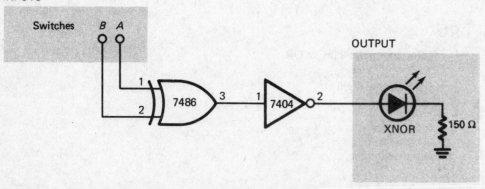

Fig. 3-19 Wiring a circuit to perform the two-input exclusive NOR (XNOR) function.

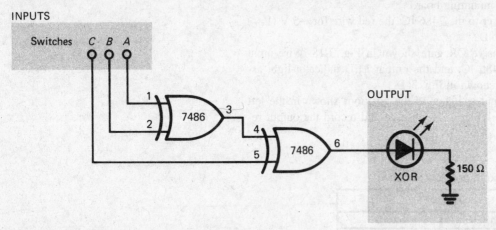

Fig. 3-20 Wiring a three-input exclusive OR gate.

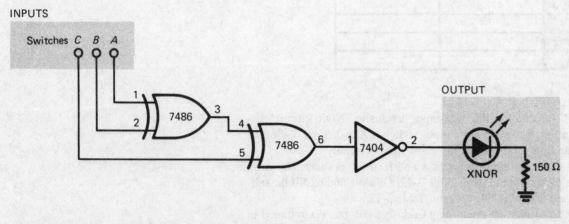

Fig. 3-21 Wiring a circuit to perform the three-input XNOR function.

8. Power ON. Place input switches *A*, *B*, and *C* in the positions shown in the left side of the truth table in Table 3-13. Observe and record the output of the circuit in the XOR column, Table 3-13.

9. Power OFF. Construct the three-input exclusive NOR circuit diagrammed in Fig. 3-21.

10. Power ON. Place input switches *A*, *B*, and *C* in the positions shown in the left side of the truth table in Table 3-13. Observe and record the output of the circuit in the XNOR column, Table 3-13.

TABLE 3-13 Truth Table for three-input XOR and XNOR Gates

INPUTS			OUTPUTS	
			Step 8	Step 10
A	*B*	*C*	XOR	XNOR
0	0	0		
0	0	1		
0	1	0		
0	1	1		
1	0	0		
1	0	1		
1	1	0		
1	1	1		

11. Power OFF. Take down the circuit, and return all equipment to its proper place.

QUESTIONS

Complete questions 1 to 9.

1. Write the Boolean expression for each of the following circuits you constructed in this experiment:
   a. Two-input XOR gate
   b. Two-input XNOR gate
   c. Three-input XOR gate
   d. Three-input XNOR gate

2. The XOR gate's output is a _____ (0, 1) when an odd number of inputs are HIGH.

3. The XNOR gate's output is a _____ (0, 1) when an odd number of inputs are HIGH.

4. If 11110000 is fed into an eight-input XOR gate, the output will be _____ (HIGH, LOW).

5. If 11110001 is fed into an eight-input XNOR gate, the output will be _____ (HIGH, LOW).

6. If 00011111 is fed into an eight-input XOR gate, the output will be _____ (HIGH, LOW).

7. If 1010 is fed into a four-input XNOR gate, the output will be _____ (HIGH, LOW).

1. a. _____
   b. _____
   c. _____
   d. _____

2. _____

3. _____

4. _____

5. _____

6. _____

7. _____

8. Refer to Fig. 3-22. Complete the XOR output column describing the four-input XOR logic function.

9. Refer to Fig. 3-22. Complete the XNOR output column describing the four-input XNOR logic function.

INPUTS				OUTPUTS	
A	B	C	D	XOR	XNOR
0	0	0	0		
0	0	0	1		
0	0	1	0		
0	0	1	1		
0	1	0	0		
0	1	0	1		
0	1	1	0		
0	1	1	1		
1	0	0	0		
1	0	0	1		
1	0	1	0		
1	0	1	1		
1	1	0	0		
1	1	0	1		
1	1	1	0		
1	1	1	1		

Fig. 3-22 Truth table for four-input XOR and XNOR gates.

3-6 LAB EXPERIMENT: USING THE NAND GATE

OBJECTIVE

To use the two-input NAND gate as a universal gate in constructing the following logic functions:

    a. Two-input AND gate
    b. Two-input OR gate
    c. Two-input NOR gate
    d. Two-input XOR gate

MATERIALS

Qty.

1   7400 two-input NAND gate IC
1   LED indicator-light assembly

Qty.

2   logic switches
1   5-V dc regulated power supply

PROCEDURE

1. Insert a 7400 IC into the mounting board.
2. Power OFF. Connect power to the 7400 IC: the red wire for +5 V (V_{CC}) and the black wire for GND.
3. Construct a two-input AND gate using 2 two-input NAND gates. See your textbook for assistance.
4. Wire the input switches A and B, a 7400 IC, and the LED output indicator-light assembly.
5. Power ON. Move the input switches A and B to the positions shown in the input part of Table 3-14. Observe and record the output in the AND gate column, Table 3-14.

TABLE **3-14**   Truth Tables

INPUTS		OUTPUTS			
		Step 5	Step 8	Step 11	Step 13
A	**B**	AND gate	OR gate	NOR gate	XOR gate
0	0				
0	1				
1	0				
1	1				

6. Power OFF. Rewire the 7400 IC to create a two-input OR gate. See the textbook.
7. Wire two input switches A and B, three sections of the 7400 IC, and the LED output indicator-light assembly.
8. Power ON. Operate the input switches A and B as shown in the input side of Table 3-14. Observe and record the output in the OR gate column.
9. Power OFF. Rewire the 7400 IC to produce a two-input NOR gate. See the textbook.
10. Wire two input switches A and B, all four sections of the 7400 IC, and the LED output indicator-light assembly.
11. Power ON. Operate the input switches A and B as shown in the input side of Table 3-14. Observe and record the output in the NOR gate column.
12. Power OFF. Rewire the 7400 IC to produce a two-input XOR gate. See the textbook.

13. Power ON. Operate input switches *A* and *B* as shown in the input side of Table 3-14. Observe and record the output in the XOR gate column.
14. Power OFF. Take down the circuit, and return all equipment to its proper place.

QUESTIONS

Complete questions 1 to 3.

1. Write the Boolean expression for each of the logic functions you constructed from NAND gates in this experiment:
   a. Two-input AND
   b. Two-input OR
   c. Two-input NOR
   d. Two-input XOR

   1. a. _____
      b. _____
      c. _____
      d. _____

2. Draw a logic symbol diagram of the following using only two-input NAND gates:
   a. Two-input AND
   b. Two-input OR
   c. Two-input NOR
   d. Two-input XOR

   2. answer below

3. Which XOR gate has fewer connections and therefore greater reliability?
   a. 7486 IC
   b. 7400 IC wired to perform the XOR function

   3. _____

3-7 DESIGN PROBLEM: GATES WITH MORE THAN TWO INPUTS

OBJECTIVES

1. To design, draw, and wire a three-input AND circuit using two-input AND gates.
2. To design, draw, and wire a four-input OR circuit using two-input OR gates.
3. To design, draw, and wire a five-input NAND circuit using 1 two-input OR gate, 1 two-input NAND gate, and 1 four-input NAND gate.
4. To design, draw, and wire a four-input XOR circuit using 3 two-input XOR gates.

MATERIALS

Qty.		Qty.	
1	7400 two-input NAND gate IC	1	7408 two-input AND gate IC
1	7420 four-input NAND gate IC	1	7432 two-input OR gate IC
5	logic switches	1	7486 two-input XOR gate IC
1	5-V dc regulated power supply	1	LED indicator-light assembly

SYSTEM DIAGRAM

Figure 3-23.

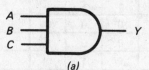

(a)

PROCEDURE

1. Design and draw a logic symbol diagram of a three-input AND gate using 2 two-input AND gates.

(b)

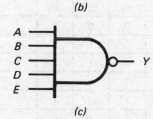

(c)

2. Insert the 7408 IC into the mounting board.
3. Power OFF. Connect power to the 7408 IC (V_{CC} and GND).
4. Construct the circuit you designed in step 1. Wire the input switches A, B, and C; the 7408 IC; and the LED output indicator-light assembly. Refer to pin diagrams in Appendix A.
5. Power ON. Operate the input switches A, B, and C as shown in the input section of the truth table in Table 3-15. Observe and record the output in the three-input AND column.
6. Power OFF. Remove the 7408 IC.
7. Design and draw a logic symbol diagram of a four-input OR gate using only two-input OR gates.

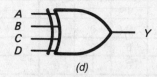

(d)

Fig. 3-23 Logic symbols for (a) three-input AND, (b) four-input OR, (c) five-input NAND, and (d) four-input XOR gates.

8. Insert a 7432 IC into the mounting board. Connect power to the IC (V_{CC} and GND).
9. Construct the circuit you designed in step 7. Wire input switches *A*, *B*, *C*, and *D*; the 7432 IC; and the LED output indicator-light assembly.
10. Power ON. Operate the input switches *A*, *B*, *C*, and *D* according to the input part of the truth table in Table 3-15. Observe and record the output in the four-input OR column.
11. Power OFF.

TABLE 3-15 Truth Tables

INPUTS					OUTPUT		
E	D	C	B	A	5-input NAND gate	4-input OR gate	3-input AND gate
0	0	0	0	0			
0	0	0	0	1			
0	0	0	1	0			
0	0	0	1	1			
0	0	1	0	0			
0	0	1	0	1			
0	0	1	1	0			
0	0	1	1	1			
0	1	0	0	0			
0	1	0	0	1			
0	1	0	1	0			
0	1	0	1	1			
0	1	1	0	0			
0	1	1	0	1			
0	1	1	1	0			
0	1	1	1	1			
1	0	0	0	0			
1	0	0	0	1			
1	0	0	1	0			
1	0	0	1	1			
1	0	1	0	0			
1	0	1	0	1			
1	0	1	1	0			
1	0	1	1	1			
1	1	0	0	0			
1	1	0	0	1			
1	1	0	1	0			
1	1	0	1	1			
1	1	1	0	0			
1	1	1	0	1			
1	1	1	1	0			
1	1	1	1	1			

12. Design and draw a logic symbol diagram of a five-input NAND gate using 7400, 7420, and 7432 ICs. Refer to pin diagrams in Appendix A.

13. Add a 7400 and a 7420 IC to the mounting board. Connect power to each IC (V_{CC} and GND).

14. Construct the circuit you designed in step 12. Wire the input switches A, B, C, D, and E; the 7400, 7420, and 7432 ICs; and the LED output indicator-light assembly.

15. Power ON. Operate the input switches A, B, C, D, and E according to the input part of the truth table in Table 3-15. Observe and record the output in the five-input NAND column.

16. Design and draw a logic symbol diagram of a four-input XOR gate using 3 two-input XOR gates.

17. Construct the circuit you designed in step 16. Wire four input switches, the 7486 quad two-input XOR gate IC, and the LED output indicator-light assembly.

18. Power ON. Operate and observe the operation of the four-input XOR circuit. Show your instructor your design. Demonstrate the operation of the circuit and be prepared to answer questions on the XOR gate.

19. Power OFF. Take down the circuit and return all equipment to its proper place.

QUESTIONS

Complete questions 1 to 7.

1. Write the Boolean expression for each of the following gates constructed in this experiment:
   a. Three-input AND
   b. Four-input OR
   c. Five-input NAND
   d. Four-input XOR

1. a. _____
   b. _____
   c. _____
   d. _____

2. Draw a logic symbol diagram of a five-input AND gate using two-input AND gates.

3. The _____ (NAND, OR) gate's unique output is a LOW, which only occurs when all inputs are HIGH.

4. The _____ (AND, OR) gate's unique output is a HIGH, which only occurs when all inputs are HIGH.

5. The _____ (AND, OR) gate's unique output is a LOW, which only occurs when all inputs are LOW.

6. The XOR gate's output is _____ (HIGH, LOW) only when an odd number of inputs are HIGH.

7. Draw a logic symbol diagram of how 7420 and 7432 ICs would be wired to produce a seven-input NAND function. Show seven inputs and a LED output indicator with a limiting resistor.

3. _____

4. _____

5. _____

6. _____

3-8 DESIGN PROBLEM: CONVERTING GATES TO OTHER LOGIC FUNCTIONS

OBJECTIVES

1. To design, draw, and wire the following logic functions using a 7404 IC and a 7408 IC:
   a. Two-input NAND gate
   b. Two-input NOR gate
   c. Two-input OR gate
2. To design, draw, and wire the following logic functions using a 7404 IC and a 7432 IC:
   a. Two-input NOR gate
   b. Two-input NAND gate
   c. Two-input AND gate
3. To design, draw, and wire a four-input NAND gate using a 7404 IC and a 7432 IC.

MATERIALS

Qty.		Qty.	
1	7404 inverter IC	1	7408 two-input AND gate IC
1	7432 two-input OR gate IC	4	logic switches
1	LED indicator-light assembly	1	5-V dc regulated power supply

PROCEDURE

1. Design and draw a logic symbol diagram for each of the following functions (use only two-input AND gates and inverters):
   a. Two-input NAND gates

   b. Two-input NOR gate

   c. Two-input OR gate

2. Insert the 7404 and 7408 ICs into the mounting board.
3. Power OFF. Connect power to both ICs (V_{CC} and GND).
4. Construct each circuit you designed. Use two input switches, the 7404 and 7408 ICs, and the LED output indicator-light assembly.
5. Power ON. Operate the input switches for each circuit as shown in the left part of the truth table in Table 3-16.
6. Observe and record the output of each circuit in the output section of Table 3-16.
7. Power OFF. Remove the 7408 IC from the mounting board.

TABLE 3-16   Truth Tables

INPUTS		OUTPUTS		
A	B	Circuit 1(a) NAND gate	Circuit 1(b) NOR gate	Circuit 1(c) OR gate
0	0			
0	1			
1	0			
1	1			

8. Design and draw a logic symbol diagram for each of the following functions (use only two-input OR gates and inverters):

   a. Two-input NOR gate

   b. Two-input NAND gate

   c. Two-input AND gate

9. Insert a 7432 IC into the mounting board, and connect power to the IC (V_{CC} and GND).

10. Construct each circuit you designed in step 8. Use two input switches, the 7404 and 7432 ICs, and the LED output indicator-light assembly.

11. Power ON. Operate the input switches A and B for each circuit as shown on the left side of the truth table in Table 3-17.

12. Observe and record the output of each circuit in the output section of Table 3-17.

TABLE 3-17   Truth Tables

INPUTS		OUTPUTS		
A	B	Circuit 8(a) NOR gate	Circuit 8(b) NAND gate	Circuit 8(c) AND gate
0	0			
0	1			
1	0			
1	1			

56

13. Design and draw a logic symbol diagram of a four-input NAND gate using only a 7404 IC and a 7432 IC. Refer to pin diagrams in Appendix A.

TABLE 3-18 Truth Table for NAND Function

INPUTS				OUTPUT
A	B	C	D	Y
0	0	0	0	
0	0	0	1	
0	0	1	0	
0	0	1	1	
0	1	0	0	
0	1	0	1	
0	1	1	0	
0	1	1	1	
1	0	0	0	
1	0	0	1	
1	0	1	0	
1	0	1	1	
1	1	0	0	
1	1	0	1	
1	1	1	0	
1	1	1	1	

14. Power OFF. Construct the four-input NAND gate you designed in step 13. Use four input switches, the 7404 and 7432 ICs, and the LED output indicator-light assembly.

15. Power ON. Operate the input switches *A*, *B*, *C*, and *D* as shown in the left side of Table 3-18. Observe and record the output in the *Y* column.

16. Have your instructor approve your designs and last circuit.

17. Power OFF. Take down the circuit, and return all equipment to its proper place.

QUESTIONS

Complete questions 1 to 8.

1. Draw a *single* logic symbol for each circuit you set up in this experiment:

a. Two-input NAND gate

b. Two-input NOR gate

c. Two-input OR gate

d. Two-input AND gate

e. Four-input NAND gate

2. Write a Boolean expression for each of the five gates (**a** to **e**) in question 1.

2. a. _____
b. _____
c. _____
d. _____
e. _____

3. Draw an alternative logic symbol sometimes used to show a NAND gate (has inverted inputs).

4. Draw an alternative logic symbol sometimes used to show a NOR gate (has inverted inputs).

5. Draw a logic symbol diagram showing how you would connect a two-input NAND gate and inverters to produce an OR function.

6. Draw a logic symbol diagram showing how you would connect a two-input NAND gate and inverters to produce a NOR function.

7. Draw a logic symbol diagram showing how you would connect a four-input NAND gate and inverters to produce the four-input AND function.

8. Draw a logic symbol diagram showing how you would connect only two-input NAND gates to produce a four-input OR function.

3-9 TROUBLESHOOTING PROBLEM: TESTING LOGIC LEVELS IN A CMOS TIMER CIRCUIT

OBJECTIVES

1. To wire and test the operation of a timer circuit using CMOS ICs.
2. To test the logic levels in the control gating circuitry of the timer using a CMOS logic probe.
3. To introduce a fault into the timer and detect the problem using logic probe readings.

MATERIALS

Qty.		Qty.	
1	74HC08 two-input AND gate CMOS IC	1	keypad (N.O. contacts)
1	74HC14 inverter CMOS IC	8	LED indicator-light assemblies
1	74HC393 binary counter CMOS IC	2	10-kΩ, ¼-W resistor
1	logic switch	1	5-V dc regulated power supply
1	free-running clock	1	logic probe

SYSTEM DIAGRAM

Troubleshooting requires that the technician know the *normal* operation of the circuit. Normal operation from the technician's or engineer's view includes various inputs and expected output(s). It also includes expected normal measurements of temperature, voltage/currents, and waveforms at various points in the circuit. Technical manuals and experience greatly aid in efficient troubleshooting. Besides technical knowledge of the circuit and testing, a technician's or engineer's powers of observation are important in troubleshooting.

A summary of the steps in troubleshooting include (1) the use of your senses (feel, look, hear, smell), (2) the use of a logic probe to check power to ICs, (3) the use of equipment manuals to determine the job of the circuit and to test unique output conditions, and (4) the use of test instruments to test input and output conditions to determine the problem.

The circuit in Fig. 3-24 is basically a timer with an 8-bit binary output. A block diagram of the system is shown in Fig. 3-24(*a*). Closing the clear input switch momentarily zeros the output to 00000000. Closing the count switch causes the control circuitry to permit the 10-Hz clock input to enter the up counter. When the count switch opens, the control circuitry blocks the clock pulses from getting through to the counter and the timer stops. The accumulated count on the output display indicates the number of *tenths of a second* the count switch was closed.

A wiring diagram for the timer circuit is detailed in Fig. 3-24(*b*). All the logic gates and counter are CMOS ICs. When the ON/OFF switch is in the ON position and when the 1 on the keypad is depressed, the counter will measure the time (in tenths of a second) that the count key is closed. The binary output needs to be translated into decimal by you, the operator. The C on the keypad will clear the output display to 00000000. The free-running clock must be set at a frequency of 10 Hz for the timer to be accurate. The ON/OFF switch can disable the timer, but it leaves the last display showing on the eight LEDs.

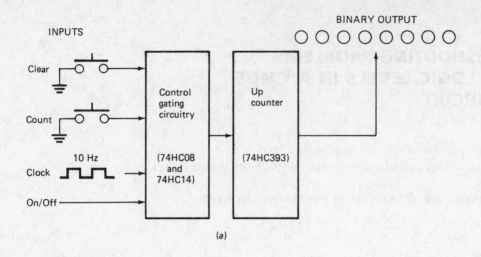

(a)

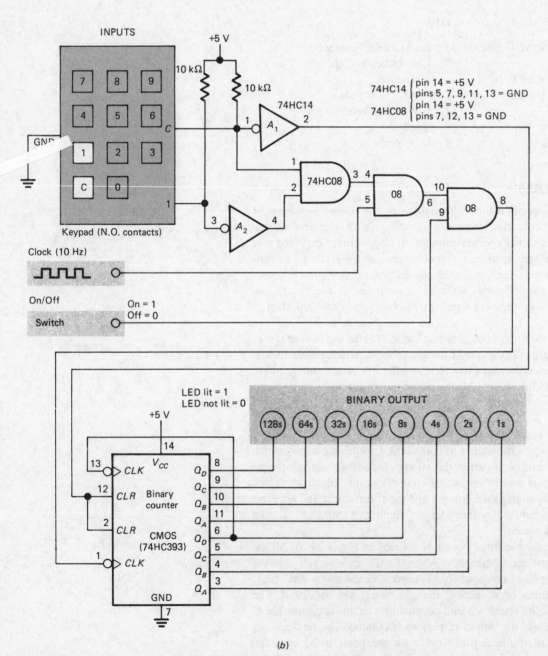

(b)

Fig. 3-24 Timer circuit using CMOS ICs. (*a*) Block diagram. (*b*) Wiring diagram.

It is suggested that the output LEDs be arranged with the LSB (1s LED) on the right and the MSB (128s LED) on the left. This will make the binary output easier to read. At this point, do not be concerned that you do not understand the exact operation of the counter IC. However, you should understand how the control circuitry operates. This is a difficult circuit at this point in the course and should be attempted only by the more advanced students.

In this problem, you will concentrate on the control gating circuitry of the timer in Fig. 3-24. You will use a logic probe to test the logic levels around the 74HC14 inverter and 74HC08 AND gate CMOS ICs.

PROCEDURE

CAUTION CMOS ICs can be damaged by static electricity. Store CMOS ICs with their pins in conductive foam or covered with aluminum foil.

1. Power OFF. Insert the 74HC08, 74HC14, and 74HC393 CMOS ICs into the mounting board.
2. Power OFF. Connect power wires to the three ICs: use red wire for +5 V and black wire for GND. Connect all unused inputs to GND [see pin numbers in Fig. 3-24(*b*)] using black wire.
3. Refer to Fig. 3-24(*b*). Power OFF. Wire the entire circuit (switch, free-running clock, keypad, ICs, and LED indicator-light assemblies). Pin numbers are shown on the wiring diagram. See your instructor if you have problems wiring this complex circuit.
4. Power ON. Check the operation of the timer using the following procedure:
   a. Move ON/OFF switch to HIGH.
   b. Press and release C on the keypad to clear display to 00000000.
   c. Press 1 on the keypad for 10 seconds. The binary display should count upward.
   d. After 10 s the display should read about decimal 100 (binary 01100100).
   e. Adjust the clock frequency to get as close to 10 Hz as possible.
   f. Move ON/OFF switch to LOW.
   g. Try pressing 1 on the keypad to use the timer. The counter is disabled and should not count upward.
5. Power OFF. Connect a logic probe to the circuit. Read the instruction manual or ask your instructor if you need help in operating the logic probe. Remember you are working with CMOS ICs.
6. Power ON. Operate the circuit for several minutes by running the timer and then clearing. Now, *touch the top* of all three ICs. Record your results on Table 3-19.
7. Power ON. Follow the testing procedure in Table 3-19 and record your results. You are testing the *normal operation* of the timer circuit. *It is important you understand the normal operation of a circuit before you can troubleshoot a faulty circuit.*

TABLE 3-19 Test Results—Normal Operation of Timer Circuit

	Test Procedure	Test Results				
		74HC08	74HC14	74HC393		
Step 1	Temperature of IC	——	——	——		WARM or COOL
Step 2	Power to ICs V_{CC} (pin 14) GND (pin 7)	—— ——	—— ——	—— ——		HIGH or LOW
Step 3	Logic levels at input devices Keypad C Keypad 1 Clock Switch (OFF position) Switch (ON position)				—— —— —— —— ——	HIGH, LOW, or pulse
Step 4	While pressing C on Keypad Outputs of 74HC393 counter Pin 1 ⎱ 74HC14 Pin 2 ⎰ Pins 2 & 12—74HC393 Pin 8—74HC08	—— 	—— ——	—— ——		HIGH or LOW
Step 5	While pressing 1 on Keypad (counting) Pin 1 ⎫ Pin 2 ⎪ Pin 5 ⎬ 74HC08 Pin 9 ⎪ Pin 8 ⎭ Pin 1—74HC393	—— —— —— —— ——		——		HIGH, LOW, or pulse
Step 6	Release 1 on Keypad (stop counting) Pin 1 ⎫ Pin 2 ⎪ Pin 5 ⎬ 74HC08 Pin 9 ⎪ Pin 8 ⎭ Pin 1 ⎱ 74HC393 Pins 2 & 12 ⎰	—— —— —— —— ——		—— ——		HIGH, LOW, or pulse

8. Power OFF. Place a jumper wire from pin 8 of the 74HC08 IC to GND. This introduces a fault into the timer system. As an alternative, your instructor may decide to introduce some unknown problem in your circuit.

9. Power ON. Run the test procedure in Table 3-20 on your faulty timer circuit. Observe and record your test results in Table 3-20.

10. Show your instructor the timer test results, and point out where the bad readings are located in Table 3-20.

11. Clear the fault (remove the jumper wire) from the timer circuit, and demonstrate operating the timer system for your instructor.

12. Power OFF. Take down the circuit, and return all equipment to its proper place. Carefully remove and properly store the CMOS ICs.

TABLE 3-20   Test Results—Faulty Timer Circuit

	Test Procedure	Test Results				
		74HC08	74HC14	74HC393		
Step 1	Temperature of IC	———	———	———		WARM or COOL
Step 2	Power to ICs V_{CC} (pin 14) GND (pin 7)	——— ———	——— ———	——— ———		HIGH or LOW
Step 3	Logic levels at input devices Keypad C Keypad 1 Clock Switch (OFF position) Switch (ON position)				——— ——— ——— ——— ———	HIGH, LOW, or pulse
Step 4	While pressing C on Keypad Outputs of 74HC393 counter Pin 1 ⎫ Pin 2 ⎬ 74HC14 Pins 2 & 12—74HC393 Pin 8—74HC08	 ———	 ——— ———	 ——— 		HIGH or LOW
Step 5	While pressing 1 on Keypad (counting) Pin 1 ⎫ Pin 2 ⎪ Pin 5 ⎬ 74HC08 Pin 9 ⎪ Pin 8 ⎭ Pin 1—74HC393	——— ——— ——— ——— ———		 ———		HIGH, LOW, or pulse
Step 6	Release 1 on Keypad (stop counting) Pin 1 ⎫ Pin 2 ⎪ Pin 5 ⎬ 74HC08 Pin 9 ⎪ Pin 8 ⎭ Pin 1 ⎫ Pins 2 & 12 ⎬ 74HC393	——— ——— ——— ——— ———		 ——— ———		HIGH, LOW, or pulse

QUESTIONS

Complete questions 1 to 11.

1. Refer to Fig. 3-24. When the switch is ON, pressing the 1 on the keypad causes the binary counter to _____.
   a. Clear the display to all 0s
   b. Count upward in binary
   c. Do nothing (counter disabled)

2. Refer to Fig. 3-24. When the switch is ON, pressing the C on the keypad causes the binary counter to _____.
   a. Clear the display to all 0s
   b. Count upward in binary
   c. Do nothing (counter disabled)

1. _____

2. _____

3. Refer to Fig. 3-24. When pins 1, 2, and 9 of the 74HC08 IC are HIGH, the clock pulses (entering pin 5) are _____ through to the counter.
   a. Permitted to pass
   b. Stopped by the gating circuit from passing

4. Refer to Fig. 3-24. When 1 is pressed on the keypad, pin 2 of the 74HC08 IC goes _____ (HIGH, LOW).

5. Refer to Fig. 3-24. When C on the keypad is *not pressed* (open switch), pin 1 of the 74HC08 IC is _____ (HIGH, LOW) because it is connected to +5 V through the 10-kΩ pull-up resistor.

6. Refer to Fig. 3-24. When 1 on the keypad is *not pressed* (open switch), pin 2 of the 74HC08 IC is _____ (HIGH, LOW).

7. Refer to Fig. 3-24. All ICs used in this timer system are (CMOS, TTL) devices.

8. Refer to Fig. 3-24. If the frequency of the clock is set at exactly 10 Hz, pressing 1 on the keypad for exactly 2 seconds will cause a binary output of _____.

9. While troubleshooting the counting circuit in Fig. 3-24(*b*), the logic probe you used for taking measurements was set for _____ (CMOS, TTL).

10. The keypad used in the counter circuit [Fig. 3-24(*b*)] consists of normally open pushbutton switches. Keypad inputs C and 1 are wired as _____ (active HIGH, active LOW) switches in this circuit and generate a LOW when pressed.

11. List four general steps in troubleshooting a digital circuit suggested in the System Diagram section of this lab.

3. _____

4. _____

5. _____

6. _____

7. _____

8. _____

9. _____

10. _____

3-10 BASIC STAMP EXPERIMENT: PROGRAMMING LOGIC FUNCTIONS

OBJECTIVES

1. To use a BASIC Stamp 2 (BS2)–based development board and personal computer to program in PBASIC several logic functions and download the program.

2. To wire input switches and LED output indicator on a BS2 development board.

3. To test the downloaded program and wired circuit to verify that it simulates a logic gate.

MATERIALS

Qty.

1	BASIC Stamp 2 development board (such as the Board of Education or HomeWork board available from Parallax)
1	BS2 module (this may be mounted on the development board)
1	PC with MS Windows
1	PBASIC editor from Parallax
1	Serial cable (for downloading to BS2 module on development board)
1	Red LED (light-emitting diode)
2	Push-button switches, N.O. contacts
1	150-Ω resistor
2	10-kΩ resistors

Note 1: BASIC Stamp 2 module, development board, PBASIC software, and both experiment and reference manuals are available from Parallax, Inc., 599 Menlo Drive, Rocklin, CA 95765. Many of the manuals and software can be downloaded free from Parallax's websites. General website: *www.parallax .com.* Telephone: 916-624-8333.

Note 2: Check with your instructor.

Older version of BS2 module. It uses a serial cable for downloading from the computer's serial port to the BS2 module.

Newer version of BS2 module. It uses a USB (universal serial bus) cable for downloading from the computer's USB port to the BS2 module.

Note 3: This lab is written for the older version of BS2 module using a serial cable.

SYSTEM DIAGRAMS

The first circuit to be wired to the BASIC Stamp 2 module (BS2 IC) is represented in Fig. 3-25. The BS2 module is functioning as a *two-input logic gate.* Input switches B and A are wired as active HIGH push-buttons connected to ports 11 and 12 (*P11* and *P12*) of the BS2 module. An *active HIGH input* switch means that *P11* is normally LOW but goes HIGH when push-button switch SW_2 is pressed or closed, connecting it directly to V_{dd} or (+) of the power supply. The PBASIC program downloaded into the BS2 IC must configure *P11* and *P12* as inputs. One red LED with limiting resistor is connected between port 1 (*P1*) of the BS2 module and V_{ss} (−) of the power supply. Port 1 (*P1*) of the BS2 module must be defined in the PBASIC program as an output. The red LED will only light when a HIGH appears at port 1 of the BS2 IC.

Programming is accomplished on a MS Windows–based personal computer using the PBASIC editor for the BS2 BASIC Stamp. The program is downloaded using a serial cable between the PC's serial port and the BASIC

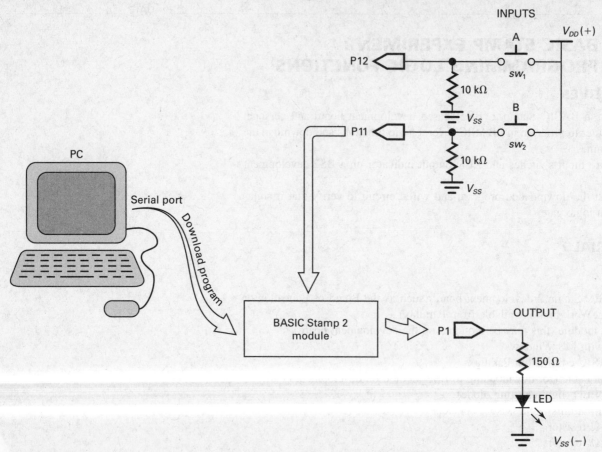

Fig. 3-25 BASIC Stamp 2 module programmed to generate two-input logic functions.

Stamp 2 development board. The BS2 module stores the program in EEPROM memory. The PBASIC program is interpreted and executed by the microcontroller on the BS2 module.

The following PBASIC program will cause the BS2 module to operate as a *two-input AND gate*. The first line of the program (**'2-input AND function**) is the title of the program. The apostrophe symbol (') at the beginning of this line identifies it as a remark statement. Remark statements are not executed by the BS2 IC but are displayed in the PBASIC listing to aid human understanding of the program. Lines 2 through 7 are statements used to declare the variables used and configure input and outputs. As an example, line 2 (**A VAR Bit 'Declare A as a variable, 1 bit**) tells the microcontroller that a variable called A will be used in the program and it will only be 1 bit long. The remark section of line 2 helps clarify the PBASIC statement in more human terms.

The **'2-input AND function** program contains a continuous program loop starting with a label (**Ckswitch:**). Labels are identified by names ending with a colon (:). Labels are places in a program that other lines of code can easily locate and jump to. The label (**Ckswitch:**) shows the beginning of the main routine, while the second label (**redLED:**) identifies the start of the subroutine, that lights the red LED.

In the **Ckswitch:** main program routine line 9 (**OUT1 = 0**) initializes output port 1 to 0. Lines 10 and 11 (**A = IN12** and **B = IN11**) assign the value (0 or 1) at the two input ports to variables A and B. Line 12 (**Y = A & B**) logically ANDs the input A and B placing the output in variable Y. IF-THEN statements are decision points in a program. Line 13 (**IF Y = 1 THEN redLED**) checks to see if Y equals 1; if the first part of the statement is true, then the BS2 executes the second part of the IF-THEN and jumps to the

66

subroutine **redLED:,** which lights the red LED. However, if the first part of the statement **(Y = 1)** is false, then the BS2 microcontroller disregards the rest of the program line and drops to the next line of code which is **GOTO Ckswitch.** The GOTO statement causes a jump back to the beginning of the main routine labeled **Ckswitch:.** The main routine continues to repeat hundreds of times per second.

```
'2-input AND function   'Title of program (Lab 3-10)

A VAR Bit               'Declare A as a variable, 1 bit
B VAR Bit               'Declare B as a variable, 1 bit
Y VAR Bit               'Declare Y as a variable, 1 bit

INPUT 11                'Declare port 11 as an input
INPUT 12                'Declare port 12 as an input
INPUT 1                 'Declare port 1 as an output (red LED)

Ckswitch:               'Label for check switch routine
  OUT1 = 0              'Initialize: port 1 at 0, red LED off
  A = IN12              'Assign value: port 12 input to variable A
  B = IN11              'Assign value: port 11 input to variable B
  Y = A & B             'Assign value: A ANDed with B to variable Y
  IF Y = 1 THEN redLED  'If Y = 1 then go to red subroutine, otherwise next line
GOTO Ckswitch           'Go to Ckswitch: begin check switch routine again


redLED:                 'Label for lighting red LED, means HIGH
  OUT1 = 1              'Output P1 goes HIGH, lights red LED
  PAUSE 100             'Pause 100 ms
GOTO Ckswitch           'Go to Ckswitch: begin check switch routine again
```

For reference, a diagram of Parallax's Board of Education (BOE) development board is shown in Fig. 3-26. A few of its features are identified on the drawing, including:

    a. Serial input (for downloading PBASIC programs from PC with serial cable)

    b. BASIC Stamp 2 module (24-pin DIP)

    c. Power connections—either 9-V battery or dc supply with barrel plug

    d. ON/OFF switch and power indicator

    e. Reset button—for restarting PBASIC programs

    f. Voltage regulator

    g. Breadboard, solderless

    h. Power supply header [V_{dd} (+) and V_{ss} (−)]

    i. BS2 input/output (I/O) pins header

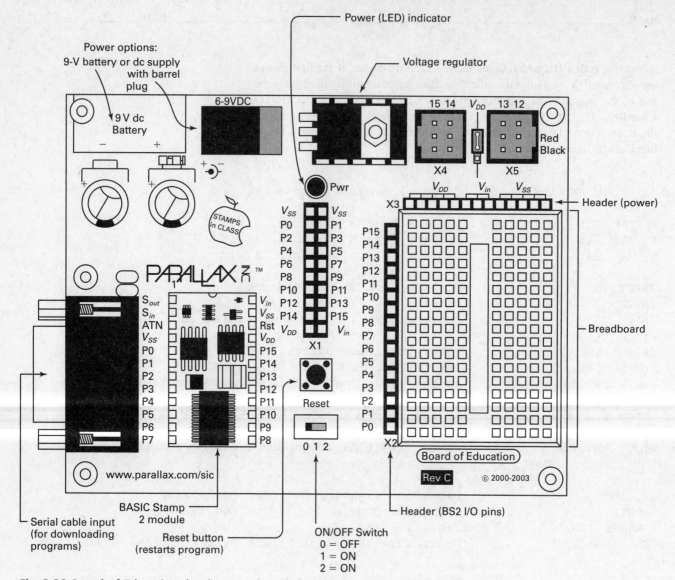

Fig. 3-26 Board of Education development board. (Used by permission of Parallax, Inc.)

The following labels appear on the figure:

Power (LED) indicator

Power options:
9-V battery or dc supply with barrel plug

Voltage regulator

6-9VDC

9 V dc Battery

15 14 V_{DD} 13 12 Red Black

X4   X5

V_{DD} V_{in} V_{SS}

STAMPS in CLASS

Pwr

Header (power)

V_{SS} V_{SS}
P0   P1
P2   P3
P4   P5
P6   P7
P8   P9
P10   P11
P12   P13
P14   P15
V_{DD} V_{in}

X1

P15
P14
P13
P12
P11
P10
P9
P8
P7
P6
P5
P4
P3
P2
P1
P0

X3

Breadboard

S_{out} V_{in}
S_{in} V_{SS}
ATN   Rst
V_{SS} V_{DD}
P0   P15
P1   P14
P2   P13
P3   P12
P4   P11
P5   P10
P6   P9
P7   P8

PARALLAX ™

Reset

0 1 2

X2

Board of Education

Rev C   © 2000-2003

www.parallax.com/sic

Serial cable input (for downloading programs)

BASIC Stamp 2 module

Reset button (restarts program)

ON/OFF Switch
0 = OFF
1 = ON
2 = ON

Header (BS2 I/O pins)

PROCEDURE

1. Refer to Fig. 3-25. Wire the circuit to a BS2 module using a development board such as Parallax's BOE (see Fig. 3-26). BS2 pins *P11* and *P12* are both used as inputs. BS2 pin *P1* is used as an output in this circuit.
2. Using an MS Windows–based PC, start the BASIC Stamp editor and write the PBASIC program titled '**2-input AND function**.
3. Connect a serial cable between the serial output port of the PC and the BOE development board.
4. Power ON (BOE board). Download the '**2-input AND function** program. The program should be running on the BOE.
5. Test the circuit by pressing combinations of the two input switches and observing the red output LED.
6. Disconnect the serial cable from the PC. Power OFF and then ON (BOE). Retest the circuit. The program was retained in EEPROM memory when the BOE was turned off and should restart when turned on again.
7. Show your instructor your circuit, and be prepared to answer questions about the circuit and the PBASIC program.

8. Using the PC and BASIC Stamp editor, rewrite the **'2-input AND function** program to perform other logic functions as assigned by your instructor. Other programs might include:
   a. Two-input OR function
   b. Two-input NAND function
   c. Two-input NOR function
   d. Two-input XOR function
   e. Two-input XNOR function
9. Reconnect the serial cable from the PC to the BOE.
10. Power ON (BOE). Download your new program, test, and show to your instructor. Be prepared to answer questions about the program.
11. Power OFF. Take down the circuit, and return all equipment to its proper place.

QUESTIONS

Answer questions 1 to 13.

1. At the heart of the BASIC Stamp 2 module is a programmable device called a _____ (microcontroller, multiplier).

2. The high-level language used to program a BS2 module is _____ (ABEL, PBASIC).

3. Refer to Fig. 3-25 and the **'2-input AND function** program. This program uses only one BS2 I/O port as an output and _____ (2, 12) I/O pins as inputs to the microcontroller module.

4. Refer to Fig. 3-25. The push-button inputs are designed as _____ (active HIGH, active LOW) switches.

5. Refer to Fig. 3-25. The red LED will light when BS2 port *P1* is driven _____ (HIGH, LOW) by the BS2 module.

6. Refer to Fig. 3-25 and the **'2-input AND function** program. When both *A* and *B* input switches are pressed, the output LED will _____ (light, not light).

7. Refer to Fig. 3-25 and the **'2-input AND function** program. When the input at port 11 is LOW and port 12 is HIGH, the output LED will _____ (light, not light).

8. Refer to Fig. 3-25 and the **'2-input AND function** program. When the input at port 11 is LOW and port 12 is LOW, the output LED will _____ (light, not light).

9. Refer to Fig. 3-26. On the Board of Education development board the BS2 module takes the form of a 24-pin DIP IC. (T or F)

10. When downloading a PBASIC program from the PC to the BS2 module, the program is stored in _____ (EEPROM, ROM) memory.

11. Refer to Fig. 3-26. On the power header on the BOE board, V_{dd} means the _____ (negative, positive) of the power supply.

12. Refer to the **'2-input AND function** program. Line 12 (**Y = A & B**) would have to be changed to _____ [**Y = ~(A ^ B), Y = ~(A & B)**] for the program to perform the two-input NAND function.

13. Refer to the **'2-input AND function** program. Line 12 (**Y = A & B**) would have to be changed to _____ [**Y = (A ^ B), Y = ~(A | B)**] for the program to perform the two-input XOR function.

1. _____
2. _____
3. _____
4. _____
5. _____
6. _____
7. _____
8. _____
9. _____
10. _____
11. _____
12. _____
13. _____

CHAPTER 4

Combining Logic Gates

TEST: COMBINING LOGIC GATES

Answer the questions in the spaces provided.

1. Write the Boolean expression for the AND-OR logic diagram shown in Fig. 4-1.

1. _____

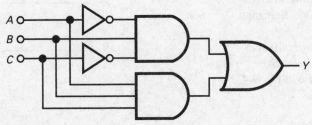

Fig. 4-1

2. The Boolean expression $A \cdot B + C \cdot D = Y$ is called the sum-of-products or _____ form.

2. _____

3. The Boolean expression $(A + B) \cdot (C + D) = Y$ is called the _____ or maxterm form.

3. _____

4. Write the Boolean expression for the OR-AND logic diagram shown in Fig. 4-2.

4. _____

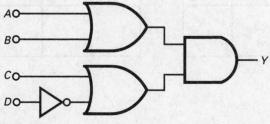

Fig. 4-2

5. Sum-of-products Boolean expressions are used to create _____ (AND-OR, OR-AND) logic circuits.

5. _____

6. Maxterm Boolean expressions are used to create _____ (AND-OR, OR-AND) logic circuits.

6. _____

7. Refer to Fig. 4-3. Write the sum-of-products Boolean expression for this truth table.

7. _____

8. Refer to Fig. 4-3. The Boolean expression $C \cdot B \cdot \overline{A} + \overline{C} \cdot B \cdot A = Y$ will produce a truth table that has HIGH outputs in what two lines?

8. _____

	INPUTS			OUTPUT
	C	B	A	Y
Line 0	0	0	0	0
Line 1	0	0	1	0
Line 2	0	1	0	0
Line 3	0	1	1	1
Line 4	1	0	0	0
Line 5	1	0	1	0
Line 6	1	1	0	0
Line 7	1	1	1	1

Fig. 4-3

9. Write the simplified minterm Boolean expression for the unsimplified expression $A \cdot B \cdot C \cdot D + A \cdot B \cdot C \cdot \overline{D} = Y$. A four-variable Karnaugh map is provided for your use in Fig. 4-4.

9. _____

10. Write the simplified minterm Boolean expression for the unsimplified expression $\overline{A} \cdot \overline{B} \cdot C \cdot D + \overline{A} \cdot B \cdot C \cdot D + A \cdot \overline{B} \cdot \overline{C} \cdot D + A \cdot \overline{B} \cdot \overline{C} \cdot \overline{D} = Y$. A four-variable Karnaugh map is provided for your use in Fig. 4-4.

10. _____

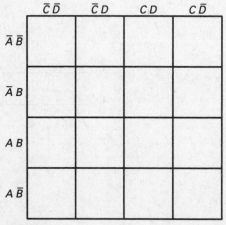
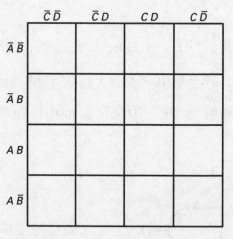
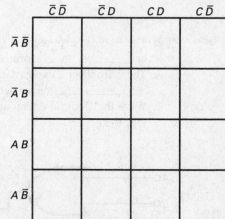

Fig. 4-4

11. Refer to Fig. 4-5. Write the *unsimplified* minterm Boolean expression for this truth table.

11. _____

12. Write the *simplified* minterm Boolean expression for the unsimplified expression in question 11. A four-variable Karnaugh map is provided for your use in Fig. 4-4.

12. _____

13. Which logic symbol diagram in Fig. 4-6 *will not* generate the truth table shown in Fig. 4-5?

13. _____

72

INPUTS				OUTPUT
A	**B**	**C**	**D**	**Y**
0	0	0	0	0
0	0	0	1	0
0	0	1	0	0
0	0	1	1	0
0	1	0	0	0
0	1	0	1	1
0	1	1	0	0
0	1	1	1	1
1	0	0	0	0
1	0	0	1	0
1	0	1	0	1
1	0	1	1	0
1	1	0	0	0
1	1	0	1	0
1	1	1	0	1
1	1	1	1	0

Fig. 4-5

14. Refer to Fig. 4-5. A one-package solution to this problem would be to use a 1-of-16 _____ IC.

14. _____

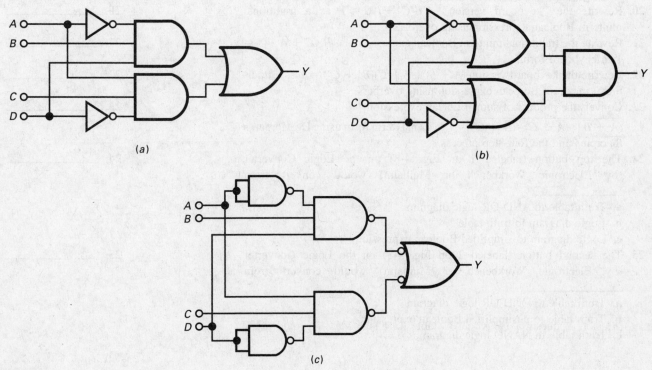

(a)

(b)

(c)

Fig. 4-6

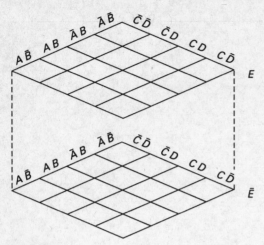

Fig. 4-7

15. Write the simplified sum-of-products Boolean expression for the unsimplified expression $\overline{A} \cdot \overline{B} \cdot C \cdot \overline{D} \cdot E + \overline{A} \cdot B \cdot C \cdot \overline{D} \cdot E + A \cdot B \cdot \overline{C} \cdot D \cdot E + A \cdot B \cdot \overline{C} \cdot D \cdot \overline{E} = Y$. A five-variable Karnaugh map is provided for your use in Fig. 4-7.

16. Write the maxterm Boolean expression for the OR-AND logic diagram shown in Fig. 4-6(*b*).

17. List the two lines of the truth table in Fig. 4-5 that would contain 1s if the Boolean expression were $A \cdot B \cdot C \cdot D + \overline{A} \cdot \overline{B} \cdot \overline{C} \cdot \overline{D} = Y$.

18. List the five lines of the truth table in Fig. 4-5 that would contain 1s if the Boolean expression were $A \cdot B \cdot \overline{C} \cdot D + \overline{A} \cdot \overline{B} = Y$.

19. Very complex logic problems involving many inputs and outputs would best be solved using:
    a. Logic gate ICs
    b. Discrete components
    c. Programmable logic devices (PLDs)

20. Rewrite the "keyboard version" $A'B'C' + AC = Y$ as a traditional minterm Boolean expression using overbars.

21. Rewrite the traditional minterm Boolean expression $\overline{A} \cdot B \cdot C + \overline{A} \cdot \overline{C} = Y$ in keyboard form.

22. Rewrite the keyboard version $(A + B')(A' + C)(B' + C') = Y$ as a traditional maxterm Boolean expression using overbars.

23. Convert the product-of-sums Boolean expression
    $(\overline{\overline{A} + \overline{B}}) \cdot (\overline{A} + C) \cdot (\overline{B} + \overline{C}) = Y$ to its maxterm form using De Morgan's theorems and the four-step process.

24. The top button (labeled 1 on Fig. 4-8) on the Logic Converter (by Electronic Workbench or Multisim) would convert from

    _____.

    a. Truth table to AND-OR logic diagram
    b. Logic diagram to truth table
    c. Logic diagram to simplified Boolean expression

25. The second button (labeled 2 on Fig. 4-8) on the Logic Converter (by Electronic Workbench or Multisim) would convert from

    _____.

    a. Truth table to AND-OR logic diagram
    b. Truth table to unsimplified Boolean expression
    c. Truth table to NAND logic diagram

15. _____

16. _____

17. _____ _____

18. _____ _____

_____ _____

_____

19. _____

20. _____

21. _____

22. _____

23. _____

24. _____

25. _____

74

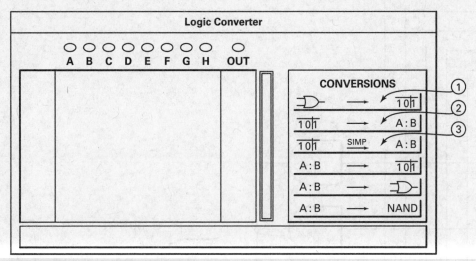

Fig. 4-8 Logic Converter problem.

26. The third button (labeled 3 on Fig. 4-8) on the Logic Converter (by Electronic Workbench or Multisim) would convert from _____.

    a. Truth table to AND-OR logic diagram
    b. Truth table to unsimplified Boolean expression
    c. Truth table to simplified Boolean expression

26. _____

27. Complex combinational logic designs can be implemented using a one-package IC called a _____ (CMOS, PLD, TTL).

27. _____

28. In lab or in school programming of PAL or GAL ICs can be achieved using development software, a PC with an output cable, and a(n) _____.

    a. DMM
    b. IC burner
    c. Logic analyzer
    d. Oscilloscope

28. _____

29. A CPLD is a complex programmable logic device good for implementing sequential logic designs while a(n) _____ (FPGA, PAL) could be used for combinational logic circuits.

29. _____

30. A PLD (such as a PAL or GAL) is a one-package solution to many logic designs, but they are extremely expensive. (T or F)

30. _____

31. The PLD pictured in Fig. 4-9 has a programmable _____ (AND, OR) array.

31. _____

32. The PLD pictured in Fig. 4-9 can implement sum-of-products Boolean expressions. (T or F)

32. _____

33. Complete the PLD fuse map shown in Fig. 4-9 that would implement the Boolean expression $\overline{A} \cdot \overline{B} \cdot C \cdot \overline{D} + \overline{A} \cdot B \cdot \overline{C} \cdot D + A \cdot B \cdot \overline{C} \cdot \overline{D} + A \cdot \overline{B} \cdot C \cdot D = Y$. Write your answer directly on Fig. 4-9. An X at an intersection means an intact fuse, while no X means a blown-open fusible link.

INPUTS

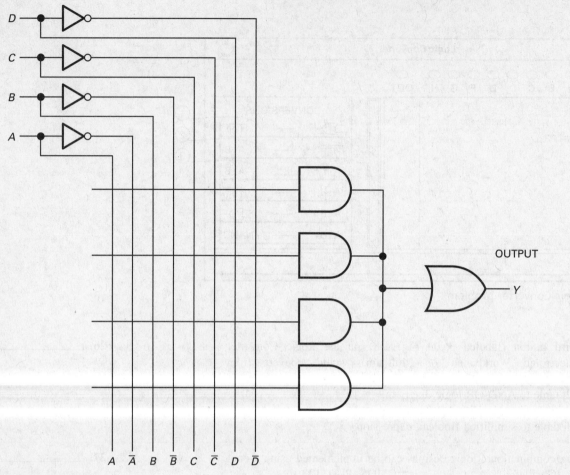

Fig. 4-9 Fuse map.

34. A programmable logic IC with PAL10L8 printed on top would use TTL technology, have 10 inputs, 8 outputs, and its outputs would be _____ (active HIGH, active LOW).

34. _____

35. Logic functions can be programmed into microcontroller-based devices such as BASIC Stamp modules. (T or F)

35. _____

36. The traditional Boolean expression $(A \cdot \overline{B} \cdot \overline{C}) + (\overline{A} \cdot C) = Y$ would commonly be rewritten as Y = (A^B^C) + (^AB) in PBASIC when using the BASIC Stamp 2 module to solve this logic function. (T or F)

36. _____

37. The BASIC Stamp 2 module always uses ports 1, 2, 3, and 4 as inputs and ports 5, 6, 7, and 8 as outputs. (T or F)

37. _____

76

4-1 LAB EXPERIMENT: DEVELOPING A LOGIC CIRCUIT

OBJECTIVES

1. To draw a logic symbol diagram from a Boolean expression.
2. To wire and operate the logic circuit using 7408 and 7432 ICs.
3. *OPTIONAL:* Electronics Workbench or Multisim. To use the EWB Logic Converter instrument to generate a logic diagram and truth table from a Boolean expression.

MATERIALS

Qty.   **Qty.**

1 7408 two-input AND gate IC   1 7432 two-input OR gate IC
3 logic switches   1 LED indicator-light assembly
1 5-V dc regulated power   1 electronic circuit simulation
   supply   program *(optional)*

PROCEDURE

1. Draw a logic circuit for the Boolean expression $A + (B \cdot C) = Y$ in Fig. 4-10 below. Use the correct logic symbols for the AND and OR gates. The input switches and output indicator are shown in Fig. 4-10 for your convenience.

INPUTS

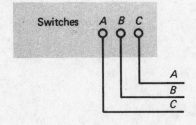

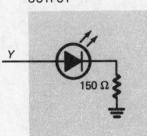

Fig. 4-10 Student logic problem, step 1.

TABLE 4-1 Truth Table for $A + (B \cdot C) = Y$

INPUTS			OUTPUTS
A	**B**	**C**	**Y**
0	0	0	
0	0	1	
0	1	0	
0	1	1	
1	0	0	
1	0	1	
1	1	0	
1	1	1	

2. Insert a 7408 and a 7432 IC into the mounting board.
3. Power OFF. Connect power to the ICs: the red wires are for +5 V (V_{CC}) and the black wires are for GND.
4. Wire the logic circuit you drew in step 1. Use three input switches, one AND gate, one OR gate, and the LED indicator-light assembly for output. Pin diagrams are found in Appendix A.
5. Power ON. Move input switches A, B, and C to each combination shown in the truth table in Table 4-1. Record the output results in the right column (LED lit = 1, LED not lit = 0).
6. Take down the circuit and return all equipment to its proper place.

7. *OPTIONAL:* Electronics Workbench or Multisim. If assigned by instructor, use EWB Logic Converter instrument to:
   a. Generate a logic diagram from the Boolean expression $A \cdot \overline{B} + A \cdot \overline{B} \cdot C = Y$ (keyboard version = $AB' + AB'C$).
   b. Generate a truth table from the same Boolean expression.
8. Show instructor your logic diagram and truth table.

QUESTIONS

Complete questions 1 to 6.

1. The Boolean expression $\overline{A} \cdot B + A \cdot \overline{B} \cdot C = Y$ is called a _____ (maxterm, minterm) or _____ -of- _____ expression.

   1. _____
      _____

2. The Boolean expression $\overline{A} \cdot B + A \cdot \overline{B} \cdot C = Y$ can be implemented with an _____ (AND-OR, OR-AND) pattern of logic gates.

   2. _____

3. Draw a logic diagram for the Boolean expression $A \cdot \overline{B} + A \cdot \overline{B} \cdot C = Y$.

4. Draw a three-variable truth table for the Boolean expression $A \cdot B + A \cdot \overline{B} \cdot C = Y$.

5. Write the keyboard version of the Boolean expression $A \cdot \overline{B} + \overline{A} \cdot B \cdot \overline{C} = Y$.

   5. _____

6. Write the keyboard version of the Boolean expression $\overline{A \cdot B + A \cdot \overline{B} \cdot \overline{C}} = Y$.

   6. _____

4-2 LAB EXPERIMENT: SIMPLIFYING LOGIC CIRCUITS

OBJECTIVES

From a given truth table, do the following:
1. Write a minterm Boolean expression.
2. Simplify the expression using a Karnaugh map.
3. Draw a logic symbol diagram from the simplified Boolean expression.
4. Wire and operate the logic circuit.
5. *OPTIONAL:* Electronics Workbench or Multisim. If assigned by instructor, use EWB circuit simulator to enter a truth table, simplify, generate a simplified Boolean expression, generate an AND-OR logic diagram, and generate a NAND-NAND logic diagram from a four-variable truth table.

MATERIALS

Qty.		Qty.	
1	7400 two-input NAND gate IC	3	logic switches
1	7404 hex inverter IC	1	LED indicator-light assembly
1	7408 two-input AND gate IC	1	5-V dc regulated power supply
1	7432 two-input OR gate IC	1	electronic circuit simulation program (*optional*)

PROCEDURE

1. Develop a minterm Boolean expression from Table 4-2.

TABLE 4-2   Truth Table

INPUTS			OUTPUT		
A	**B**	**C**	**Y**	**AND-OR circuit**	**NAND circuit**
				Y	**Y**
0	0	0	0		
0	0	1	1		
0	1	0	0		
0	1	1	1		
1	0	0	0		
1	0	1	0		
1	1	0	0		
1	1	1	1		

2. From the unsimplified Boolean expression you wrote in question 1, plot the 1s in the Karnaugh map in Fig. 4-11.
3. Draw loops on the Karnaugh map in Fig. 4-11.
4. Write the simplified Boolean expression (minterm form) for Fig. 4-11 by eliminating variables.

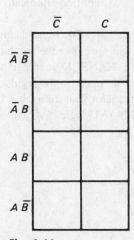

Fig. 4-11

5. Draw a logic circuit for your simplified Boolean expression in step 4. Use input switches, inverters, AND gates, an OR gate, and an LED output indicator-light assembly.

6. Power OFF. Wire the AND-OR logic circuit drawn in step 5. Find pin diagrams in Appendix A.
7. Power ON. Move input switches, A, B, and C to each combination shown in the truth table in Table 4-2. Record the outputs in the AND-OR Circuit column, Table 4-2.
8. Power OFF. Remove the ICs from the mounting board.
9. Redraw your logic circuit for the simplified Boolean expression in step 4 using NAND gates only. Use input switches, NAND gates, and an LED output indicator-light assembly.

10. Power OFF. Insert the 7400 IC into the mounting board, and connect the power (V_{CC} and GND).
11. Wire the logic circuit using three input switches, the 7400 IC, and an LED output indicator-light assembly.
12. Power ON. Operate the NAND logic circuit according to the truth table in Table 4-2. Record the outputs in the NAND Circuit column, Table 4-2.
13. Take down the circuit and return all equipment to its proper place.
14. *OPTIONAL:* Electronics Workbench or Multisim. If assigned by instructor, use the EWB circuit simulator to:
    a. From Fig. 4-12, output Y, fill in truth table on EWB's Logic Converter instrument.
    b. Simplify Boolean expression using EWB's Logic Converter.
    c. Generate a simplified Boolean expression using EWB's Logic Converter.
    d. Generate an AND-OR logic diagram using EWB's Logic Converter (diagram will appear in the regular workspace).
    e. Generate a NAND logic diagram using EWB's Logic Converter (diagram will appear in the regular workspace).
15. Show your instructor your truth table, Boolean expression, AND-OR logic diagram, and NAND logic diagram.

INPUTS				OUTPUT	OUTPUT
A	B	C	D	Y	Z
0	0	0	0	1	0
0	0	0	1	1	0
0	0	1	0	1	1
0	0	1	1	1	0
0	1	0	0	0	0
0	1	0	1	0	0
0	1	1	0	0	1
0	1	1	1	0	0
1	0	0	0	0	0
1	0	0	1	1	0
1	0	1	0	0	1
1	0	1	1	0	0
1	1	0	0	0	1
1	1	0	1	1	1
1	1	1	0	0	1
1	1	1	1	0	1

Fig. 4-12 Truth table.

QUESTIONS

Complete questions 1 to 8.

1. In this experiment it was found that the _____ (AND-OR, NAND) circuit used *fewer ICs* to perform the logic function in Table 4-2.

1. _____

2. Which method of simplifying a Boolean expression was used in this experiment (steps 1–4)?
   a. Karnaugh map
   b. Tabular method
   c. Venn diagram
   d. Veitch diagram

2. _____

3. From the truth table in Fig. 4-13, do the following:
   a. Write the unsimplified Boolean expression.
   b. Record five 1s in the Karnaugh map in Fig. 4-14.
   c. Loop the adjacent groups of 1s in Fig. 4-14.
   d. From the looping in Fig. 4-14, eliminate variables.
   e. Write the simplified minterm Boolean expression.
   f. Draw a logic diagram of the simplified Boolean expression (use AND, OR, and NOT gates).

3. **a.** _____

_____

e. _____

   g. Redraw the AND-OR circuit in *f* to form a NAND circuit.

INPUTS			OUTPUT
A	B	C	Y
0	0	0	1
0	0	1	1
0	1	0	1
0	1	1	1
1	0	0	0
1	0	1	0
1	1	0	0
1	1	1	1

Fig. 4-13 Truth table.

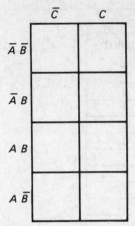

Fig. 4-14 Karnaugh map.

4. Refer to question 3. Which logic circuit would use *fewer ICs*?
   a. AND-OR circuit
   b. NAND circuit

4. _____

5. Computer circuit simulation programs are likely to use the _____ (Karnaugh map, Quine-McCluskey tabular) method of simplifying Boolean expressions.

5. _____

6. Using circuit simulation software or Karnaugh mapping, simplify and write the simplified minterm Boolean expression for output Z of the truth table in Fig. 4-12.

6. _____

7. Draw the AND-OR logic diagram from the simplified Boolean expression in question 6 for the logic function described in output Z of the truth table in Fig. 4-12.

8. Draw the NAND logic diagram based on the AND-OR logic diagram developed in question 7.

4-3 MULTISIM EXPERIMENT: LOGIC SIMPLIFICATION

OBJECTIVES

1. Using the Logic Converter instrument from Electronics Workbench or Multisim, convert a five-variable truth table to its simplified Boolean expression and then to an AND-OR logic diagram.
2. Using a trainer, wire the five-input AND-OR logic diagram generated by the Logic Converter and test its performance.

MATERIALS

Qty.   **Qty.**

- Logic Converter instrument   5   input logic switches
  from Electronics Workbench   1   LED indicator-light assembly
  or Multisim software
- personal computer (PC)
- various 7400 series ICs
  (selected by the student)

INFORMATION

The Logic Converter is a virtual instrument available in the Electronics Workbench or Multisim software package. A screen image of the Logic Converter is sketched in Fig. 4-15. The possible conversions are abbreviated on buttons on the right. Activating the conversion buttons allow conversion back and forth from truth table, Boolean expression, and logic symbol diagram. A very valuable logic simplification button is also available (third button down). The Logic Converter instrument can generate AND-OR, OR-AND, and NAND logic circuits.

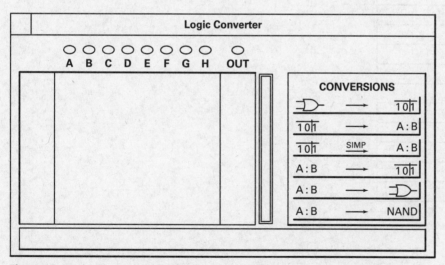

Fig. 4-15 Logic Converter screen (by Electronics Workbench or Multisim).

PROCEDURE

1. Using the Logic Converter instrument from Electronics Workbench or Multisim, fill in the data from the output *Y* column of the five-variable truth table in Table 4-3.
2. Using the Logic Converter instrument, convert the truth table to a simplified Boolean expression. Write the simplified Boolean expression below.

_____

Table 4-3 Truth Table

INPUTS					OUTPUT	
A	B	C	D	E	Y	AND-OR hardware circuit
0	0	0	0	0	0	
0	0	0	0	1	1	
0	0	0	1	0	0	
0	0	0	1	1	1	
0	0	1	0	0	0	
0	0	1	0	1	1	
0	0	1	1	0	0	
0	0	1	1	1	1	
0	1	0	0	0	0	
0	1	0	0	1	0	
0	1	0	1	0	0	
0	1	0	1	1	0	
0	1	1	0	0	0	
0	1	1	0	1	0	
0	1	1	1	0	0	
0	1	1	1	1	0	
1	0	0	0	0	1	
1	0	0	0	1	1	
1	0	0	1	0	0	
1	0	0	1	1	0	
1	0	1	0	0	0	
1	0	1	0	1	0	
1	0	1	1	0	0	
1	0	1	1	1	0	
1	1	0	0	0	1	
1	1	0	0	1	1	
1	1	0	1	0	0	
1	1	0	1	1	0	
1	1	1	0	0	0	
1	1	1	0	1	0	
1	1	1	1	0	0	
1	1	1	1	1	0	

3. Using the Logic Converter instrument, convert the simplified Boolean expression to a five-input AND-OR logic diagram. Draw the AND-OR logic diagram below.

4. Test your five-input AND-OR logic circuit using actual ICs. Select the appropriate AND, OR, and inverter ICs from Appendix A. If you need help, see your instructor.

5. Power OFF on the trainer. Wire the five-input AND-OR logic circuit from step 3. Connect power to each IC (+5 V and GND) even though these connections may not be shown on the logic diagram.

6. Power OFF. Finish wiring using five-input switches, AND, OR, and inverter ICs, and a single LED indicator-light assembly.

7. Power ON. Test your five-input AND-OR logic circuit. Fill in the AND-OR Hardware Circuit output column in Table 4-3.

8. Show your operating AND-OR hardware circuit and the results to your instructor. Be prepared to answer selected questions on your circuit.

9. Power OFF. Take down the circuit, and return all equipment to its proper place.

QUESTIONS

Complete questions 1 to 6.

1. List the six conversions available when using the Logic Converter instrument.

   a. _____

   b. _____

   c. _____

   d. _____

   e. _____

   f. _____

2. Entering a sum-of-products Boolean expression into the Logic Converter instrument and activating the Boolean expression-to-logic diagram button (fifth button down in Fig. 4-15) will generate an _____ (AND-OR, OR-AND) schematic.

2. _____

3. Entering a product-of-sums Boolean expression into the Logic Converter instrument and activating the Boolean expression-to-logic diagram button (fifth button down in Fig. 4-15) will generate an _____ (AND-OR, OR-AND) schematic.

4. Logic simplification using the Logic Converter instrument is considered to be quicker, easier, and more accurate than the paper-and-pencil methods. (T or F)

5. Simplify the following five-variable Boolean expression using the Logic Converter instrument: $AB'CD'E + AB'CD'E' + AB'C'DE + ABC'DE + A'BC'DE + A'B'C'DE$.

6. Draw the AND-OR logic diagram for the simplified Boolean expression you developed in question 5.

3. _____

4. _____

5. _____

4-4 LAB EXPERIMENT: DATA SELECTORS

OBJECTIVES

1. To wire and operate a 1-of-16 data selector.
2. To observe the logic level of a "floating" TTL input.

MATERIALS

Qty.   **Qty.**

1   7404 inverter IC   1   74150 1-of-16 data selector
6   logic switches   IC
1   5-V dc regulated power   2   LED indicator-light
    supply   assemblies

SYSTEM DIAGRAM

The circuit you will connect for testing the 74150 data selector is shown in Fig. 4-16. Notice the three separate groups of inputs. The single enable input will simply turn on the unit. Note that it takes a logical 0 to enable the IC. The data select inputs choose which input will be connected to output W of the 74150 data selector. The data inputs at the upper left are connected to whatever data (0 or 1) you want transmitted to the output. The small inverter bubble at output W of the 74150 means that data come out in inverted form. The 7404 inverter complements the data output W, so output Y equals the correct data input.

Looking carefully at Fig. 4-16, note that the data inputs (0 to 15) are not connected to anything. We say that they are "floating." You will find that these

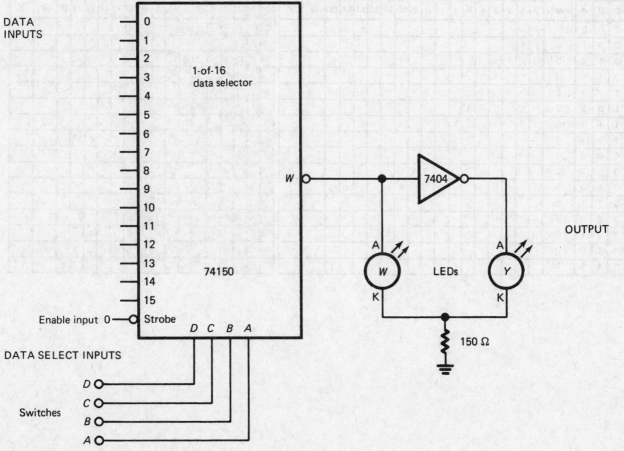

Fig. 4-16 Wiring a 74150 1-of-16 data selector IC.

inputs will float at a logical 1 in TTL ICs. We say that *the data inputs float HIGH*. During part of the experiment you will connect each data input to an input switch.

Think of both the enable and data select inputs as *controls*. These controls select when and which data are transferred to the output of the 74150 IC.

PROCEDURE

1. Insert the 7404 and 74150 ICs into the mounting board. Connect power to both ICs (V_{CC} and GND).
2. Power OFF. Wire the circuit in Fig. 4-16. Use five input switches, the 7404 and 74150 ICs, and two LED output indicator-light assemblies. Temporarily leave data inputs 0 to 15 disconnected (floating HIGH). Pin diagrams are in Appendix A.
3. Power ON. Set the data select switches (D, C, B, and A) as shown in the truth table in Table 4-4. Apply a logical 1 to the correct data input using an input switch. Observe the outputs W and Y. Record the 32 outputs in columns W and Y, Table 4-4.
4. Repeat step 3, but apply a logical 0 to the correct data input using an input switch. Observe the outputs W and Y. Record the 32 outputs in the W_1 and Y_1 columns, Table 4-4.
5. Power OFF. Leave the 74150 IC mounted for use in the next experiment.

TABLE 4-4 Truth Table for 74150 TTL IC

D	C	B	A	Enable	0	1	2	3	4	5	6	7	8	9	10	11	12	13	14	15	W	Y	0	1	2	3	4	5	6	7	8	9	10	11	12	13	14	15	W_1	Y_1
0	0	0	0	0	1																		0																		
0	0	0	1	0		1																		0																	
0	0	1	0	0			1																		0																
0	0	1	1	0				1																		0															
0	1	0	0	0					1																		0														
0	1	0	1	0						1																		0													
0	1	1	0	0							1																		0												
0	1	1	1	0								1																		0											
1	0	0	0	0									1																		0										
1	0	0	1	0										1																		0									
1	0	1	0	0											1																		0								
1	0	1	1	0												1																		0							
1	1	0	0	0													1																		0						
1	1	0	1	0														1																		0					
1	1	1	0	0															1																		0				
1	1	1	1	0																1																		0			

QUESTIONS

Complete questions 1 to 10.

1. In Fig. 4-17 identify the following:
   a. Pin numbers 1 to 24
   b. Control inputs, enable and data select (A, B, C, and D)
   c. Inputs, data inputs (0 to 15)
   d. Output W
   e. Power, V_{CC} and GND

74150

(Top view)

Fig. 4-17

2. What *input control* on the 74150 data selector might be considered a "main switch" to turn on the unit?

2. _____

3. The enable (strobe) input of the 74150 data selector is enabled with a logical _____.

3. _____

4. The small bubble at output W on the 74150 data selector means what?

4. _____

5. With the data select inputs of the 74150 IC at binary 0101, what data input is connected to output W?

5. _____

6. To transfer a bit of information from data input 13 to output W of the 74150 IC, the following must be true:
   a. Enable is at _____.
   b. Data select inputs are $D =$ _____, $C =$ _____, $B =$ _____, and $A =$ _____.

6. a. _____
   b. _____
   _____
   _____
   _____

7. If the input settings on the 74150 IC are data select (0111), enable (0), and data inputs (floating), then what is the output at W?

7. _____

8. When a data input is floating on the 74150 IC, it is at logical _____.

8. _____

9. When a data input on the 74150 IC is floating, we say it floats _____ (HIGH, LOW).

9. _____

10. A data selector is also called a _____, according to manufacturers' data manuals.

10. _____

4-5 DESIGN PROBLEM: SOLVING GATING PROBLEMS WITH DATA SELECTORS

OBJECTIVES

1. To draw the simplest NAND logic circuit for a given four-variable truth table.
2. To wire and operate the NAND logic circuit.
3. To wire a 74150 data selector to solve the same logic problem.

MATERIALS

Qty.

1 7404 hex inverter IC
1 7420 four-input NAND gate IC
4 logic switches
1 5-V dc regulated power supply

Qty.

1 74150 1-of-16 data selector IC
1 7410 three-input NAND gate IC
1 LED indicator-light assembly

PROCEDURE

1. Write the minterm Boolean expression for the truth table in Table 4-5.

   _____

   _____

2. Plot seven 1s on the Karnaugh map in Fig. 4-18.
3. Loop adjacent groups of 1s on the Karnaugh map in Fig. 4-18.
4. Write the simplified minterm Boolean expression for the truth table in Fig. 4-18.

   _____

5. Draw an AND-OR logic circuit for the Boolean expression in question 4.

6. Redraw the AND-OR logic circuit using NAND gates and inverters.

7. Insert the 7404, 7410, and 7420 ICs into the mounting board. Connect power to each IC (V_{CC} and GND).

TABLE 4-5 Truth table

| INPUTS | | | | OUTPUT | | |
D	C	B	A	Y	NAND circuit	Data selector circuit
0	0	0	0	1		
0	0	0	1	0		
0	0	1	0	0		
0	0	1	1	1		
0	1	0	0	0		
0	1	0	1	0		
0	1	1	0	1		
0	1	1	1	0		
1	0	0	0	1		
1	0	0	1	0		
1	0	1	0	0		
1	0	1	1	1		
1	1	0	0	0		
1	1	0	1	1		
1	1	1	0	1		
1	1	1	1	0		

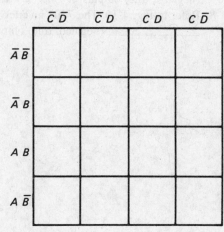

Fig. 4-18

8. Power OFF. Wire the NAND logic circuit you drew in step 6. Use four input switches; the 7404, 7410, and 7420 ICs; and an LED indicator-light assembly. Pin diagrams are in Appendix A.

9. Power ON. Operate input switches A, B, C, and D according to the input side of Table 4-5. Record the outputs in the NAND Circuit column of Table 4-5.

10. Power OFF. Take down the NAND circuit.

11. Mount the 74150 IC and connect power (V_{CC} and GND).

12. Wire the 74150 data selector as shown in Fig. 4-16, Lab Experiment 4-4. Omit the LED output indicator W.

13. Connect data inputs 0 to 15 to logical 1 or 0 according to column Y in Table 4-5 (logical 0 = GND, logical 1 = +5 V).

14. Enable input to logical 0.

15. Power ON. Operate data select input switches D, C, B, and A according to the truth table in Table 4-5. Record the outputs in the Data Selector Circuit column, Table 4-5.

16. Power OFF. Take down the circuit and return all equipment to its proper place.

QUESTIONS

Complete questions 1 to 8.

1. This logic problem was solved faster and cheaper using a(n)
   a. Data selector
   b. AND-OR logic gate circuit
   c. NAND logic gate circuit

    1. _____

2. If a data input on a 74150 data selector were left disconnected, that input would be considered to be at a logical _____.

    2. _____

3. To save switches, data inputs (74150) at a logical 0 are connected to _____.

    3. _____

4. To save switches, data inputs (74150) at a logical 1 are connected to _____.

    4. _____

5. Inside the single 74150 IC are _____. (Refer to the manufacturer's data manual.)
   a. 3 to 4 gates
   b. 10 to 15 gates
   c. 25 or more gates

    5. _____

6. Refer to Fig. 4-16. If the strobe or enable input is HIGH, the 74150 IC is _____ (disabled, enabled) and output W will be _____ (HIGH, LOW).

    6. _____
        _____

7. Refer to Fig. 4-16. The output W of the 74150 IC is a _____ (negated, true) output.

    7. _____

8. Refer to Fig. 4-16. The 7404 inverter is used in this circuit to display the _____ (negated, true) output at output indicator Y.

    8. _____

4-6 DESIGN PROBLEM: USING CMOS TO SOLVE A FIVE-VARIABLE LOGIC PROBLEM

OBJECTIVES

1. To design, draw, construct, and test a logic circuit having five variables.
2. To implement the circuit using CMOS NAND gates.
3. *OPTIONAL:* Electronics Workbench or Multisim. To use the Logic Converter of EWB to convert from a truth table to Boolean expression to an AND-OR logic diagram.

MATERIALS

Qty.   **Qty.**

1   74HC00 quad two-input   1   LED indicator-light
    NAND gate CMOS IC   assembly
1   74HC30 eight-input NAND gate   1   5-V dc regulated power supply
    CMOS IC   1   electronic circuit simulation
5   logic switches   program *(optional)*

SYSTEM DIAGRAM

The block diagram in Fig. 4-19 illustrates the logic circuit you will design. The block diagram shows five logic switch inputs and a single output indicator coming from the logic circuit. The customary input logic switches and LED output indicator-light assembly will work. Your system will be powered by a 5-V dc regulated power supply. The logic circuit will be implemented with CMOS ICs. It is suggested that NAND logic be used in the circuit. The 74HC00 (two-input NAND gate) and 74HC30 (eight-input NAND gate) CMOS ICs can be used in your logic circuit.

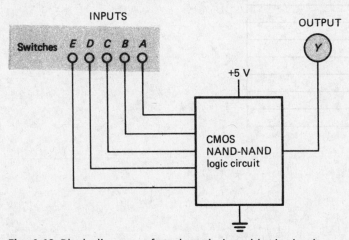

Fig. 4-19 Block diagram of student-designed logic circuit.

PROCEDURE

1. Design a logic circuit that will solve the problem posed in Table 4-6. The procedure is:

   a. Write an unsimplified minterm Boolean expression for the truth table.

   _____

   _____

   b. On Fig. 4-20 plot nine 1s on a five-variable Karnaugh map. Loop adjacent 1s in groups of two, four, or eight.

TABLE 4-6 Truth Table for Design Problem

INPUTS					OUTPUT	
A	B	C	D	E	Y	NAND circuit Y
0	0	0	0	0	0	
0	0	0	0	1	0	
0	0	0	1	0	0	
0	0	0	1	1	0	
0	0	1	0	0	0	
0	0	1	0	1	0	
0	0	1	1	0	0	
0	0	1	1	1	0	
0	1	0	0	0	0	
0	1	0	0	1	0	
0	1	0	1	0	0	
0	1	0	1	1	0	
0	1	1	0	0	0	
0	1	1	0	1	0	
0	1	1	1	0	0	
0	1	1	1	1	1	
1	0	0	0	0	1	
1	0	0	0	1	1	
1	0	0	1	0	1	
1	0	0	1	1	1	
1	0	1	0	0	0	
1	0	1	0	1	0	
1	0	1	1	0	0	
1	0	1	1	1	0	
1	1	0	0	0	1	
1	1	0	0	1	1	
1	1	0	1	0	1	
1	1	0	1	1	1	
1	1	1	0	0	0	
1	1	1	0	1	0	
1	1	1	1	0	0	
1	1	1	1	1	0	

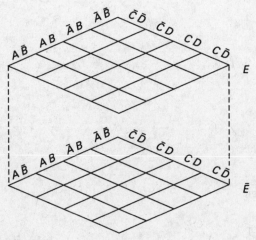

Fig. 4-20

94

c. From the Karnaugh map, write the simplified Boolean expression in minterm form.

_____

d. Draw an AND-OR logic diagram from the simplified Boolean expression.

e. Redraw the circuit as a NAND-NAND logic diagram.

CAUTION CMOS ICs can be damaged from static electricity. Store CMOS ICs with their pins in conductive foam or covered with aluminum foil.

2. Wire the NAND-NAND logic circuit. It is suggested you use 74HC00 quad two-input NAND gate and 74HC30 eight-input NAND gate ICs. Use five input switches and an LED indicator-light assembly.

3. Power ON. Test your CMOS logic circuit. Write the results in Table 4-6 under the column labeled NAND Circuit.

4. Have your instructor approve your Karnaugh map, logic diagram, and circuit. Your instructor may have you implement this logic using TTL ICs.

5. Power OFF. Take down the circuit and return all equipment to its proper place.

6. *OPTIONAL:* Electronics Workbench or Multisim: If assigned by instructor, use EWB circuit simulator to:

   a. From Table 4-6, fill in truth table on the simulator's Logic Converter instrument.

   b. Simplify Boolean expression using the simulator's logic converter.

   c. Generate an AND-OR logic diagram using the simulator's Logic Converter.

7. Show instructor your truth table, simplified Boolean expression, and AND-OR logic diagram.

QUESTIONS

Answer questions 1 and 2.

1. Using a five-variable Karnaugh map or the Logic Converter instrument from Electronics Workbench or Multisim, write the simplified Boolean expression for *ABCDE* + *ABCDE'* + *AB'C'D'E* + *ABC'D'E* + *A'BC'D'E* + *A'B'C'D'E* = *Y*. A Karnaugh map is provided in Fig. 4-21.

1. _____

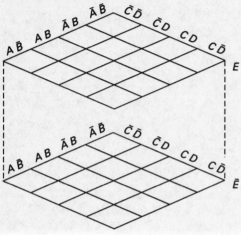

Fig. 4-21

2. From the simplified sum-of-products Boolean expression you developed in question 1, draw an AND-OR logic symbol diagram for this logic problem.

96

4-7 DESIGN PROBLEM: A DECODER USING A PLD

OBJECTIVES

1. To design (using a fuse map) a hexadecimal-to-seven-segment display decoder using a programmable logic device (PLD).
2. *OPTIONAL:* If you have the correct PLD software (such as CUPL or ABEL), design a hex-to-seven-segment decoder.

MATERIALS

Qty.

- Fuse map
- *OPTIONAL:* CUPL or ABEL software packages for designing PLDs

INFORMATION

Seven-segment LED displays are easy-to-use output devices designed to show either decimal or hexadecimal numbers. You will do a "paper-and-pencil" design of a decoder using a fuse map. The block diagram in Fig. 4-22 suggests how to use hex-to-seven-segment display decoder to translate a hexadecimal input to an output that lights the appropriate LED segments on the seven-segment display reading directly in hex.

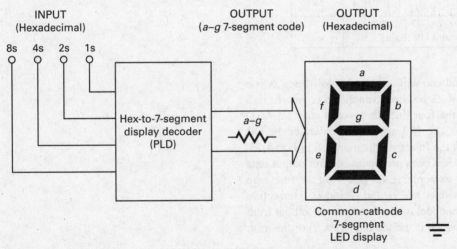

Fig. 4-22 Block diagram of a hex-to-seven-segment display decoder using a PLD.

A truth table for a hexadecimal-to-seven-segment decoder is detailed in Table 4-7. The left column lists the hexadecimal number followed by the 4-bit binary equivalent (input column). Listed next are seven output columns for the seven segments on an LED display (labeled *a* through *g*). The right column shows how the hexadecimal numbers will be formed on a seven-segment LED display.

A simplified programmable logic device (PLD) is sketched in Fig. 4-23 in the form of a fuse map. This unit has four inputs, and a single output appears like an AND-OR logic circuit. The inputs to the AND gates are programmable with fuses in this example. An *X* at an intersection means an intact fuse whereas no *X* means a blown fuse. It is assumed that this unit comes with all

TABLE 4-7 Truth Table for Hex-to-Seven-Segment Decoder

Hex	INPUTS				OUTPUTS							Display
	A	B	C	D	Decoder							
	8s	4s	2s	1s	a	b	c	d	e	f	g	
0	0	0	0	0	1	1	1	1	1	1	0	0
1	0	0	0	1	0	1	1	0	0	0	0	1
2	0	0	1	0	1	1	0	1	1	0	1	2
3	0	0	1	1	1	1	1	1	0	0	1	3
4	0	1	0	0	0	1	1	0	0	1	1	4
5	0	1	0	1	1	0	1	1	0	1	1	5
6	0	1	1	0	1	0	1	1	1	1	1	6
7	0	1	1	1	1	1	1	0	0	0	0	7
8	1	0	0	0	1	1	1	1	1	1	1	8
9	1	0	0	1	1	1	1	0	0	1	1	9
A	1	0	1	0	1	1	1	0	1	1	1	A
B	1	0	1	1	0	0	1	1	1	1	1	b
C	1	1	0	0	0	0	0	1	1	0	1	c
D	1	1	0	1	0	1	1	1	1	0	1	d
E	1	1	1	0	1	0	0	1	1	1	1	E
F	1	1	1	1	1	0	0	0	1	1	1	F
					0 = segment OFF (no light)							
					1 = segment ON (light)							

fuses intact from the manufacturer, and you will selectively blow fuses as you burn the PLD. The fuse map in Fig. 4-23 has been simplified compared to a real AND-OR logic diagram in that the four lines that would enter an AND gate are shown as a single line. Also, the 16 lines that would enter the OR gate are condensed to a single line on the fuse map diagrammed in Fig. 4-23.

The fuse map shown in Fig. 4-23 has been programmed with output data from column a, Table 4-7. As a first example, AND gate 1 on the fuse map in Fig. 4-23 has been programmed with $\overline{A}$ and $\overline{B}$ and $\overline{C}$ and $\overline{D}$ intersection fuses intact, while all others are blown open. In this example, the output from AND gate 1 will be $\overline{A} \cdot \overline{B} \cdot \overline{C} \cdot \overline{D}$, which is the top line (Hex 0) of the truth table, Table 4-7.

Consider a second example: AND gate 2 on the fuse map in Fig. 4-23 does not need to be programmed because line 2 in the truth table (Table 4-7) has an output of 0. All fuses are left intact, which has no effect on the output of the OR gate.

Consider a third example: AND gate 16 on the fuse map in Fig. 4-23 has been programmed with A and B and C and D intersection fuses intact, while all others are blown open. In this example, the output from AND gate 16 will be $A \cdot B \cdot C \cdot D$, which is the bottom line (Hex F) of the truth table, Table 4-7.

The fuse map shown in Fig. 4-23 generates the output required to drive only segment a of the seven-segment display. The PLD needed to drive all of the segments of a display would be much more complicated.

NAME ______________________________ DATE ______________

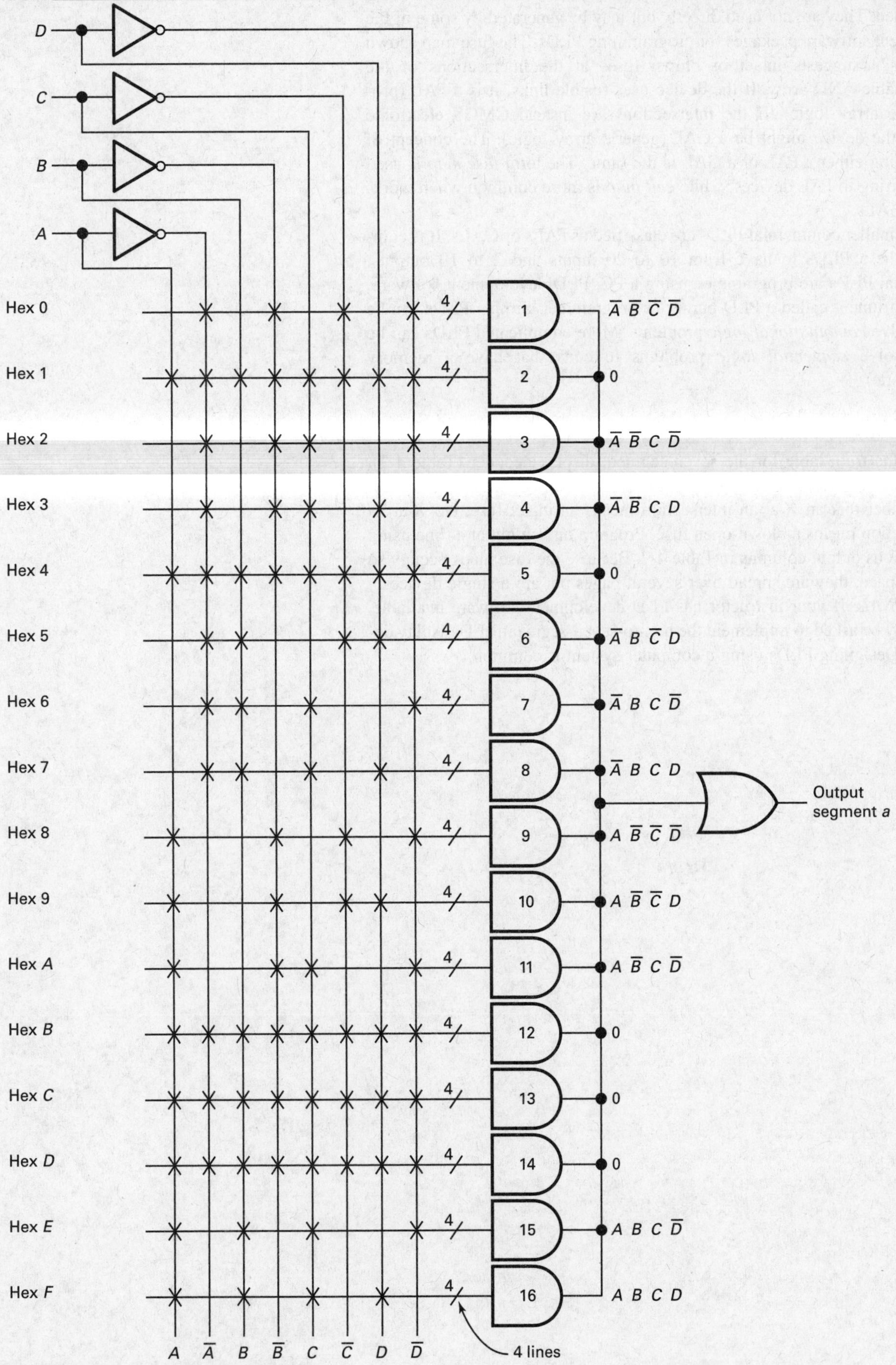

Fig. 4-23 Fuse map–solving problem.

The fuse map is a graphic method to help visualize how a PLD is being programmed. They are not used directly but may be generated by some of the development software packages for programming PLDs. The fuse map shown in Fig. 4-23 suggests intact or blown fuses at the intersections of the programmable AND array. If the device uses fusible links, it is a PAL (programmable array logic). If the intersections are instead CMOS electronic switches, the device might be a GAL (generic array logic). The concept of programming either a PAL or a GAL is the same. The term *fuse map* is used when referring to PAL devices, while *cell map* is more common when working with GALs.

Many smaller commercial PLDs are classified as PALs or GALs. It is common for these PLDs to have from 10 to 20 inputs and 2 to 10 outputs. Commercial PLDs are programmed using a PC, PLD development software, and an instrument called a PLD burner or programmer. Simple PLDs can be used to solve *combinational logic* problems. More complicated PLDs can be used to solve *sequential logic* problems (circuits that have a memory characteristic).

PROCEDURE

1. Using the truth table for the seven-segment display decoder (Table 4-7), complete the programming of the PLD fuse map shown in Fig. 4-24. Remember that an *X* at an intersection means an intact fuse. No *X* at an intersection means a blown-open fuse. Program one output at a time using the various output columns in Table 4-7. Because the fuse maps occupy so much space, they are spread over several pages but are a single device.
2. *OPTIONAL:* If your instructor has PLD development software available, you may be asked to implement the hex-to-seven-segment LED display decoder. Designing PLDs using a computer system is common.

100

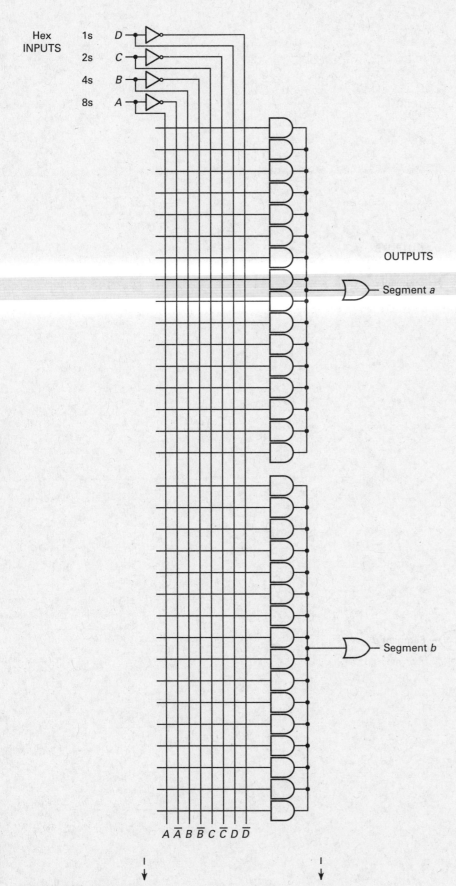

Fig. 4-24(a) Fuse map.

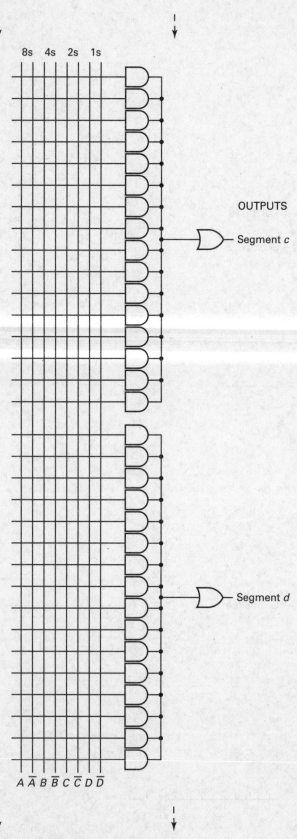

8s 4s 2s 1s

OUTPUTS

Segment c

Segment d

$A\ \overline{A}\ B\ \overline{B}\ C\ \overline{C}\ D\ \overline{D}$

Fig. 4-24(*b*) Fuse map (*continued*).

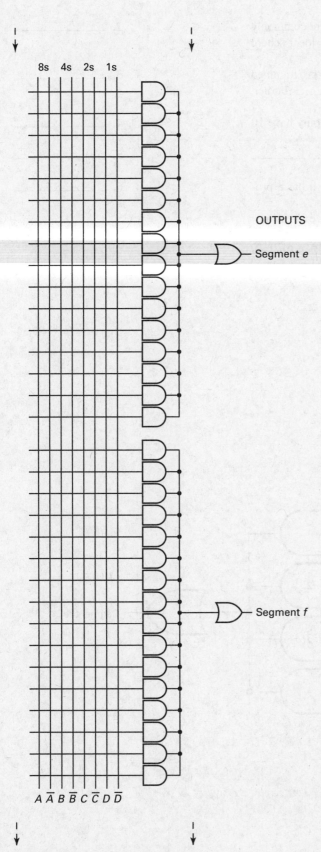

Fig. 4-24(c) Fuse map (*continued*)

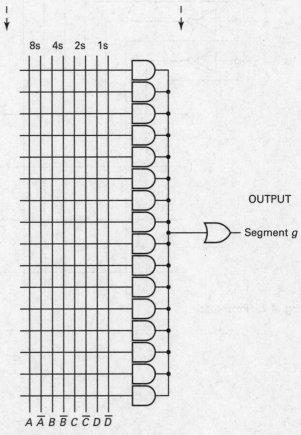

Fig. 4-24(d) Fuse map (*continued*)

QUESTIONS

Answer questions 1 to 7 below.

1. A fuse map _____ (is, is not) used in programming commercial programmable logic devices such as PALs and GALs.

2. Real programmable logic devices such as PALs and GALs are commonly programmed _____ (at the manufacturer, in the local school or lab).

3. Commercial PLDs (such as PALs and GALs) are commonly programmed using development software, a PC, and a PLD _____ (burner, scanner).

4. Smaller commercial PLDs (such as PALs and GALs) commonly have 10 to 20 inputs and _____ (a single, many) outputs.

5. The simple PLD in this problem was used to solve a _____ (combinational, sequential) logic problem.

6. The PLD in Fig. 4-23 is like a commercial PAL device and it has a programmable _____ (AND, OR) array.

7. Some PLDs can be programmed using simplified Boolean expressions. Complete the fuse map in Fig. 4-25 for the segment *e* output from the hex-to-seven-segment decoder from Table 4-7. Use the simplified Boolean expression $\overline{B} \cdot \overline{D} + A \cdot B + C \cdot \overline{D} + A \cdot C =$ segment *e*.

1. _____

2. _____

3. _____

4. _____

5. _____

6. _____

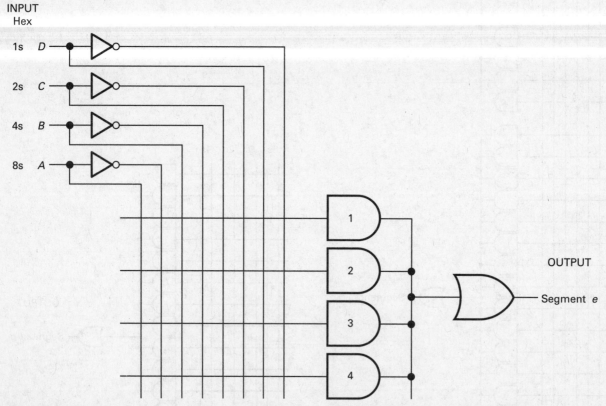

Fig. 4-25 Fuse map.

4-8 BASIC STAMP EXPERIMENT: BINARY-TO-SEVEN-SEGMENT DISPLAY DECODER

OBJECTIVES

1. To use a BASIC Stamp 2 (BS2)–based development board and personal computer to program a binary-to-seven-segment decoder/driver.

2. To wire and test the decoder and verify its proper operation.

MATERIALS

1   BASIC Stamp 2 development board (such as the Board of Education or HomeWork board available from Parallax)

1   BS2 module (this may be mounted on the development board)

1   PC with MS Windows

1   PBASIC text editor from Parallax

1   Serial cable (for downloading to BS2 module on development board) or USB cable

3   Push-button switches, N.O. contacts

3   10-kΩ resistor's

1   Seven-segment LED display, common anode (mounted on DB1000 display board by Dynalogic Concepts)

7   150-Ω resistor's (mounted on DB1000 display board)

Note: BASIC Stamp 2 module, development board, PBASIC software, and both experiment and reference manuals are available from Parallax, Inc., 599 Menlo Drive, #100, Rocklin, CA 95765. Many of the manuals and software can be downloaded free from Parallax's websites. General website: *www.parallax.com.* Telephone: 916-624-8333.

SYSTEM DIAGRAMS

The diagram in Fig. 4-26 shows three active HIGH input switches connected to ports *P8*, *P9*, and *P10* of the BASIC Stamp 2 module. Ports *P1* through *P7* are outputs that connect to the various segments of the LED display. A segment on the LED display will only light when its port goes LOW. For instance, if *only* output ports *P1*, *P2*, and *P3* are LOW, then only segments *a*, *b*, and *c* will light, forming the decimal number 7.

The input switches represent one of three bits of a binary number. Input switch *C* (*P8*) is the LSB or 1s bit. Input switch *A* (*P10*) is the MSB or 4s bit. For instance, to enter binary 011 would mean pressing both input switches *B* and *C* (activating ports *P8* and *P9* with 1s).

Programming is accomplished on a MS Windows–based personal computer using the PBASIC editor for the BS2 BASIC Stamp. The program is downloaded using a serial cable between the PC's serial port and the BASIC Stamp 2 development board. The BS2 module stores the program in EEPROM memory. The PBASIC program is interpreted and executed by the microcontroller on the BS2 module.

The following PBASIC program will cause the BS2 module to operate as a 3-bit binary-to-seven-segment LED display decoder/driver. The first line of the program (**'Decoder binary-to-7 segment**) is the title of the program. The apostrophe symbol (') at the beginning of this line identifies it as a remark statement. Remark statements are not executed by the BS2 IC but are displayed in the PBASIC listing to aid human understanding of the program. Lines 2 through 21 are statements used to declare the variables used and configure input and outputs. As an example, line 2 (**A VAR Bit 'Declare A as variable, 1 bit**) tells the microcontroller that a variable called *A* will be used

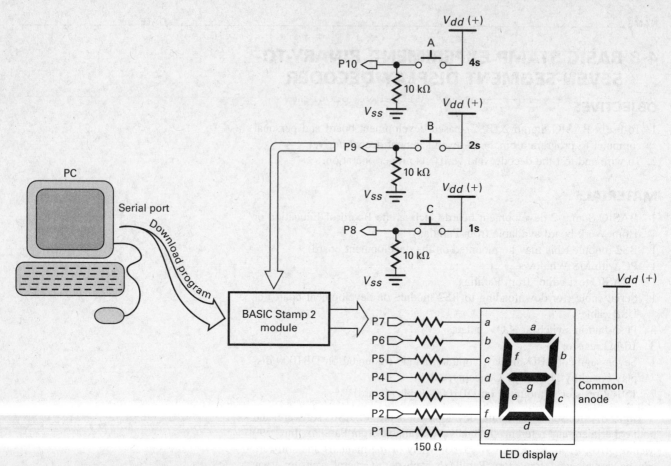

Fig. 4-26 BASIC Stamp 2 module programmed as a 3-bit binary-to-seven-segment LED display decoder/driver.

in the program and it will be only 1 bit long. The remark section of line 2 helps clarify the PBASIC statement in more human terms. Another example, line 12 (**INPUT 8**) informs the microcontroller that port 8 will be used as an input in this decoder program.

The **'Decoder binary-to-7 segment** program contains a continuous program loop starting with a label (**CkAllSwit:**). Labels are identified by names ending with a colon (:). Labels are places in a program that other lines of code can easily locate and jump to. The label (**CkAllSwit:**) shows the beginning of the main routine. Each of seven subroutines also begins with labels such as line 54 (**Segg:**) or line 72 (**Sega:**).

In the **CkAllswit:** main program routine, lines 23 through 29 initialize output ports 1 through 7 (ports are set HIGH which turns off all LEDs). Lines 30, 31, and 32 (**A = IN10** and **B = IN9** and **C = IN8**) assign the value (0 or 1) at the three input ports to variables *A*, *B*, and *C*.

```
'{$STAMP BS2}

'Decoder binary-to-7 segment 'Title of program                            L1

A  VAR Bit                    'Declare A as variable, 1 bit                L2
B  VAR Bit                    'Declare B as variable, 1 bit                L3
C  VAR Bit                    'Declare C as variable, 1 bit                L4
Y1 VAR Bit                    'Declare Y1 as variable, 1 bit               L5
Y2 VAR Bit                    'Declare Y2 as variable, 1 bit               L6
Y3 VAR Bit                    'Declare Y3 as variable, 1 bit               L7
Y4 VAR Bit                    'Declare Y4 as variable, 1 bit               L8
Y5 VAR Bit                    'Declare Y5 as variable, 1 bit               L9
Y6 VAR Bit                    'Declare Y6 as variable, 1 bit               L10
Y7 VAR Bit                    'Declare Y7 as variable, 1 bit               L11

INPUT 8                       'Declare port 8 as an input                  L12
INPUT 9                       'Declare port 9 as an input                  L13
INPUT 10                      'Declare port 10 as an input                 L14
OUTPUT 1                      'Declare port 1 as output Y1 (segment g)     L15
OUTPUT 2                      'Declare port 2 as output Y2 (segment f)     L16
OUTPUT 3                      'Declare port 3 as output Y3 (segment e)     L17
OUTPUT 4                      'Declare port 4 as output Y4 (segment d)     L18
OUTPUT 5                      'Declare port 5 as output Y5 (segment c)     L19
OUTPUT 6                      'Declare port 6 as output Y6 (segment b)     L20
OUTPUT 7                      'Declare port 7 as output Y7 (segment a)     L21

CkAllSwit:                    'Label for main routine                      L22
  OUT1 = 1                    'Initialize port 1 at 1, segment g off       L23
  OUT2 = 1                    'Initialize port 2 at 1, segment f off       L24
  OUT3 = 1                    'Initialize port 3 at 1, segment e off       L25
  OUT4 = 1                    'Initialize port 4 at 1, segment d off       L26
  OUT5 = 1                    'Initialize port 5 at 1, segment c off       L27
  OUT6 = 1                    'Initialize port 6 at 1, segment b off       L28
  OUT7 = 1                    'Initialize port 7 at 1, segment a off       L29
  A = IN10                    'Assign value at input port 10 to variable A L30
  B = IN9                     'Assign value at input port 9 to variable B  L31
  C = IN8                     'Assign value at input port 8 to variable C  L32
  Y1 = (~A&~B)|(A&B&C)        'Assign value (0 or 1) of expression to Y1   L33
  IF Y1 = 0 THEN Segg         'If Y1=0 then light segment g (P1= 0)        L34
  P2:                         'Label                                       L35
  Y2 = (~A&B)|(~A&C)|(B&C)    'Assign value (0 or 1) of expression to Y2   L36
  IF Y2 = 0 THEN Segf         'If Y2=0 then light segment f (P2= 0)        L37
  P3:                         'Label                                       L38
  Y3 = (A&~B)|C.              'Assign value (0 or 1) of expression to Y3   L39
  IF Y3 = 0 THEN Sege         'If Y3=0 then light segment e (P3= 0)        L40
  P4:                         'Label                                       L41
  Y4 = (~A&~B&C)|(A&~B&~C)|   'Assign value (0 or 1) of expression to Y4   L42
  (A&B&C)
  IF Y4 = 0 THEN Segd         'If Y4=0 then light segment d (P4= 0)        L43
  P5:                         'Label                                       L44
  Y5 = (~A&B&~C)              'Assign value (0 or 1) of expression to Y5   L45
  IF Y5 = 0 THEN Segc         'If Y5=0 then light segment c (P5= 0)        L46
  P6:                         'Label                                       L47
  Y6 = (A&~B&C)|(A&B&~C)      'Assign value (0 or 1) of expression to Y6   L48
  IF Y6 = 0 THEN Segb         'If Y6=0 then light segment b (P6= 0)        L49
```

```
P7:                                      'Label                                            L50
Y7 = (~A&~B&C)|(A&~B&~C)                  'Assign value (0 or 1) of expression to Y7        L51
IF Y7 = 0 THEN Sega                       'If Y7=0 then light segment a (P7= 0)             L52
GOTO CkallSwit                            'Go to start of main routine                     L53

Segg:                                    'Label Segg subroutine                            L54
OUT1 = 0                                 'Port 1 = 0, lights segment g                     L55
GOTO P2                                  'Go back to label P2                              L56

Segf:                                    'Label                                            L57
OUT2 = 0                                 'Port 2 = 0, lights segment f                     L58
GOTO P3                                  'Go back to label P3                              L59

Sege:                                    'Label                                            L60
OUT3 = 0                                 'Port 3 = 0, lights segment e                     L61
GOTO P4                                  'Go back to label P4                              L62

Segd:                                    'Label                                            L63
OUT4 = 0                                 'Port 4 = 0, lights segment d                     L64
GOTO P5                                  'Go back to label P5                              L65

Segc:                                    'Label                                            L66
OUT5 = 0                                 'Port 5 = 0, lights segment c                     L67
GOTO P6                                  'Go back to label P6                              L68

Segb:                                    'Label                                            L69
OUT6 = 0                                 'Port 6 = 0, lights segment b                     L70
GOTO P7                                  'Go back to label P7                              L71

Sega:                                    'Label                                            L72
OUT7 = 0                                 'Port 7 = 0, lights segment a                     L73
GOTO Ckallswit                           'Go back to beginning of main routine             L74
```

The truth table in Fig. 4-27 details the inputs and outputs for the binary-to-seven-segment decoder. Notice that a LOW (logical 0) is required to light a segment of a common-anode seven-segment LED display. When a segment on the LED lights, we say the microcontroller module is *driving* the display.

Boolean expressions are given for the seven outputs (*a–g*) in Fig. 4-28. These expressions have been simplified using either software or Karnaugh mapping techniques. Traditional Boolean expressions are listed on the left of

Decimal	INPUTS			OUTPUTS							Display
	A	B	C	0 = light			1 = no light				
	4s	2s	1s	a	b	c	d	e	f	g	
0	0	0	0	0	0	0	0	0	0	1	0
1	0	0	1	1	0	0	1	1	1	1	1
2	0	1	0	0	0	1	0	0	1	0	2
3	0	1	1	0	0	0	0	1	1	0	3
4	1	0	0	1	0	0	1	1	0	0	4
5	1	0	1	0	1	0	0	1	0	0	5
6	1	1	0	0	1	0	0	0	0	0	b
7	1	1	1	0	0	0	1	1	1	1	7

Fig. 4-27 Truth table for 3-bit binary-to-seven-segment LED display decoder.

Segment	Boolean Expression	PBASIC Assignment Statement	Line Number
a	$\bar{A}\cdot\bar{B}\cdot C + A\cdot\bar{B}\cdot\bar{C} = Y$	Y7 = (~A&~B&C) | (A&~B&~C)	L51
b	$A\cdot\bar{B}\cdot C + A\cdot B\cdot\bar{C} = Y$	Y6 = (A&~B&C) | (A&B&~C)	L48
c	$\bar{A}\cdot B\cdot\bar{C} = Y$	Y5 = ~A&B&~C	L45
d	$\bar{A}\cdot\bar{B}\cdot C + A\cdot\bar{B}\cdot\bar{C} + A\cdot B\cdot C = Y$	Y4 = (~A&~B&C) | (A&~B&~C) | (A&B&C)	L42
e	$A\cdot\bar{B} + C = Y$	Y3 = (A&~B) | C	L39
f	$\bar{A}\cdot B + \bar{A}\cdot C + B\cdot C = Y$	Y2 = (~A&B) | (~A&C) | (B&C)	L36
g	$\bar{A}\cdot\bar{B} + A\cdot B\cdot C = Y$	Y1 = (~A&~B) | (A&B&C)	L33

Fig. 4-28 Binary-to-seven segment decoder Boolean expressions and PBASIC assignment statements.

the chart for each output. The PBASIC assignment statement for each segment is shown at the right. For reference, the line numbers of the assignment statements are listed at the far right. These line numbers refer to the PBASIC program '**Decoder binary-to-7 segment.**

Consider line 33 of the '**Decoder binary-to-7 segment** program. The PBASIC statement **Y1 = (~A&~B) | (A&B&C)** assigns either a 0 or a 1 to variable *Y1* based on column *g* of the truth table in Fig. 4-27. Remember that variables *A*, *B*, and *C* are either 0 or 1 based on the condition of input switches connected to ports P10, P9, and P8 of the BASIC Stamp 2 module (Fig. 4-26).

IF-THEN statements are decision points in a program. The BASIC Stamp 2 module evaluates line 34 (**IF Y1 = 0 THEN Segg**). If *Y1* = 1, the program will drop to the next line of code, which is line 35. However, if *Y1* = 0 then the program will jump to the subroutine labeled **Segg:** which will light segment *g* of the seven-segment LED display. Notice that the **Segg:** subroutine will force port P1 to go LOW, causing segment *g* of the display to light (line 55). Line 56 of the subroutine will cause the program to jump back to the main routine (to **P2:**). The main routine is executed many times per second. The three input switches are monitored, and the appropriate segments of the seven-segment LED display light.

For reference, a diagram of Parallax's Board of Education development board is drawn in Fig. 4-29. A few of its features are identified on the drawing including:

a. Serial input (for downloading PBASIC programs from PC with serial cable)
b. BASIC Stamp 2 module (24-pin DIP)
c. Power connections—either 9-V battery or dc supply with barrel plug
d. ON/OFF switch and power indicator
e. Reset button—for restarting PBASIC programs
f. Voltage regulator
g. Breadboard, solderless
h. Power supply header [V_{dd} (+) and Vss (−)]
i. BS2 input/output (I/O) pins header

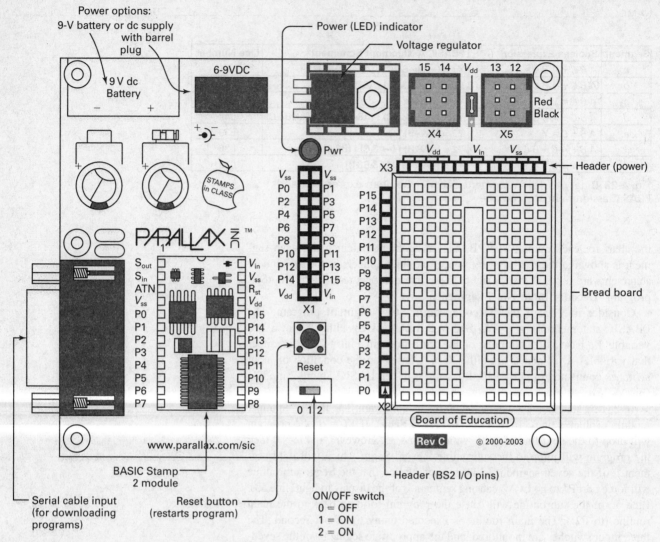

Fig. 4-29 Board of Education development board. (Used by permission of Parallax, Inc.)

Labels in figure:

Power options:
9-V battery or dc supply
with barrel plug

Power (LED) indicator

Voltage regulator

6-9VDC

9 V dc Battery

15 14 V_{dd} 13 12

Red
Black

X4   X5

STAMPS in CLASS

V_{dd} V_{in} V_{ss}

X3

Header (power)

Pwr

V_{ss} V_{ss}
P0   P1
P2   P3
P4   P5
P6   P7
P8   P9
P10   P11
P12   P13
P14   P15
V_{dd} V_{in}

X1

P15
P14
P13
P12
P11
P10
P9
P8
P7
P6
P5
P4
P3
P2
P1
P0

X2

Bread board

PARALLAX INC™

S_{out} V_{in}
S_{in} V_{ss}
ATN R_{st}
V_{ss} V_{dd}
P0   P15
P1   P14
P2   P13
P3   P12
P4   P11
P5   P10
P6   P9
P7   P8

Reset

0 1 2

www.parallax.com/sic

Board of Education

Rev C   © 2000-2003

BASIC Stamp 2 module

Header (BS2 I/O pins)

Serial cable input (for downloading programs)

Reset button (restarts program)

ON/OFF switch
0 = OFF
1 = ON
2 = ON

PROCEDURE

1. Refer to Fig. 4-26. Wire the circuit to a BS2 module using a development board such as Parallax's BOE (see Fig. 4-29). BS2 pins *P8*, *P9*, and *P10* are used as inputs. BS2 pins *P1*, *P2*, *P3*, *P4*, *P5*, *P6*, and *P7* are used as outputs to drive LED segments in this circuit. The seven 150-Ω limiting resistors and seven-segment LED common-anode display may be mounted on a display board.

2. Using an MS Windows–based PC, start the BASIC Stamp editor and write the PBASIC program titled '**Decoder binary-to-7 segment.**

3. Connect a serial cable between the serial output port of the PC and the BOE development board. Your instructor may have you use a USB cable.

4. Power ON (BOE board). Download the '**Decoder binary-to-7 segment** program. The program should be running on the BOE.

5. Test the circuit by pressing combinations of the three input switches and observing the seven-segment LED display.

6. Disconnect the serial cable from the PC. Power OFF and then ON (BOE). Retest the circuit. The program was retained in EEPROM memory when the BOE was turned off and should restart when turned on again.

7. Show your instructor your circuit, and be prepared to answer questions about the circuit and the PBASIC program.

8. Power OFF. Take down the circuit, and return all equipment to its proper place.

QUESTIONS

Answers questions 1 to 13.

1. At the heart of the BASIC Stamp 2 module is a programmable device called a _____ (microcontroller, multiplier).

2. The high-level language used to program a BS2 module is _____ (FORTRAN, PBASIC).

3. Refer to Fig. 4-26 and the **'Decoder binary-to-7 segment** program. List the three input ports and seven output ports.

4. Refer to Fig. 4-26. The push-button inputs are designed as _____ (active HIGH, active LOW) switches, which means a logical 1 is applied to the input port when the switch is activated.

5. Refer to Fig. 4-26, Fig. 4-27, and the **'Decoder binary-to-7 segment** program. If all input switches (*A*, *B*, and *C*) are activated, the seven-segment LED display will read the decimal number _____ because segments *a*, *b*, and *c* light.

6. Refer to Fig. 4-26, Fig. 4-27, and the **'Decoder binary-to-7 segment** program. When only input switch *B* is activated, the decimal number 2 displays because segments _____ (answer with 5 letters) light.

7. When downloading a PBASIC program from the PC to the BS2 module, the program is stored in _____ (EEPROM, ROM) memory.

8. Refer to Fig. 4-29. On the power header on the BOE board, V_{ss} means the _____ (negative, positive) of the power supply.

9. Refer to the truth table in Fig. 4-27 and the chart in Fig. 4-28. What is the simplified Boolean expression for segment *f*? What is the PBASIC assignment statement for segment *f*?

10. Refer to line 49 of the **'Decoder binary-to-7 segment** program. If $y6 = 1$, which line of code is executed next?

11. Refer to the **'Decoder binary-to-7 segment** program. What is the label that identifies the subroutine that lights segment *a* of the seven-segment LED display?

12. The main routine in the **'Decoder binary-to-7 segment** program begins with the label _____ at line _____. The main routine ends with the PBASIC statement _____.

13. PBASIC code such as **P7:** is called a _____ (label, remark) and is a location that can be found by GOTO statements.

1. _____

2. _____

3. _____

4. _____

5. _____

6. _____

7. _____

8. _____

9. _____
   _____
   _____

10. _____

11. _____

12. _____
    _____
    _____

13. _____

CHAPTER 5

IC Specifications and Simple Interfacing

TEST: IC SPECIFICATIONS AND SIMPLE INTERFACING

Answer the questions in the spaces provided.

1. Interfacing may be defined as the simultaneous transmission of two or more signals over a single path. (T or F)

   1. _____

2. Applying 1.4 V to a TTL input is interpreted by the IC as a(n) _____ logic level.
   - **a.** HIGH
   - **b.** LOW
   - **c.** Undefined

   2. _____

3. A TTL output of 0.2 V is considered a(n) _____ output.
   - **a.** HIGH
   - **b.** LOW
   - **c.** Undefined

   3. _____

4. Applying 1 V to a 4000 series CMOS input (10-V power supply) is interpreted by the IC as a(n) _____ logic level.
   - **a.** HIGH
   - **b.** LOW
   - **c.** Undefined

   4. _____

5. Applying 2.8 V to a 74HC00 series CMOS input (5-V power supply) is interpreted by the IC as a(n) _____ logic level.
   - **a.** HIGH
   - **b.** LOW
   - **c.** Undefined

   5. _____

6. A "typical" HIGH output voltage for a standard TTL gate would be about _____ volts.
   - **a.** 0.1
   - **b.** 0.8
   - **c.** 1.9
   - **d.** 3.5

   6. _____

7. A "typical" HIGH output voltage for a CMOS gate (10-V power supply) would be about _____ volts.
   - **a.** 0.1
   - **b.** 3
   - **c.** 5
   - **d.** 7
   - **e.** 10

   7. _____

8. Applying 3 V to a 74HCT00 series CMOS input (5-V power supply) is interpreted by the IC as a(n) _____ logic level.
   - **a.** HIGH
   - **b.** LOW
   - **c.** Undefined

   8. _____

9. Unwanted voltages induced in the connecting wires and PC board traces in a digital system that might affect its operation are called
   - **a.** Digital junk
   - **b.** Noncoherent flux
   - **c.** Noise
   - **d.** Nonlinear induction
   - **e.** Clutter

   9. _____

10. Excellent noise immunity is a characteristic of the _____ logic family.
   - **a.** CMOS
   - **b.** DDL
   - **c.** TTL
   - **d.** RTL

   10. _____

11. The switching threshold is _____ for standard TTL logic gate inputs.
   a. About 0.0 V   d. Exactly 2.4 V
   b. Exactly 0.6 V   e. About 5.0 V
   c. Between 0.8 and 2.0 V

11. _____

12. The fan-out of standard TTL is 10 when driving other standard TTL gates. (T or F)

12. _____

13. The fan-out of the 4000 series CMOS family is considered to be about
   a. 0.001   c. 1   e. 50
   b. 0.1   d. 10

13. _____

14. The _____ family of ICs has the least input loading.
   a. CMOS
   b. TTL

14. _____

15. The FAST-TTL family of ICs has very
   a. Low propagation delays and low speed
   b. High propagation delays and high speed
   c. Low propagation delays and high speed
   d. High propagation delays and low speed

15. _____

16. It is recommended that the pins of CMOS ICs be stored in conductive foam to guard against static electricity. (T or F)

16. _____

17. The V_{SS} pin on a 4000 series CMOS IC is connected to a(n) _____ output of the power supply.
   a. Alternating current   c. Negative (GND)
   b. Positive   d. Neutral

17. _____

18. The _____ series CMOS subfamily features the highest speed and very good drive capabilities and operates at low voltages (3 V or less).
   a. 74ALVC00   c. 74ALS00
   b. 4000   d. 74AC00

18. _____

19. Refer to Fig. 5-1. Component R_1 is called a _____ resistor.
   a. Limiting   c. Pull-up
   b. Pull-down   d. Load

19. _____

20. Refer to Fig. 5-1. Closing SW_1 causes the input of the inverter to go _____, which results in the LED output _____.
   a. HIGH, going out   c. LOW, going out
   b. LOW, lighting   d. HIGH, lighting

20. _____ _____

21. Refer to Fig. 5-1. With SW_1 open, a _____ appears at the input of the inverter, causing the output LED to _____.
   a. HIGH, go out   c. LOW, go out
   b. LOW, light   d. HIGH, light

21. _____ _____

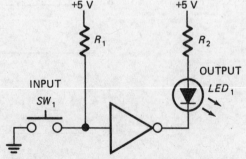

Fig. 5-1 Interface circuit for questions 19 through 21.

22. Refer to Fig. 5-2. The cross-coupled NAND gates are being used to _____ the input switch.

22. _____

   a. Decode   **c.** Debounce
   b. Lubricate   **d.** Multiplex

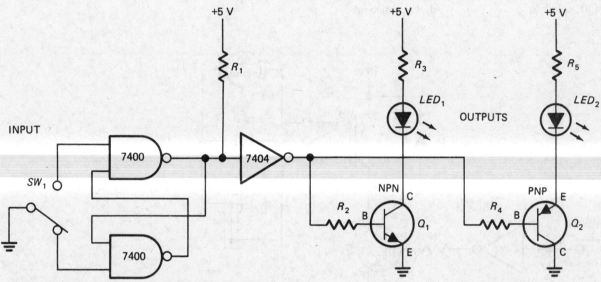

Fig. 5-2 Interface circuit for questions 22 through 24.

23. Refer to Fig. 5-2. When SW_1 is in the down position as shown, the input to the inverter is

23. _____

   a. HIGH
   b. LOW

24. Refer to Fig. 5-2. With a LOW input at the bottom of R_1, the output of the inverter is _____, transistor _____ is turned on, and output indicator _____ is lit.

24. _____ _____
_____

   a. LOW, Q_2, LED_2 **c.** HIGH, Q_1, LED_1
   b. HIGH, Q_2, LED_2 **d.** LOW, Q_1, LED_1

25. A TTL output can drive a CMOS input with the addition of a(n)

25. _____

   a. Switch debouncer   **c.** Input capacitor
   b. Pull-up resistor   **d.** Output inductor

26. A 74HC00 series CMOS gate can drive an LS-TTL input with no additional parts. (T or F)

26. _____

27. A 4000 series CMOS output can drive a standard TTL input with the addition of a

27. _____

   a. Buffer (such as 4049)   **c.** Switch debouncer
   b. Pull-up resistor   **d.** Pull-down resistor

28. Open-collector TTL gates require the use of _____ at the outputs.

28. _____

   a. Debouncers
   b. Pull-up resistors
   c. Latches
   d. Pull-down resistors

29. The _____ series of ICs is designed to serve as an interface between TTL and CMOS logic elements.
 a. 7400
 b. 4000
 c. 74HCT00
 d. 74C00

29. _____

30. Refer to Fig. 5-3. With SW_1 open, the transistor is turned _____ and the _____ is activated and operates.
 a. Off, motor   c. On, motor
 b. Off, solenoid   d. On, solenoid

30. _____ _____

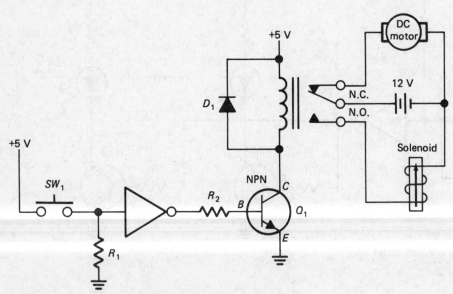

Fig. 5-3 Interface circuit for questions 30 through 33.

31. Refer to Fig. 5-3. Component R_1 is called a _____ resistor.
 a. Limiting
 b. Voltage-boosting
 c. Clamping
 d. Pull-down

31. _____

32. Refer to Fig. 5-3. With SW_1 closed, the output of the inverter goes
 a. HIGH
 b. LOW

32. _____

33. Refer to Fig. 5-3. With SW_1 closed, the transistor is turned _____ and the _____ is activated and operates.
 a. Off, motor
 b. Off, solenoid
 c. On, motor
 d. On, solenoid

33. _____ _____

34. Refer to Fig. 5-4. The purpose of the 4N25 device is to electrically _____ the digital circuitry from the higher-voltage and noise-producing motor circuit.
 a. Debounce
 b. Connect
 c. Isolate

34. _____

35. Refer to Fig. 5-4. The 4N25 device is called a(n) _____.
 a. SCR
 b. Triac
 c. Laser detector
 d. Optoisolator

35. _____

116

36. Refer to Fig. 5-4. Pressing switch SW_1 will cause the DC motor to
_____.
  a. Rotate
  b. Not rotate

36. _____

37. Refer to Fig. 5-4. When switch SW_1 is pressed a HIGH appears at pin 2
of the 4N25, the LED does not light, the 4N25's phototransistor is turned
off, the base of transistor Q_1 goes positive, transistor Q_1 turns on, and the
dc motor rotates. (T or F)

37. _____

38. Refer to Fig. 5-4. When switch SW_1 is released, a LOW appears at pin 2
of the 4N25, the LED lights, the 4N25's phototransistor is turned on, the
base of transistor Q_1 goes LOW, and transistor Q_1 and the dc motor are
deactivated. (T or F)

38. _____

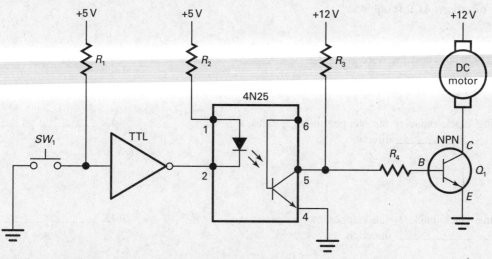

Fig. 5-4 Interface circuit for questions 34 through 39.

39. Refer to Fig. 5-4. With switch SW_1 open the output of the inverter is LOW
and is said to be _____, which is flowing through resister R_2
and the LED inside the 4N25 optoisolator.
  a. Sinking current
  b. Sourcing current

39. _____

40. Refer to Fig. 5-5. The pulse generator will vary the _____,
causing the servo motor to adjust the angular position of the output shaft.
  a. Pulse width from about 1 to 2 ms
  b. Pulse amplitude from about 1 to 10 V

40. _____

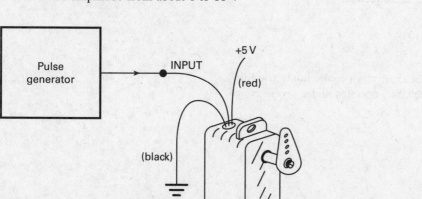

Fig. 5-5 Pulse generator for question 40.

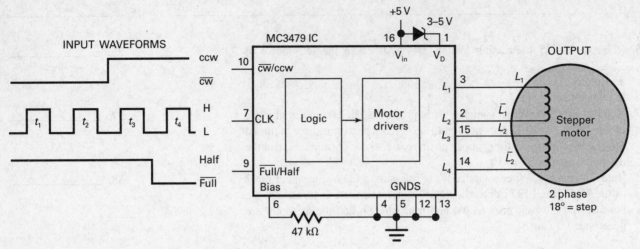

Fig. 5-6 Interface circuit for questions 41 through 45.

41. Refer to Fig. 5-6. The MC3479 IC is best described as a _____.　　41. _____
    a. Servo motor driver
    b. Optocoupler
    c. Stepper motor driver

42. Refer to Fig. 5-6. During clock pulse t_1 the stepper motor rotates a　　42. _____ _____
    _____ in the _____ direction.
    a. Half step, CW
    b. Half step, CCW
    c. Full step, CW
    d. Full step, CCW

43. Refer to Fig. 5-6. During clock pulse t_2 the stepper motor rotates a　　43. _____ _____
    _____ in the _____ direction.
    a. Half step, CW
    b. Half step, CCW
    c. Full step, CW
    d. Full step, CCW

44. Refer to Fig. 5-6. During clock pulse t_3 the stepper motor rotates a　　44. _____ _____
    _____ in the _____ direction.
    a. Half step, CW
    b. Half step, CCW
    c. Full step, CW
    d. Full step, CCW

45. Refer to Fig. 5-6. During clock pulse t_4 the stepper motor rotates a　　45. _____ _____
    _____ in the _____ direction.
    a. Half step, CW
    b. Half step, CCW
    c. Full step, CW
    d. Full step, CCW

46. The Hall-effect sensor is a _____ device commonly used in　　46. _____
    automobiles because it is rugged, is reliable, operates under severe con-
    ditions, and is inexpensive.
    a. Light-activated
    b. Magnetically activated
    c. Thermal-activated

47. Refer to Fig. 5-7. If the Hall-effect IC uses *bipolar switching,* then first increasing the magnetic S pole will turn the switch _____, while second changing to the N pole and increasing the magnetic field will turn the Hall-effect switch _____.
  a. Off, on
  b. On, off
  c. Toggle, toggle

48. Refer to Fig. 5-7. The output of just the Hall-effect sensor (see point *A* inside the IC) is _____ in nature, while the output of the switch at pin 3 is classified as _____.
  a. Analog, analog
  b. Analog, digital
  c. Digital, analog

47. _____ _____

48. _____ _____

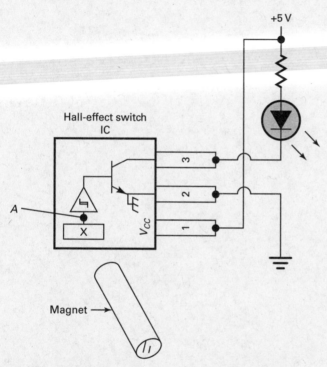

Fig. 5-7 Hall-effect switch for questions 47 and 48.

49. Refer to Fig. 5-5. A BASIC Stamp microcontroller module could be programmed to act as the PWM pulse generator to rotate the servo motor. (T or F)

50. The PBASIC statement PULSOUT 14, 1000 would generate 14 positive pulses, each 1000 seconds long. (T or F)

49. _____

50. _____

5-1 LAB EXPERIMENT: INTERFACING SWITCHES WITH TTL

OBJECTIVES

1. To wire and test active LOW and active HIGH push-button switches.
2. To observe the use of pull-up and pull-down resistors.
3. To wire and test the effect of switch bounce on a TTL counter circuit.
4. To add switch-debouncing circuitry to the keypad for normal operation of the counter circuit.

MATERIALS

Qty.		Qty.	
1	7404 hex inverter TTL IC	1	33-Ω, $\frac{1}{4}$-W resistor
1	7414 hex Schmitt trigger inverter TTL IC	1	330-Ω, $\frac{1}{4}$-W resistor
1	74192 decade counter TTL IC	2	10-kΩ, $\frac{1}{4}$-W resistors
1	keypad (or two N.O. push-button switches)	1	47-μF electrolytic capacitor
		1	5-V dc regulated power supply
4	LED indicator-light assemblies	1	logic probe (or dc voltmeter)

SYSTEM DIAGRAMS

Figure 5-8(*a*) is a wiring diagram for interfacing a N.O. push-button switch with a TTL inverter. With the switch open (key not depressed), the 10-kΩ *pull-up resistor* makes sure the input of the 7404 inverter is HIGH. When the push button is pressed, the input of the inverter is grounded by the switch. This is called an active LOW switch. That is, the input is driven LOW when the switch is activated.

An active HIGH switch is diagrammed in Fig. 5-8(*b*). With the switch open, the 330-Ω *pull-down resistor* holds the TTL input LOW. The pull-down resistor's value is fairly low to supply the higher current required by a standard TTL LOW input.

The schematic in Fig. 5-9 is a *faulty* circuit. It will be used to *demonstrate switch bounce*. Pressing C on the keyboard clears the output display to 0000. Pressing and releasing the 1 key should deliver *just one pulse* to the count-up (clock) input of the 74192 TTL counter IC. You will observe that because of switch bounce, the very fast TTL counter will detect and count from one to four or more pulses per keyboard entry. The normal binary counting sequence should be 0000, 0001, 0010, 0011, 0100, 0101, 0110, 0111, 1000, 1001, and then 0000, 0001, and so forth.

An experimental switch-debouncing circuit has been added to the counter circuit in Fig. 5-10. The addition of the capacitor and Schmitt trigger inverters should eliminate most switch bounce. The two Schmitt trigger inverters have no logical function but do serve to square up the waveform that enters the count-up (clock) input of the counter. The inverters' hysteresis characteristics also tend to ignore minor switch bouncing. The capacitor absorbs the HIGH-LOW-HIGH-LOW variations caused by switch bounce. The switch-debouncing circuit in Fig. 5-10 is considered experimental in that it has a high parts count and is not totally reliable. During testing, mechanical switches with metal contacts tested well in this circuit. However, a few bounces were detected with a Mylar keypad. Mylar film keypads use conductive ink instead of metal contacts and cause more switch bounce.

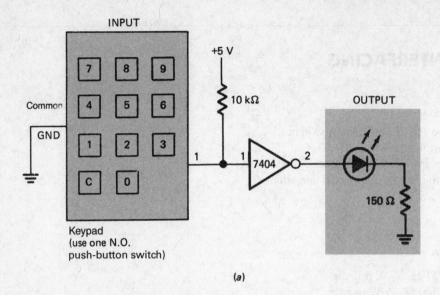

(a)

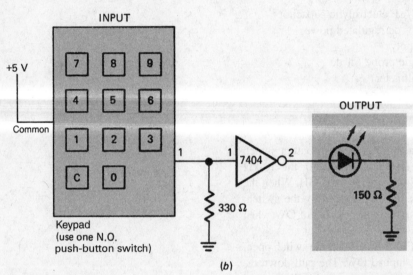

Fig. 5-8 Push-button interface circuits. (a) Active LOW switch. (b) Active HIGH switch.

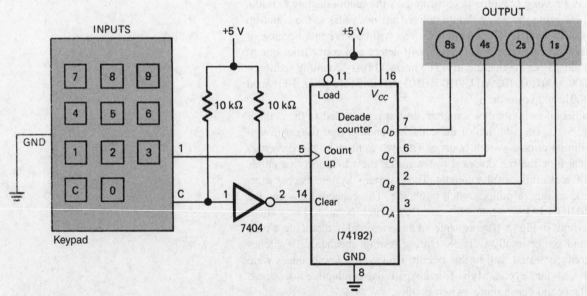

Fig. 5-9 Faulty pulse-counting circuit (without switch debouncing).

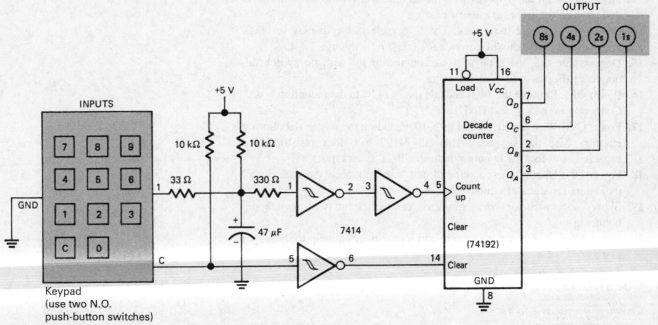

Fig. 5-10 Pulse-counting circuit with switch debouncing.

PROCEDURE

1. Insert the 7404 IC into a mounting board.
2. Power OFF. Connect power to the 7404 IC: the red wire for +5 V (V_{CC}) and the black wire for GND.
3. Refer to Fig. 5-8(*a*). Wire the entire circuit (keypad, resistor, 7404 IC, and LED indicator-light assembly). Note only the GND (or common) and N.O. push-button switch 1 are being used on the keypad. See your instructor if you have questions about your keypad.
4. Power ON. The LED should be OFF. The pull-up resistor is holding the input of the inverter HIGH. The output of the inverter should be LOW.
5. Press the 1 key and observe the results. The LED should light.
6. Connect a logic probe to the power supply of this circuit. Check the input and output readings to see if they correspond to your expectations. Why is this called an *active LOW switch?*

_____

7. Remove the 10-kΩ pull-up resistor. Power ON. Operate the circuit and observe the results. Does it work? It probably works because the input to the 7404 inverter is "floating HIGH when the input switch is open."
8. With a logic probe (or dc voltmeter) check the logic level at the input of the 7404 inverter (switch open). What are the results? The input is probably "floating" in the undefined region which is commonly interpreted as a HIGH input by TTL. *Floating inputs can cause trouble.*
9. Power OFF. Rewire the circuit as in Fig. 5-8(*b*). Note that a 330-Ω pull-down resistor is added and the common connection on the keypad goes to +5 V.
10. Power ON. Operate the circuit and observe the results.
11. Remove the pull-down resistor. Operate the circuit and observe the results. Does the circuit work properly without the pull-down resistor in place?

_____

12. Power OFF. Add the 74192 IC to the mounting board. Connect power (V_{CC} and GND).

13. Wire the faulty counter circuit in Fig. 5-9. Use the keypad, 7404 and 74192 ICs, two resistors, and four LED output indicator-light assemblies. Physically align the displays with the 8s indicator on the left and the 1s indicator on the right for ease of reading.
14. Power ON. Operate the circuit. The C key clears the display to 0000. Pressing the 1 key should *advance the count by only one.* Try this.
15. Demonstrate the operation of your counter circuit with no switch debouncing (Fig. 5-9) to your instructor.
16. Power OFF. Remove the 7404 and add the 7414 IC to the mounting board. Connect power (V_{CC} and GND).
17. Power OFF. Wire the circuit in Fig. 5-10, including the switch-debouncing circuitry. Use the keypad, 7414 and 74192 ICs, four resistors, one capacitor, and four LED output indicator-light assemblies.
18. Power ON. Operate the counting circuit, carefully observing the output indicators for signs of contact bounce.
19. Show your instructor your completed counter circuit with switch debouncing.
20. Power OFF. Take down the circuit, and return all equipment to its proper place.

QUESTIONS

Complete questions 1 to 11.

1. Refer to Fig. 5-8(*a*). Pressing the switch places a _____ (HIGH, LOW) at the input of the inverter and the LED _____ (goes out, lights).

   1. _____ _____

2. Refer to Fig. 5-8. Which of these two circuits is considered an active HIGH switch?

   2. _____

3. Refer to Fig. 5-8(*b*). The 330-Ω component is called a(n) _____ resistor.

   3. _____

4. Removing the pull-up resistor in Fig. 5-8(*a*) causes the input to the inverter to "float." Floating TTL inputs are usually interpreted as a _____ (HIGH, LOW) by the IC.

   4. _____

5. Refer to the faulty counter circuit in Fig. 5-9. What is the problem with this circuit?

   5. _____
   _____

6. Refer to Figs. 5-9 and 5-10. What is the purpose of the added components in the counter circuit in Fig. 5-10?

   6. _____
   _____
   _____

7. Refer to Fig. 5-10. The hysteresis symbol in the middle of the 7414 symbols shows that these are _____ trigger inverters.

   7. _____

8. Refer to Fig. 5-10. Pressing the C key sends a _____ (HIGH, LOW) signal to the clear input of the 74192 counter IC, clearing the outputs to _____.

   8. _____ _____

9. The keypad used in this experiment contained normally _____ (closed, open) pushbutton switches.

   9. _____

10. What is the normal counting sequence of the 74192 counter IC in Fig. 5-10?

    10. _____
    _____
    _____
    _____

11. Refer to Fig. 5-8(*a*). Removing the 10 kΩ pull-up resistor causes the input to the 7404 TTL IC to "float HIGH" which can sometimes cause problems. (T or F)

    11. _____

5-2 LAB EXPERIMENT: INTERFACING LEDS WITH TTL AND CMOS

OBJECTIVES

1. To wire and test several simple TTL-to-LED interface circuits.
2. To wire and test a CMOS-to-LED interface using driver transistors.

MATERIALS

Qty.

1   7404 hex inverter TTL IC
1   74HC14 hex inverter CMOS IC
1   2N3904 general-purpose NPN transistor
1   2N3906 general-purpose PNP transistor
1   logic switch

Qty.

1   red LED
1   green LED
2   150-Ω, $\frac{1}{4}$-W resistors
1   330-Ω, $\frac{1}{4}$-W resistor
1   680-Ω, $\frac{1}{4}$-W resistor
2   33-kΩ, $\frac{1}{4}$-W resistors
1   5-V dc regulated power supply

SYSTEM DIAGRAMS

Figure 5-11(a) diagrams a simple LED output indicator. The 7404 TTL IC drives the LED directly. The current draw is greater than specification for standard TTL, but the circuit works as a simple output indicator in noncritical applications.

Another simple LED output indicator circuit is diagrammed in Fig. 5-11(b). This circuit will indicate a green light for a LOW or red light for a HIGH, or both LEDs will light when the output of the gate is in the undefined region. The disadvantage of this display is that the standard TTL inverter must drive the LEDs directly. Another disadvantage is that the voltage profile does not exactly match the output voltage profile for TTL.

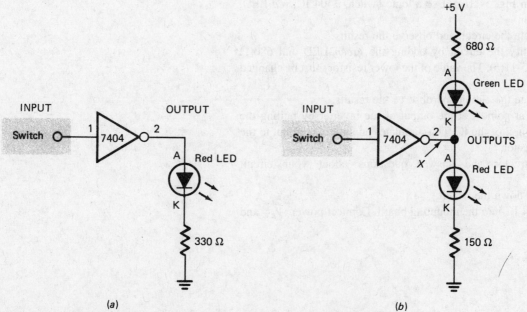

(a)   (b)

Fig. 5-11 Interfacing LEDs to standard TTL. (a) Driving a single LED direct. (b) Driving HIGH-LOW indicators.

The output indicator circuit in Fig. 5-12 operates the same as the previous HIGH-LOW indicator. However, it does not load the inverter, and its voltage profile is very close to matching TTL output voltage characteristics. The HIGH-LOW indicator in Fig. 5-12 can be driven by either TTL or CMOS ICs. The HIGH-LOW indicator is a simple logic probe.

When the *output* of the inverter in Fig. 5-11(b) is LOW, it is said the IC is *sinking* current to ground. However, when the *output* of the inverter in Fig. 5-11(b) is HIGH, it is said that the IC is *sourcing* current. TTL and FACT ICs usually can sink more current than they can source.

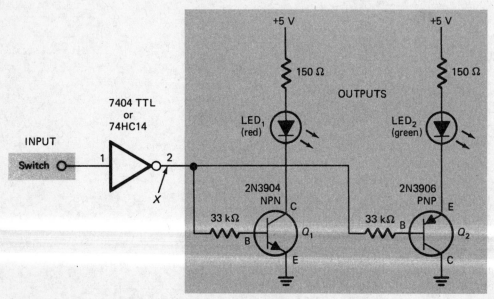

Fig. 5-12 Interfacing LEDs with CMOS and TTL using transistor drivers.

PROCEDURE

1. Insert the 7404 IC into the mounting board.
2. Power OFF. Connect power (V_{CC} and GND) to the IC.
3. Wire the circuit in Fig. 5-11(a). Use a logic switch, 7404 IC, *red* LED, and 330-Ω resistor.
4. Power ON. Operate the circuit and observe the results.
5. Power OFF. Modify the circuit by adding the *green* LED and 680-Ω resistor as in Fig. 5-11(b). The value of the lower resistor must be changed to 150-Ω.
6. Power ON. Operate the circuit and observe the results.
7. Break the circuit at point X at the output of the inverter by pulling the wire loose from pin 2 of the IC. This simulates an undefined input to the HIGH-LOW indicator.
8. Show your instructor and answer questions about your circuit [Fig. 5-11(b)].
9. Power OFF. Take down the circuit.
10. Insert the 74HC14 IC into the mounting board. Connect power (V_{CC} and GND).

NAME _____ DATE _____

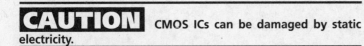

CAUTION CMOS ICs can be damaged by static electricity.

11. Wire the circuit in Fig. 5-12. Use an input switch, a 74HC14 IC, two transistors (NPN and PNP), two LEDs (green and red), and four resistors. See Appendix A for IC pin diagrams and transistor basing diagrams.
12. Power ON. Operate the circuit and observe the results.
13. Break the circuit at point X (see Fig. 5-12) at the output of the inverter by pulling the wire loose from pin 2 of the IC. This simulates an undefined input to the HIGH-LOW indicator.
14. Show your instructor and answer questions about your interfacing circuit (Fig. 5-12).
15. Power OFF. Take down the circuit, and return all equipment to its proper place. The pins of CMOS ICs should be stored in conductive foam.

QUESTIONS

Complete questions 1 to 10.

1. Refer to Fig. 5-11(a). The LED lights when the output of the inverter goes _____ (HIGH, LOW).

1. _____

2. Refer to Fig. 5-11(b). The _____ (green, red) LED lights when the output of the inverter goes LOW.

2. _____

3. Refer to Fig. 5-11(b). When the *input* to the inverter is LOW, the _____ (green, red) LED lights.

3. _____

4. Refer to Fig. 5-11(b). If the output of the inverter is floating at an undefined logic level, which LED(s) light?

4. _____

5. Refer to Fig. 5-12. With the output of the inverter LOW, transistor _____ (Q_1, Q_2) turns ON and the _____ (green, red) LED lights.

5. _____ _____

6. Refer to Fig. 5-12. With the *input* to the inverter LOW, the _____ (green, red) LED lights.

6. _____

7. List at least one disadvantage of the HIGH-LOW indicator in Fig. 5-11(b) compared with the circuit in Fig. 5-12.

7. _____
_____

8. Refer to Fig. 5-11(b). When the *output* of the inverter is LOW, the green LED lights and the 7404 IC is said to be _____ (sinking, sourcing) the current to ground.

8. _____

9. Refer to Fig. 5-11(b). When the *output* of the inverter is HIGH, the red LED lights and the 7404 IC is said to be _____ (sinking, sourcing) the current.

9. _____

10. Refer to Fig. 5-12. The output section of this circuit is a simple test instrument called a _____ (frequency counter, logic probe).

10. _____

5-3 LAB EXPERIMENT: INTERFACING TTL AND CMOS INTEGRATED CIRCUITS

OBJECTIVES

1. To design, draw, and wire a TTL-to-HC-CMOS interface circuit.
2. To design, draw, and wire an HC-CMOS-to-standard TTL interface circuit.
3. To observe how some "faulty" designs may work properly most of the time even if they violate some "worst-case-conditions" design rules.

MATERIALS

Qty.		Qty.	
1	74HC00 two-input NAND gate CMOS IC	3	logic switches
1	7404 hex inverter TTL IC	1	LED indicator-light assembly
1	7432 two-input OR gate TTL IC	2	1-kΩ, $\frac{1}{4}$-W resistors
1	74LS244 octal buffer/driver TTL IC	1	5-V dc regulated power supply

SYSTEM DIAGRAM

Two logic diagrams are sketched in Fig. 5-13. All of the ICs are powered by a single 5-V power supply. Your task is to design an appropriate interface in the area shown for both circuits. Section 5-6 and the table in Fig. 5-6(*b*) in the companion textbook *Digital Electronics, Principles and Applications*, 8th ed., will be useful in this design work.

Many of the design rules in digital electronics are based on *worst-case conditions*. For instance, a "floating TTL input should be tied HIGH through a pull-up resistor" is the rule, but 95 percent of the time an unconnected TTL input will float HIGH. Many of the rules seem overly strict because they must guarantee satisfactory operation 100 percent of the time (not just 95 percent of the time). This idea will be reinforced in the experiment when, near the end, you will wire the circuits in Fig. 5-13 directly (with no interfacing) and find they *will probably work*. Remember that many design rules and specifications consider the worst possible conditions of temperature, high frequency, noise, poor-quality power supply, and so forth.

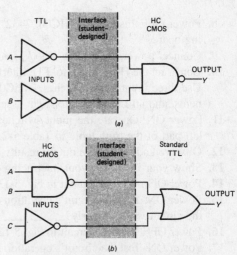

Fig. 5-13 Student interface design problems. (*a*) TTL-to-HC-CMOS interface problem. (*b*) HC-CMOS-to-TTL interface problem.

PROCEDURE

1. Design and draw a wiring diagram for the TTL-to-HC-CMOS interfacing circuit outlined in Fig. 5-13(*a*). Suggested parts include the 7404 and 74HC00 ICs and several resistors.

2. Power OFF. Insert the 7404 and 74HC00 ICs into the mounting board.

CAUTION CMOS ICs can be damaged by static electricity.

3. Connect power to both ICs (V_{CC} and GND).
4. Construct the TTL-to-HC-CMOS interface circuit you designed in step 1. Use two input logic switches, 7404 and 74HC00 ICs, interface components, and an LED indicator-light assembly.
5. Power ON. Operate the input switches to the circuit as shown in the left part of the truth table in Table 5-1.
6. Observe and record the output results in the right column, Table 5-1.
7. Power OFF. Design and draw a wiring diagram for the HC-CMOS-to-standard-TTL interfacing circuit outlined in Fig. 5-13(b). Suggested parts include the 74HC00, 7432, and 74LS244 ICs.

TABLE 5-1 TTL-to-HC-CMOS Circuit

INPUTS		OUTPUT
B	A	Y
0	0	
0	1	
1	0	
1	1	

8. Power OFF. Insert the 74HC00, 7432, and 74LS244 ICs into the mounting board.
9. Connect power to all three ICs (V_{CC} and GND).
10. Construct the HC-CMOS-to-TTL interface circuit you designed in step 7. Use three input logic switches, 74HC00 and 7432 ICs, interface components, and an LED indicator-light assembly.
11. Power ON. Operate the input switches for each circuit as shown in the left part of the truth table in Table 5-2.
12. Observe and record the output results in the right column, Table 5-2.
13. Show your instructor your interfacing circuit designs.
14. Power OFF. Wire the circuit in Fig. 5-13(a) with *no interfacing*.
15. Power ON. Test the circuit's operation and compare with Table 5-1. Does the circuit work properly?_____
16. Power OFF. Wire the circuit in Fig. 5-13(b) with *no interfacing*.
17. Power ON. Test the circuit's operation and compare with Table 5-2. Does the circuit work properly?_____
18. Power OFF. Take down the circuit, and return all equipment to its proper place. The pins of CMOS ICs should be stored in conductive foam.

TABLE 5-2 HC-CMOS-to-Standard TTL Circuit

INPUTS			OUTPUT
C	B	A	Y
0	0	0	
0	0	1	
0	1	0	
0	1	1	
1	0	0	
1	0	1	
1	1	0	
1	1	1	

QUESTIONS

Complete questions 1 to 5.

1. Refer to Fig. 5-13(a). This circuit generates the truth table for a two-input _____ gate.

1. _____

2. Refer to Fig. 5-13(b). This circuit generates the truth table for a three-input _____ gate.

2. _____

3. Refer to Fig. 5-13(b). The interface between the HC-CMOS and standard TTL consisted of _____ (buffer ICs, pull-up resistors).

3. _____

4. Refer to Fig. 5-13(a). The interface between the TTL and HC-CMOS consisted of _____ (buffer ICs, pull-up resistors).

4. _____

5. Design rules are very restrictive because they are based on _____ conditions.

5. _____

130

5-4 LAB EXPERIMENT: INTERFACING CMOS WITH BUZZERS, RELAYS, AND MOTORS

OBJECTIVES

1. To wire and test a CMOS logic circuit interfaced with a relay, motor, and buzzer.

2. To observe and study the operation of the interface circuit.

MATERIALS

Qty.

1	74HC00 two-input NAND gate CMOS IC
1	74LS244 octal buffer/driver TTL IC
1	2N3904 general-purpose NPN transistor
2	1N4001 silicon diodes
1	keypad (three N.O. switches)
3	100-kΩ, $\frac{1}{4}$-W resistors

Qty.

1	2.2-kΩ, $\frac{1}{4}$-W resistor
1	relay, 5-V coil, 2-A (or more) contacts
1	piezo (or electronic) buzzer
1	dc electric motor (voltage choice of instructor)
1	5-V dc regulated power supply
1	dc power supply (to match dc motor)

SYSTEM DIAGRAM

Figure 5-14 is a wiring diagram for a circuit that demonstrates CMOS logic gates controlling electric devices such as relays, motors, and buzzers. Note that the dc motor circuit is isolated from the 5-V logic circuit. Your instructor

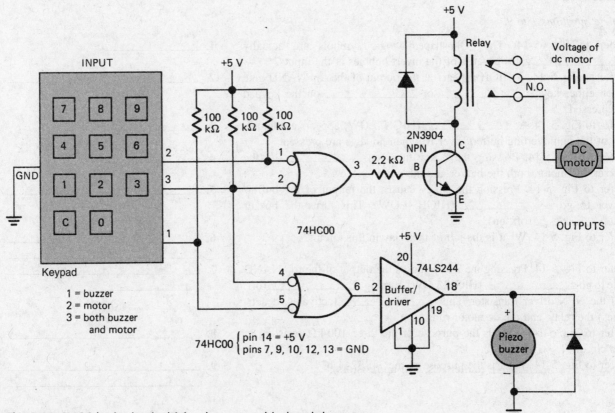

Fig. 5-14 CMOS logic circuit driving buzzer and isolated dc motor.

will give you the motor and suggest the voltage to be used in the isolated circuit. The 74LS244 IC is a TTL buffer/driver which has adequate output capacity to drive most piezo or electronic buzzers. The diodes across the relay coil and buzzer suppress transient voltages that may be induced in the system. The NPN driver transistor serves as the interface between the NAND gate and the relay coil.

PROCEDURE

1. Insert the 74HC00 and 74LS244 ICs into the mounting board.

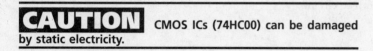

CAUTION CMOS ICs (74HC00) can be damaged by static electricity.

2. Power OFF. Connect power (V_{CC} and GND) to the two ICs. Connect common (GND) of keypad to GND of 5-V power supply. *Do not* connect power to the isolated dc motor at this time.
3. Wire the circuit in Fig. 5-14. Note the extra GND connections, on both ICs. Wire the ICs, keypad, resistors, NPN transistor, relay and diode, buzzer and diode, and electric motor.
4. Power ON. Connect a separate power supply (or battery) to the *isolated* dc motor circuit. Your instructor will suggest the correct voltage of power supply.
5. Operate the circuit and observe the results.
6. Show your instructor your wiring and answer questions on the circuit.
7. Power OFF. Turn off the isolated dc motor power supply. Turn off the 5-V power supply for the logic circuit.
8. Power OFF. Take down the circuit, and return all equipment to its proper place. The pins of CMOS ICs should be stored in conductive foam.

QUESTIONS

Complete questions 1 to 9.

1. Refer to Fig. 5-14. The "OR-shaped" logic symbols are actually _____ gates because of the invert bubbles at the inputs.
2. Refer to Fig. 5-14. A HIGH appears at the output of the top NAND gate when either keys _____ or _____ on the keypad are pressed.
3. Refer to Fig. 5-14. A _____ (HIGH, LOW) appears at the output of the noninverting buffer/driver IC when no keys are pressed.
4. Refer to Fig. 5-14. Pressing the 3 key turns on _____ (the buzzer, the motor, both the buzzer and the motor).
5. Refer to Fig. 5-14. Pressing the 3 key causes the output of the buffer/driver to go _____ (HIGH, LOW). This turns the buzzer _____ (off, on).
6. Refer to Fig. 5-14. What is the job of the relay in this circuit?

7. Refer to Fig. 5-14. Pressing the 2 key causes the output of the top NAND gate to go _____ (HIGH, LOW), turning _____ (off, on) the NPN driver transistor. This _____ (activates, deactivates) the relay, and the dc motor runs.
8. Refer to Fig. 5-14. What is the purpose of the three 100-kΩ resistors in this circuit?
9. Refer to Fig. 5-14. What is the purpose of the 74LS244 IC?

1. _____
2. _____ _____
3. _____
4. _____
5. _____ _____ _____
6. _____ _____ _____
7. _____ _____ _____
8. _____
9. _____ _____

5-5 LAB EXPERIMENT: USING AN OPTOISOLATOR IN INTERFACING

OBJECTIVES

1. To wire and operate a TTL logic circuit interfaced to a buzzer and dc motor.
2. To use an optoisolator to separate the logic circuit from the higher voltage and noisy dc motor circuit.
3. To observe and study the operation of the interface circuit.

MATERIALS

Qty.

1 7400 two-input NAND gate IC
1 7408 two-input AND gate IC
1 4N25 optoisolator IC
1 TIP31A NPN power transistor
1 1N4001 silicon diode
1 keypad (three N.O. switches)
3 10-kΩ resistors
1 4.7-kΩ resistor
1 2.2-kΩ resistor

Qty.

1 piezo (or electronic) buzzer
1 dc motor (voltage choice of instructor)
1 dc power supply (to match dc motor)
1 5-V dc regulated power supply

SYSTEM DIAGRAM

Figure 5-15 is a wiring diagram for a circuit that demonstrates TTL logic gates controlling electric devices such as motors and buzzers. The three pull-up resistors hold the inputs of the AND and NAND gates HIGH until the active LOW keypad activates a gate input. A TTL gate has enough capacity to drive the piezo buzzer directly so no special interfacing is required. The diode across the buzzer suppresses transient voltages that may be induced in the system.

Note in Fig. 5-15 that the dc motor is isolated from the 5-V logic circuit using the *4N25 optoisolator*. The top TTL AND gate is sourcing current for

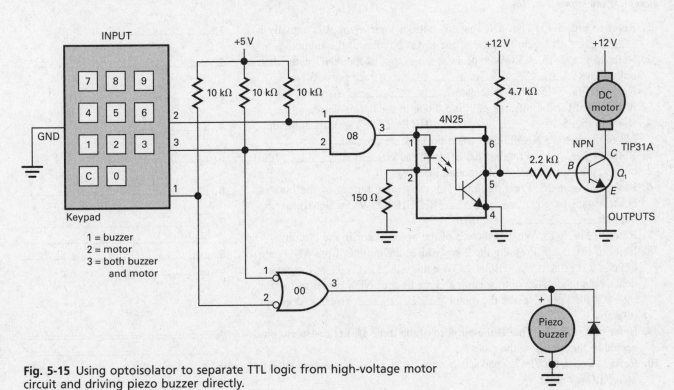

Fig. 5-15 Using optoisolator to separate TTL logic from high-voltage motor circuit and driving piezo buzzer directly.

the infrared-emitting diode inside the 4N25 optoisolator. When the output of the AND gate goes HIGH, the 4N25's LED will light, activating (turning on) the phototransistor driving the base of the NPN power transistor LOW, thus turning off the motor. However, if the output of the AND gate goes LOW, the 4N25 optoisolator is deactivated and the base of the power transistor (Q_1) is pulled positive by the 4.7-kΩ pull-up resistor. The positive voltage at the base of the NPN power transistor will turn on Q_1 and the dc motor will rotate.

Your instructor will give you information on which dc motor to use and its voltage and current characteristics. The schematic in Fig. 5-15 suggests the motor circuit operate at 12-V dc. The dc motor is driven (turned on or turned off) by the NPN power transistor (Q_1). The collector current capacity of the TIP31A NPN power transistor is about 3 A. If the dc motor you use draws more current, your instructor may have you substitute a power transistor with a higher current rating.

PROCEDURE

1. Insert the three ICs into the mounting board.
2. Power OFF. Connect power (V_{CC} and GND) to the ICs. Connect common (GND) of keypad to GND of 5-V power supply. *Do not* connect power to the isolated dc motor at this time.
3. Wire the circuit in Fig. 5-15. Wire the ICs, keypad, resistors, optoisolator, NPN power transistor, dc motor, piezo buzzer, and diode.
4. Power ON. Connect a separate power supply (or battery) to the *isolated* dc motor circuit.
5. Operate the circuit and observe the results.
6. Show your instructor your wiring and answer questions on the circuit.
7. Power OFF. Turn off the isolated dc motor power supply. Turn off the 5-V power supply for the logic circuit and buzzer.
8. Power OFF. Take down the circuit, and return all equipment to its proper place.

QUESTIONS

Complete questions 1 to 10.

1. Refer to Fig. 5-15. The "OR-shaped" bottom logic symbol is actually a _____ gate because of the invert bubbles at the inputs.

1. _____

2. Refer to Fig. 5-15. A LOW appears at the output of the AND gate when either keys _____ or _____ are pressed.

2. _____ _____

3. Refer to Fig. 5-15. The AND gate is _____ (sinking, sourcing) current for the infrared-emitting diode in the optoisolator.

3. _____

4. Refer to Fig. 5-15. A _____ (HIGH, LOW) appears at the output of the bottom NAND gate when no keys are pressed.

4. _____

5. Refer to Fig. 5-15. Pressing the 3 key turns on _____ (the buzzer, the motor, both the buzzer and motor).

5. _____

6. Refer to Fig. 5-15. Pressing the 3 key causes the output of the bottom NAND gate to go _____ (HIGH, LOW), which turns the buzzer _____ (on, off).

6. _____ _____

7. Refer to Fig. 5-15. What is the job of the optoisolator in the circuit?

7. _____
   _____

8. Refer to Fig. 5-15. Pressing the 2 key causes the output of the AND gate to go _____ (HIGH, LOW); this _____ (activates, deactivates) the 4N25 optoisolator, and the base of NPN power transistor (Q_1) is pulled positive and the motor _____ (rotates, does not rotate).

8. _____ __ _____
   _____

9. Refer to Fig. 5-15. What is the purpose of the three 10-kΩ resistors connected to the inputs of the gates?

9. _____
   _____

10. Refer to Fig. 5-15. The keypad acts as an _____ (active HIGH, active LOW) input.

10. _____

5-6 LAB EXPERIMENT: INTERFACING WITH A STEPPER MOTOR

OBJECTIVES

Option 1: Driving a bipolar stepper motor with a MC3479 IC.

1. To wire and test a bipolar stepper motor.
2. To observe and record the control sequences used to drive the bipolar stepper motor.
3. To test the bipolar stepper motor in the full- and half-step mode; the clockwise (CW) and counterclockwise (CCW) rotation mode; and try to operate as a continuous rotation motor.

Option 2: Driving a unipolar stepper motor with a UCN5804 IC.

4. To wire and test a unipolar stepper motor.
5. To test the unipolar stepper motor in the full- and half-step mode; the CW and CCW rotation mode; and try to operate as a continuous rotation motor.

MATERIALS

Qty:

Option 1

1   MC3479 bipolar stepper motor driver IC (or a plug-in substitute IC)
1   bipolar stepper motor, four-wire, 5-V, less thaa 350-mA/coil
1   4.3-V zener diode
1   47-kΩ, $\frac{1}{4}$-W resistor
4   LED indicator-light assemblies
1   5-V dc regulated power supply
1   clock (single-positive)
1   clock (free-running)
2   logic switches

Option 2

1   UCN5804 BiMOS unipolar stepper motor translator/driver IC
1   unipolar stepper motor, six-wire, 6 V to 9 V, less than 250 mA/coil
2   10-kΩ, $\frac{1}{4}$-W resistors
2   50-Ω to 75-Ω, l-W resistors (use parallel 150-Ω, $\frac{1}{2}$-W resistors)
4   1N5817 Schottky diodes
4   LED indicator-light assemblies
1   5-V dc regulated power supply
1   +12-V dc power supply
1   clock (single-positive)
1   clock (free-running)
2   logic switches
1   VOM or DMM

Your instructor will inform you which option you will perform.

SYSTEM DIAGRAMS

Option 1: The wiring diagram in Fig. 5-16 shows the MC3479 IC driving a four-wire bipolar stepper motor. The MC3479 (or a plug-in substitute) driver IC is a one-package solution to interfacing with smaller four-wire bipolar stepper motors. The MC3479 IC is designed to drive a stepper positioning motor in application such as disk drives and robotics. The MC3479 IC contains a logic section that allows either CW or CCW rotation of the stepper motor. The logic section also generates the proper *control sequence* to operate the stepper motor in either the full-step or half-step mode. The MC3479 IC contains internal circuitry that will allow the chip to drive smaller motors with up to 350 mA per coil. The driver IC will operate on a wide range of voltages. Each

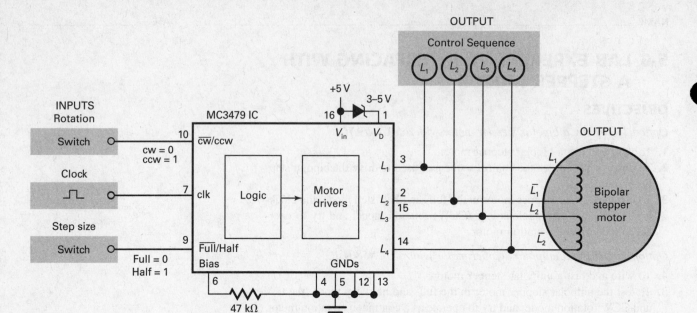

Fig. 5-16 Interfacing using the MC3479 IC to translate logic and drive a bipolar stepper motor.

positive clock pulse entering the CLK input causes the stepper motor to advance either a full-step or half-step. The exact angle that the motor moves (steps) is dependent on the design of the stepper motor and on whether the driver is in the half-step or full-step mode.

To operate the stepper as a continuous rotation motor, you will increase the frequency of the input clock pulses. In the lab you may do this by connecting to a free-running clock. As the frequency increases, the speed of rotation will increase. Depending on the design of the stepper motor, there will be a point as the speed increases where the motor will freeze. This is because the inertia of the rotor will not allow it to be stepped at this speed or beyond.

Your instructor will relay information on the bipolar stepper motor you will use in the lab. Important information about the stepper motor includes the colors of the wires for $L_1/\overline{L}_1$ and $L_2/\overline{L}_2$ coils. Record these colors directly on the schematic in Fig. 5-16. In the lab, if the CW and CCW control seems backward, try reversing the leads on just one of the coils.

A commercial PC board that has a mounted four-wire bipolar stepper motor may be used. This unit features the bipolar stepper motor, MC3479 (or substitute) interface IC and associated parts all prewired with solderless breadboard for external inputs. This handy product (SMB-1000) is available from Dynalogic Concepts at 1-800-246-4907 (www.dynalogic concepts.com).

Option 2: The wiring diagram in Fig. 5-17 shows the UCN5804 BiMOS translator/driver IC driving a six-wire unipolar stepper motor. The UCN5804 driver IC is a one-package solution to interfacing with smaller six-wire unipolar stepper motors. The UCN5804 IC contains a logic section that allows either CW or CCW rotation of the stepper motor. The logic section also generates the proper *control sequence* to operate the stepper motor in either the full-step or half-step mode. The UCN5804 IC contains internal circuitry that will allow the chip to drive smaller motors with up to 1.5 A. The driver IC will operate on a wide range of voltages. Each positive clock pulse entering the STEP input causes the stepper motor to advance either a full-step or half-step. The exact angle that the motor moves (steps) is dependent on the design of the unipolar stepper motor.

136

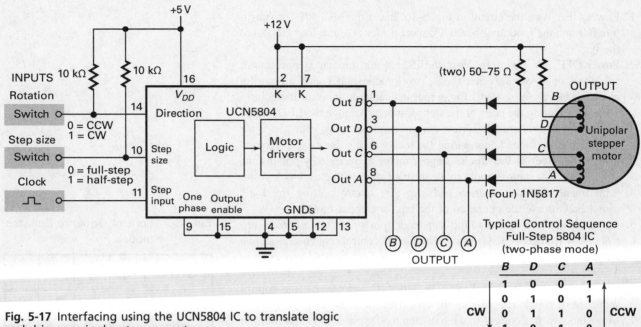

Fig. 5-17 Interfacing using the UCN5804 IC to translate logic and drive a unipolar stepper motor.

The UCN5804 BiMOS translator/driver IC combines two manufacturing technologies in a single chip. The CMOS logic section is combined with the high-current and high-voltage bipolar outputs. Notice in Fig. 5-17 the unusual power supply connections. Two different voltages (+5 V and +12 V) are connected to the UCN5804 translator/driver IC. The UCN5804 IC can be considered a hybrid circuit in that it contains both digital circuitry (logic section) and analog circuitry (stepper motor drivers). The 5804 unipolar driver features *active LOW outputs* which must sink larger currents than normal ICs. The heavy output currents are why the chip has four ground pins (pins 4, 5, 12 and 13). The four 1N5817 Schottky diodes protect the 5804 IC from dangerous voltage spikes generated by the motor coils.

Unipolar stepper motors vary in the resistance of their coils and the maximum current they can tolerate. In Fig. 5-17, the two 50- to 75-Ω resistors limit the current to safe levels from the +12-V power supply for many smaller 6-V to 9-V unipolar stepper motors. The values of the two low-resistance limiting resistors may need to be adjusted depending the specifications of the stepper motor. Your instructor will advise you if the resistance values of the limiting resistors need to be changed.

The two 10-kΩ resistors near the inputs (Direction and Step size) in Fig. 5-17 are pull-up resistors. The input switches control the direction of rotation and step size. Each positive clock pulse causes the UCN5804 IC to change the control sequence and step the unipolar stepper motor. A typical control sequence for full-stepping the unipolar stepper motor is shown in Fig. 5-17.

PROCEDURE

Option 1: Driving a bipolar stepper motor.

1. Power OFF. Wire the circuit in Fig. 5-16. Insert the MC3479 (or substitute IC) into the mounting board. Connect power (V_{in} and four GNDs) to the IC.
2. Power OFF. See Fig. 5-16. Wire the IC, resistor, bipolar stepper motor (watch color of wires), zener diode, two logic switches, and a one-shot clock (+ pulse). Wire four LED output indicators to monitor the control sequence entering the coils of the stepper motor. Arrange the LEDs in the order shown to record the correct control sequence. Ask your instructor if you have questions about wiring the four-wire bipolar stepper motor.
3. Power ON. Operate the bipolar stepper motor by repeatedly pulsing the clock input. The motor should rotate in small steps with each clock pulse.
4. Experiment with the *rotation* and *step size* controls using the logic switches. Observe the operation of the bipolar stepper motor.
5. Move the *step size* control to full-step mode (pin 9 = 0) and *rotation* control to CW (pin 10 = 0). Carefully observe the control sequence as shown on the LED indicator-light assemblies. Record the control sequence in Table 5-3.
6. Change the *rotation* control to CCW. Pulse IC and observe the control sequence. It should progress up the chart (see Table 5-3).
7. Move the *step size* control to the half-step mode (pin 9 = 1) and *rotation* control to CW (pin 10 = 0). Carefully observe the control sequence as shown on the LEDs. Record the control sequence in Table 5-4. The control sequence (half-steps) repeats, so try several cycles to make sure you have recorded it correctly.
8. Change *rotation* control to CCW. Pulse IC and observe the control sequence. It should progress up the chart (see Table 5-4).
9. Move the step size control to full-step mode. Carefully count the number of steps it takes for the motor to rotate one revolution. Record this.
10. Experiment with feeding a higher-frequency square wave (from the free-running clock) into the CLK input of the IC, and observe the rotation of the stepper. At higher frequencies the stepper motor will stop or freeze.
11. Show your instructor your wiring, control sequences, how many steps per revolution, and be prepared to answer questions on the circuit operation.
12. Power OFF. Take down the circuit, and return all equipment to its proper place.

Option 2: Driving a unipolar stepper motor.

13. Power OFF. Wire the circuit in Fig. 5-17. Insert the UCN5804 into the mounting board. Connect power (+5 V, +12 V, and four GNDs) to the IC.
14. Power OFF. See Fig. 5-17. Wire the IC, resistors, unipolar stepper motor (watch color of wires), two logic switches, four Schottky diodes, and a one-shot clock (+ pulse). Wire four LED output indicators to monitor the control sequence entering the coils of the stepper motor. Arrange the LEDs in the order shown to record the correct control sequence. Ask your instructor if you have questions about wiring the six-wire unipolar stepper motor.
15. Power ON. Operate the unipolar stepper motor by repeatedly pulsing the clock input. The motor should rotate in small steps with each clock pulse.
16. Experiment with the *rotation* and *step size* controls using the logic switches. Observe the operation of the unipolar stepper motor.
17. Move the *step size* control to full-step mode (pin 10 = 0) and *rotation* control to CW (pin 14 = 1). Carefully observe the control sequence as shown on the LED indicator-light assemblies. Record the control sequence.

TABLE 5-3 Control sequence (full-step mode)

STEP	L_1	L_2	L_3	L_4
1	1	0	1	0
2				
3				
4				
1				
2				

TABLE 5-4 Control sequence (half-step mode)

STEP	L_1	L_2	L_3	L_4
1	1	0	1	0
2				
3				
4				
5				
6				
7				
8				
1				
2				
3				
4				

18. Change *rotation* control to CCW. Pulse IC and observe the control sequence.

19. Move the *step size* control to the half-step mode (pin l0 = 1) and *rotation* control to CW (pin 14 = 1). Carefully observe the control sequence as shown on the LEDs. Record the control sequence. The control sequence (half-steps) repeats, so try several cycles to make sure you have recorded it correctly.

20. Change *rotation* control to CCW. Pulse IC and observe the control sequence.

21. Move the step size control to full-step mode. Carefully count the number of steps it takes for the motor to rotate one revolution. Record this. Calculate the step rotation of the stepper motor (in degrees) by dividing 360° by the number of steps in one revolution of the output shaft.

22. Experiment with feeding a higher-frequency square wave (from the free-running clock) into the *step* input of the IC, and observe the rotation of the stepper. At higher frequencies the stepper motor will stop or freeze.

23. Feed a lower-frequency square wave (about 1 to 5 Hz) into the *step* input of the IC, and measure the voltage drop across one of the 75-Ω limiting resistors. Record the voltage drop. Using Ohm's law, calculate the current through the *AC* or *BD* coil of the unipolar stepper motor. Record the calculated current.

24. Show your instructor your wiring, control sequences, how many steps per revolution, and your calculations for coil current, and be prepared to answer questions on the circuit operation.

25. Power OFF. Take down the circuit, and return all equipment to its proper place.

QUESTIONS

Complete questions 1 to 19.

1. List the two functional blocks inside either the MC3479 IC or the UCN5804 IC.
    1. _____
    _____

2. List the two *control inputs* to either the MC3479 IC or the UCN5804 IC.
    2. _____
    _____

3. How is the MC3479 IC or UCN5804 IC triggered to step the motor?
    3. _____
    _____

4. Changing the *direction of rotation* input from CW to CCW or CCW to CW causes the logic section of either IC (3479 or 5804) to cycle downward or upward through the control sequence. (T or F)
    4. _____

5. The *full-step control sequence* generated by either the MC3479 IC or UCN5804 IC repeats after 13 unique codes. (T or F)
    5. _____

6. The *half-step control sequence* generated by either the MC3479 IC or UCN5804 IC repeats after _____ (2, 8) unique codes.
    6. _____

7. If the MC3479 IC or UCN5804 IC step-size control is set at *full-step*, what dictates the exact angle that the rotor of the stepper motor will rotate on each step?
    7. _____

8. A stepper motor with just four leads entering the unit is a _____ (bipolar, unipolar)-type motor.
    8. _____

9. A common unipolar stepper motor may have _____ (6, 16) leads entering the motor.
    9. _____

10. Calculate the *step size (in degrees)* of a stepper motor that steps 24 times to make the output shaft rotate one revolution (360°).
    10. _____

11. Calculate the *number of steps* it takes for a stepper motor having a step size of 7.5° to rotate the output shaft one revolution.
    11. _____

12. Refer to Fig. 5-17. The outputs of the UCN5804 IC (pins 1, 3, 6, and 8) are described as _____ (active HIGH, active LOW) outputs.

12. _____

13. Refer to Fig. 5-17. The two 10-kΩ resistors are called _____ (pull-up, push-down) resistors.

13. _____

14. Refer to Fig. 5-17. The two 75-kΩ resistors are called _____ current-limiting, pull-down) resistors.

14. _____

15. Refer to Fig. 5-17. The four 1N5817 Schottky diodes help protect the 5804 IC from voltage spikes that may be generated by the stepper motor coils. (T or F)

15. _____

16. Refer to Fig. 5-17. The UCN5804 IC has both +5-V and +12-V power supplies connected to the single chip, suggesting it is manufactured using _____ (BiMOS, RTL) technology.

16. _____

17. The MC3479 IC was used to drive a bipolar stepper motor. Which coil diagram in Fig. 5-18 is most commonly associated with a bipolar stepper motor?

17. _____

18. The six-wire stepper motor coil diagram shown in Fig. 5-18(b) is most commonly associated with a _____ (bipolar, unipolar) stepper motor.

18. _____

19. Refer to Fig. 5-18(c). The eight-wire stepper motor coil diagram is sometimes referred to as a "universal" stepper motor because it can be wired as a bipolar or unipolar unit. (T or F)

19. _____

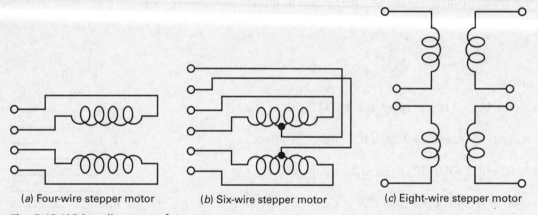

(a) Four-wire stepper motor   (b) Six-wire stepper motor   (c) Eight-wire stepper motor

Fig. 5-18 Wiring diagrams of stepper motors.

5-7 LAB EXPERIMENT: USING A HALL-EFFECT SWITCH

OBJECTIVES

1. To test the operation of two Hall-effect switches and observe both bipolar and unipolar switching.
2. To use a Hall-effect switch to drive a CMOS binary counter IC and observe its bounce-free operation.

MATERIALS

Qty.		Qty.	
1	3132 bipolar Hall-effect switch IC	1	light-emitting diode (LED)
1	3141 unipolar Hall-effect switch IC	1	150-Ω, $\frac{1}{4}$-W resistor
		1	33-kΩ, $\frac{1}{4}$-W resistor
1	74HC393 binary counter CMOS IC	8	LED output indicator lights
		1	5-V regulated power supply
1	permanent magnet	1	logic probe

SYSTEM DIAGRAM

Hall-effect sensors and switches are commonly used in automobiles and industrial and computer equipment. Hall-effect sensors and switches are magnetically activated devices. Hall-effect sensors and switches are rugged and reliable, operate under harsh environmental conditions, and are inexpensive. In switch form, the Hall-effect device produces a bounce-free output.

The *3132 bipolar Hall-effect switch IC* by Allegro MicroSystems is shown driving an LED in Fig. 5-19. The 3132 IC has a built-in NPN open-collector transistor that can sink 25 mA continuously. The 3132 IC will turn *on* (collector of transistor will drop LOW) when a south pole of a magnet approaches the face of the IC. The 3132 IC uses *bipolar switching*, which means if the Hall-effect switch is turned *on* with a south pole then it is turned *off* with a north magnetic pole.

The *3141 unipolar Hall-effect switch IC* by Allegro MicroSystems is shown driving an LED in Fig. 5-20. The 3141 IC has a built-in NPN open-collector transistor that can sink 25 mA continuously. The 3141 Hall-effect switch uses *unipolar switching* unlike the 3132 IC, which used bipolar switching. The 3141 unipolar Hall-effect switch will turn *on* (collector of transistor will drop LOW) when a south pole of a magnet approaches the face of the IC. The 3141 unipolar switch will turn *off* when the magnet moves away from the IC or if the magnetic field is removed from the area of the Hall-effect sensor. In summary, the 3141 IC switches ON with a south magnetic field present and OFF when the magnetic field is removed.

One of the many advantages of Hall-effect switches such as the 3132 and 3141 ICs is that they are bounce-free. Both ICs contain Schmitt trigger circuitry, which causes the output to toggle from LOW to HIGH to LOW. This feature will be demonstrated by using a Hall-effect switch to trigger an 8-bit binary counter. The counter circuit is drawn in Fig. 5-21. The 74HC393 CMOS binary counter IC has clear (CLR) inputs to clear the count to binary 00000000. The 74HC393 is an up counter with a count sequence of 00000000, 00000001, 00000010, 00000011, up to 11111111. If the Hall-effect switch is working properly (bounce-free) the counter should not skip counts.

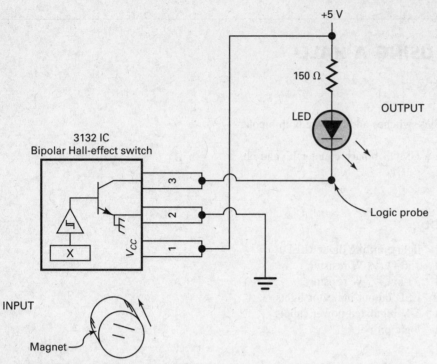

Fig. 5-19 Bipolar 3132 Hall-effect switch driving an LED. (3132 IC is courtesy of Allegro MicroSystems.)

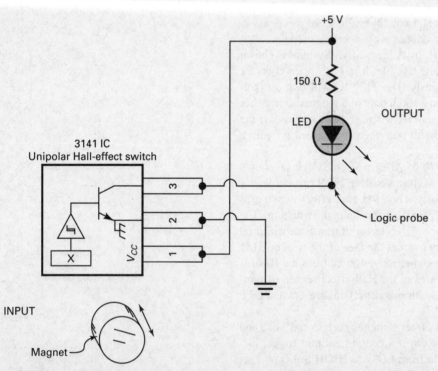

Fig. 5-20 Unipolar 3141 Hall-effect switch driving an LED. (3141 IC is courtesy of Allegro MicroSystems.)

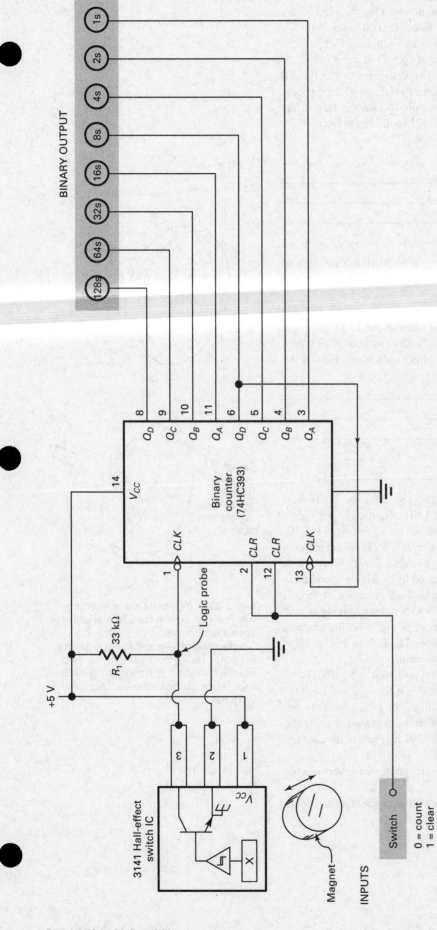

Fig. 5-21 Hall-effect switch driving a binary counter. (3141 IC is courtesy of Allegro MicroSystems.)

PROCEDURE

1. Power OFF. Connect power to the IC. The pin numbers for the 3132 IC are read when viewing the IC from the printed side (branded side). Wire the LED and resistor to the output of the Hall-effect switch as in Fig. 5-19.

2. Power OFF. Attach a logic probe to the output (pin 3) of the 3132 IC.

3. Power ON. Move either pole of a magnet perpendicular to the printed side of the 3132 Hall-effect switch, and observe the LED and logic probe. Reverse the magnet, and move the opposite pole toward the Hall-effect switch. Which pole of the magnet turns the 3132 bipolar Hall-effect switch *on*? Which pole of the magnet turns the switch *off*?

_____

_____

_____

_____

4. Power OFF. Wire the unipolar Hall-effect switch (3141 IC) detailed in Fig. 5-20. The pin numbers for the 3141 are read when viewing the printed side (branded side) of the IC.

5. Power ON. Move the S pole of a magnet perpendicular to the printed side of the 3141 unipolar Hall-effect switch, and observe the LED and logic probe. Remove the magnet from near the Hall-effect switch. Which pole of the magnet turns the 3141 unipolar Hall-effect switch *on*? How is the 3141 unipolar Hall-effect switch turned *off*?

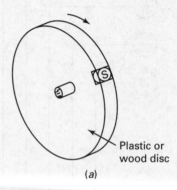

(a)

_____

_____

_____

_____

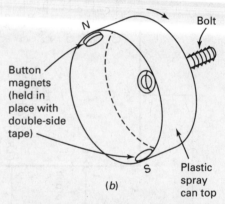

6. Power OFF. Connect power (V_{cc} and GND), and wire the Hall-effect switch, logic switch, 74HC393 IC, and eight LED output-indicator lights. See Fig. 5-21. Attach a logic probe to the output (pin 3) of the 3141 IC.

7. To aid in rapid counting, you can construct a simple wheel-like device such as shown in Fig. 5-22(a) with the magnet glued in place. With a bolt through the middle hole the wheel can be placed in a variable-speed electric drill and rotated close to the face of the Hall-effect sensor. A second variation is shown in Fig. 5-22(b) with both S and N poles available around the circumference. This variation uses a "spray can top" with the magnets adhered inside the rim with double-sided tape. The unit in Fig. 5-22(b) can be used for either unipolar or bipolar switching.

8. Power ON. Clear the counter to 00000000 (clear switch to HIGH and back LOW). Using the S pole of a magnet, move it forward and back toward the printed face of the Hall-effect switch, causing the counter to operate. Observe the action at the output of the Hall-effect switch using the logic probe. Observe the triggering of the clock (CLK) input to the counter. Observe the output of the 74HC393 counter.

9. Demonstrate the counter to your instructor, and prepare to answer selected questions on the circuit.

10. Power OFF. Take down the circuit, and return all equipment to its proper place.

Fig. 5-22 Assemblies for mounting magnets to be rotated by variable speed drill or small dc motor. (*a*) Single magnet—good for triggering unipolar Hall-effect switch. (*b*) Two magnets—good for triggering either bipolar or unipolar Hall-effect switches.

144

QUESTIONS

Complete questions 1 to 13.

1. Hall-effect sensors and switches are _____ (magnetically, thermally) activated devices.

2. The 3132 Hall-effect IC features _____ (bipolar, unipolar) switching.

3. The 3132 Hall-effect IC features a driver transistor with a(n) _____ (open-collector, totem-pole) output.

4. Refer to Fig. 5-19. When a S pole of a magnet approaches the 3132 IC, the output (pin 3) will go _____ (HIGH, LOW) and the LED will _____ (light, not light).

5. Refer to Fig. 5-19. When a N pole of a magnet approaches the 3132 IC, the output (pin 3) will go _____ (HIGH, LOW) and the LED will _____ (light, not light).

6. The 3141 Hall-effect IC features _____ (bipolar, unipolar) switching.

7. Refer to Fig. 5-20. When a S pole of a magnet approaches the 3141 IC, the output (pin 3) will go _____ (HIGH, LOW) and the LED will _____ (light, not light).

8. Refer to Fig. 5-20. How is the 3141 Hall-effect switch turned off?

_____

_____

_____

9. Hall-effect switches such as the 3132 and 3141 ICs used in this experiment are considered to be bounceless. (T or F)

10. Refer to Fig. 5-21. The clear input switch (to pins 2 and 12 of the 74HC393 IC) needs to be _____ (HIGH, LOW) for the counter to count upward.

11. Refer to Fig. 5-21. Component R_1 is commonly called a(n) _____ (input, pull-up) resistor and pulls the output of the 3141 IC HIGH when the Hall-effect switch is OFF.

12. Refer to Fig. 5-21. The 74HC393 appears to count upward on the _____ (HIGH-to-LOW, LOW-to-HIGH) transition of signal generated by the Hall-effect switch.

13. Refer to Fig. 5-21. Explain the movement of the magnet required to get the 3141 Hall-effect switch to toggle and the counter to count.

_____

_____

_____

1. _____

2. _____

3. _____

4. _____
   _____

5. _____
   _____

6. _____

7. _____
   _____

9. _____

10. _____

11. _____

12. _____

5-8 LAB EXPERIMENT: INTERFACING WITH A SERVO MOTOR

OBJECTIVES

1. To wire and observe the action of a hobby servo motor driven by an experimental pulse-width modulation circuit.
2. To measure the pulse widths entering the servo motor and the trigger-pulse frequency using an oscilloscope.

MATERIALS

Qty.		Qty.	
1	555 timer IC	1	330-kΩ, $\frac{1}{4}$-W resistor
1	74121 monostable multivibrator IC	1	50-kΩ, potentiometer
1	7404 inverter IC	1	0.033-μF capacitor
1	hobby servo motor	1	0.1-μF capacitor
1	10-kΩ, $\frac{1}{4}$-W resistor	1	5-V regulated power supply
1	27-kΩ, $\frac{1}{4}$-W resistor	1	oscilloscope

SYSTEM DIAGRAM

The fundamentals of driving either a stepper motor or a hobby servo motor are represented by the block diagrams in Fig. 5-23. Recall that a stepper motor rotates a discrete amount for each input pulse. For instance, the stepper motor in Fig. 5-23(a) will rotate five steps counterclockwise (CCW). If the step size designed into the stepper motor were 3.6° per step, then the rotation shown in Fig. 5-23(a) would be 18° (5 × 3.6° = 18°). Recall that the interface circuit also allows for changing the rotation direction of the stepper motor.

Hobby servo motors of the type used in this experiment use *pulse-width modulation (PWM)* to adjust the angular position of the output shaft. This idea is portrayed using the block diagram in Fig. 5-23(b). The inputs to this pulse-width modulator are short trigger pulses with a frequency of about 40 to 60 Hz and a variable resistance. The PWM circuit is designed to vary the width of the output pulse from about 1 to 2 ms. The hobby servo will center the shaft when a 1.5-ms pulse repeatedly enters the input of the motor. When the pulse width decreases to about 1.0 ms, the shaft will rotate clockwise (CW) about 90°. When the pulse width increases to about 2.0 ms, the shaft will rotate counterclockwise to a position about 90° CCW from center. The duration of the pulse emitted by the PWM circuit is controlled by the operator. The operator changes the variable resistance to adjust the angular position of the servo motor's output shaft.

You will find some hobby servo motors that rotate in the opposite direction from that shown in Fig. 5-23. Instead of rotating CCW as the pulse width increases, some servos will rotate in the opposite direction.

A detailed diagram for an experimental PWM circuit that will control a hobby servo motor is drawn in Fig. 5-24. The duration of the pulse entering the input lead of the hobby servo determines the angular position of the motor's shaft. Most hobby servos have internal stops that allow positions in a range of about 180°.

The 74121 one-shot multivibrator IC is at the heart of the PWM circuit. The short positive trigger pulses are generated by the 555 timer IC wired as a free-running multivibrator (MV). Each trigger pulse entering input B of the 74121 causes the PWM to emit a pulse having a duration of from 1 to 2 ms. The duration of the 74121 IC's output pulse is varied by the operator changing

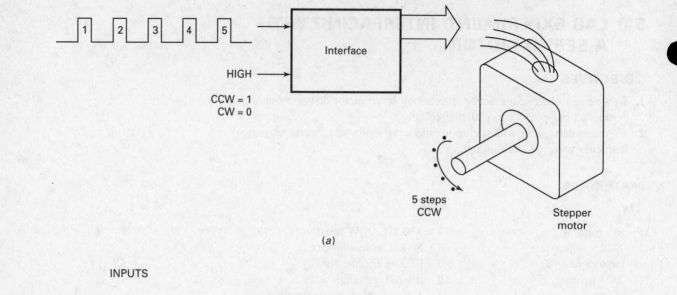

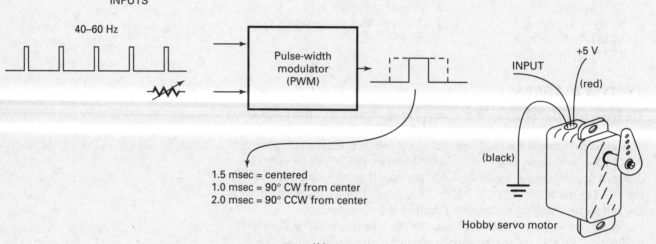

Fig. 5-23 Block diagram for Lab Experiment 5-8. (*a*) Driving a stepper motor. (*b*) Driving a hobby servo motor. (*Note:* Some hobby servos rotate in the opposite direction of the motor featured in this figure.)

the value of potentiometer R_3. If the resistance of R_3 is near LOW, the pulse width is narrower (about 1 ms). As the resistance of R_3 increases, the pulse width becomes wider to a maximum of about 2 ms. From Fig. 5-23(*b*) you will see how the hobby servo motor responds to various pulse widths. A pulse width of about 1 ms forces the shaft to rotate to its most CW position. A pulse width of 2 ms causes the shaft of the servo to rotate to its most CCW position.

The 7404 inverter has been added to the PWM circuit in Fig. 5-24 to invert the output of the free-running MV so the trigger pulses are positive. The output of a free-running MV is about 40 Hz.

PROCEDURE

1. Power OFF. See Fig. 5-24. Insert the three TTL ICs in the mounting board, and connect power (V_{cc} and GND).
2. Power OFF. Using the schematic diagram in Fig. 5-24, wire the experimental PWM circuit to the hobby servo motor.
3. Power ON. Vary potentiometer R_3 and observe the action of the servo motor.
4. Power OFF. Attach an oscilloscope to the output Q of the 74121 IC (pin 6).

148

5. Power ON. Measure the pulse widths when the servo motor is (1) rotated fully CW, (2) centered, and (3) rotated fully CCW. Record your observations.

_____

_____

_____

6. Power OFF. Connect the oscilloscope to the input *B* pin (pin 5) of the 74121 IC.

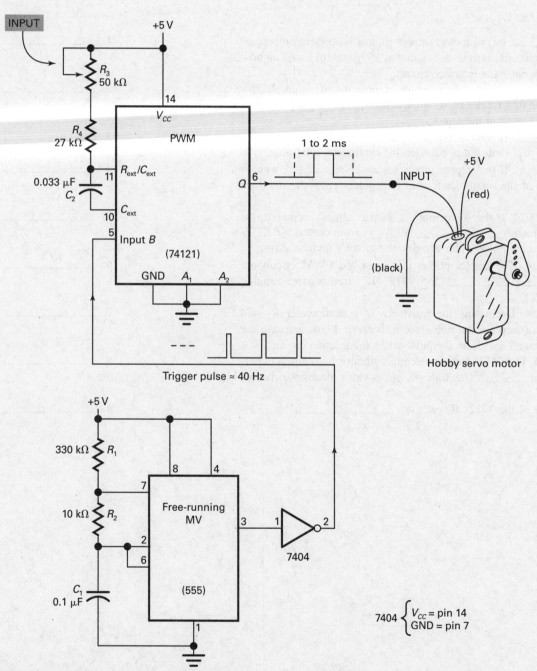

Fig. 5-24 Experimental pulse-width modulation circuit controlling a hobby servo motor.

7. Power ON. Observe the shape and measure the frequency of the trigger pulse input (f in Hz $= 1/t$ in seconds). Record your observations.

_____

_____

_____

8. Demonstrate the hobby servo control circuit to your instructor, and prepare to answer selected questions about the circuit.
9. Power OFF. Take down the circuit, and return all equipment to its proper place.

QUESTIONS

Complete questions 1 to 9.

1. The _____ (servo motor, stepper motor) is an electromechanical device that rotates in discrete steps (such as 15° per step) based on how many input pulses enter the interface circuit.

2. The _____ (hobby servo motor, stepper motor) is an electromechanical device that rotates to an angular position based on the pulse-width entering the input of the motor.

3. The _____ (hobby servo motor, dc motor) is well suited for angular positioning of a shaft but is not used for continuous rotation.

4. Refer to Fig. 5-23(*a*). If the stepper motor has a step size of 15°, what is the total rotation of the output shaft after five pulses enter the interface circuit?

5. Refer to Fig. 5-23(*b*). If the servo motor's shaft is already centered, the output shaft will rotate _____ (90° CW from center, 90° CCW from center) when the pulse width entering the motor's input is 2 ms.

6. Refer to Fig. 5-24. The trigger pulses that enter the PWM circuit are generated by the _____ (555, 74121) IC wired as a free-running multivibrator (MV).

7. Refer to Fig. 5-24. Decreasing the resistance of potentiometer R_3 will _____ (decrease the pulse width to about 1 ms, increase the pulse width to about 2 ms) from the pulse-width modulation circuit.

8. Refer to Fig. 5-24. The 74121 monostable multivibrator IC is wired to perform as a _____ (multiplexer, pulse-width modulator) in this control circuit.

9. The manufacturer of the 74121 IC calls it a _____ _____.

1. _____

2. _____

3. _____

4. _____

5. _____

6. _____

7. _____

8. _____

9. _____ _____

150

5-9 DESIGN PROBLEM: CONTROL A STEPPER MOTOR FROM A KEYPAD

OBJECTIVES

1. To design the logic required to control direction of rotation and half- or full-step of either a bipolar or unipolar stepper motor.
2. To design two logic circuits: one for the direction of rotation and a second for half- or full-step.
3. To combine the two logic circuits into one with four common inputs and two outputs.
4. At the direction of your instructor, to implement the logic circuit and one of the two stepper motors.

MATERIALS

Qty.

1 MC3479 stepper motor driver IC and bipolar stepper motor (4 wire) *or* UCN5804 stepper motor driver IC and unipolar stepper motor (6 wire) associated parts shown in either Fig. 5-26 (bipolar stepper motor option) *or* associated parts shown in Fig. 5-27 (unipolar stepper motor option)

For logic block (other solutions possible)

1 7404 hex inverter IC
1 7408 two-input NAND gate IC
2 7420 four-input NAND gate ICs
1 electronic circuit simulation program

SYSTEM DIAGRAM

Figure 5-25 is a block diagram of the system required to solve this problem. The idea is to be able to press a 1 through 4 on the keypad and have the stepper motor respond with the correct direction of rotation and either half- or full-step.

The truth tables in Table 5-5 define the logic problem. Note that the shaded areas on the truth tables are of importance in this problem. For instance, under column A (key 1) only the eighth line down on the truth table shows column A at logical 0 while the others are at logic 1. This would be the input conditions to the logic block when key 1 is activated (active LOW). When key 1 is pressed, the $\overline{\text{CW}}$/CCW output must go LOW while the $\overline{\text{Full}}$/Half output must go HIGH as shown on the truth table.

Detailed wiring diagrams are given in Fig. 5-26 (bipolar stepper motor) and Fig. 5-27 (unipolar stepper motor). Your instructor will decide if you wire an entire system (as in Fig. 5-26 or 5-27) or just do the design work for the control logic section.

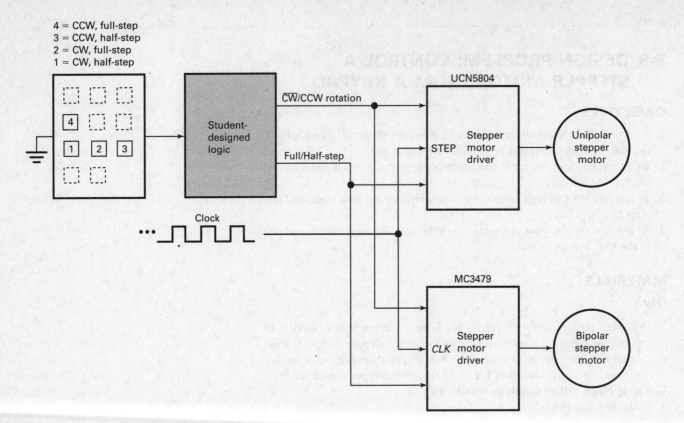

Fig. 5-25 Block diagram of system for keypad control of either a unipolar or bipolar stepper motor.

TABLE 5-5 Truth Table for Control Logic Problem

Key 1	Key 2	Key 3	Key 4	OUTPUT	OUTPUT
A	B	C	D	$\overline{CW}$/CCW	$\overline{Full}$/half
0	0	0	0	0	0
0	0	0	1	0	0
0	0	1	0	0	0
0	0	1	1	0	0
0	1	0	0	0	0
0	1	0	1	0	0
0	1	1	0	0	0
0	1	1	1	0	1
1	0	0	0	0	0
1	0	0	1	0	0
1	0	1	0	0	0
1	0	1	1	0	0
1	1	0	0	0	0
1	1	0	1	1	1
1	1	1	0	1	0
1	1	1	1	0	0

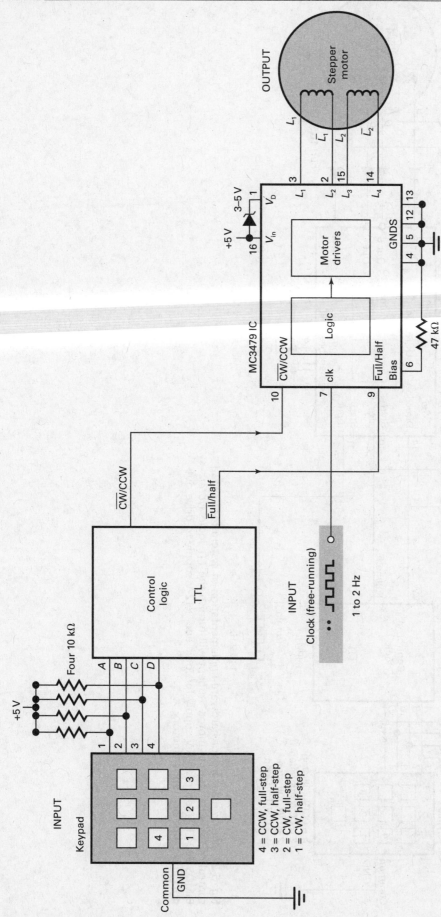

Fig. 5-26 Wiring diagram for hardware implementation of system for keypad control of bipolar stepper motor. Student to furnish the TTL control logic.

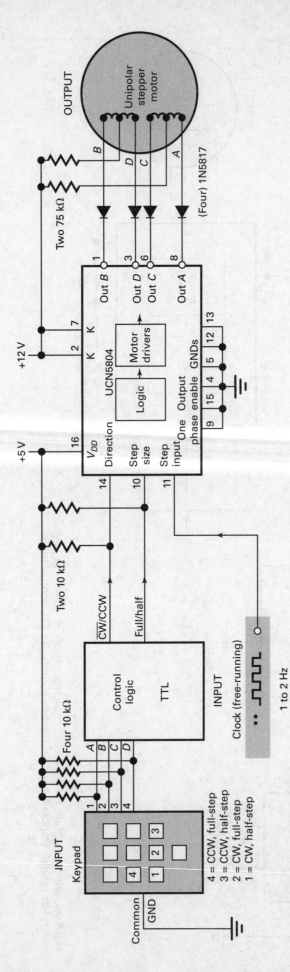

Fig. 5-27 Wiring diagram for hardware implementation of system for keypad control of unipolar stepper motor. Student will furnish the TTL logic. *Note:* BiMOS 5804 IC features +5-V supply for logic and +12-V supply to drive outputs and unipolar stepper motor.

154

PROCEDURE

1. Your instructor will direct you as to which option to use during the design phase.

Option 1: Use Electronic Workbench's or Multisim's logic converter instrument to enter truth tables, generate simplified Boolean expressions, and generate a logic diagram for each of the two circuits.

Option 2: Without the aid of a computer, list a Boolean expression for the truth table, simplify each Boolean expression, and draw a logic diagram for each of the two circuits.

2. Manually, combine the two circuits into one logic circuit with four inputs (*A*, *B*, *C*, and *D*) and two outputs ($\overline{CW}$/CCW and $\overline{Full}$/Half).

3. Your instructor will direct you as to which option to use during the testing and implementation phase.

Option 3: Wire the entire circuit using keypad, ICs, resistors and zener diode, stepper motor, 5-V power supply, and free-running clock. Show the instructor your working circuit, your control logic design, and answer questions about the circuit.

Option 4: Test only your control logic design (four inputs and two outputs) using a circuit simulation program. Show the instructor your circuit and answer questions about the circuit.

5-10 BASIC STAMP EXPERIMENT: DRIVING A SERVO MOTOR

OBJECTIVES

1. To use a BASIC Stamp 2 (BS2)–based development board and personal computer to program to test a servo motor.
2. To follow the operation of the program that controls the servo motor on the PC monitor using the DEBUG command.

MATERIALS

Qty.

1   BASIC Stamp 2 development board (such as the Board of Education, HomeWork Board or NX-1000 board available from Parallax)
1   BS2 module (this may be mounted on the development board)
1   PC with MS Windows
1   PBASIC text editor from Parallax
1   serial cable (for downloading to BS2 module on development board) or USB cable
1   hobby servo motor

Note: BASIC Stamp 2 module, development board, PBASIC software, and both experiment and reference manuals are available from Parallax, Inc., 599 Menlo Drive, #100, Rocklin, CA 95765. Many of the manuals and software can be downloaded free from Parallax's website. General website: *www.parallax.com.* Telephone: 916-624-8333.

SYSTEM DIAGRAMS

Figure 5-28 shows a hobby servo motor connected to port 14 of the BASIC Stamp 2 microcontroller module. The BS2 module will generate pulses to

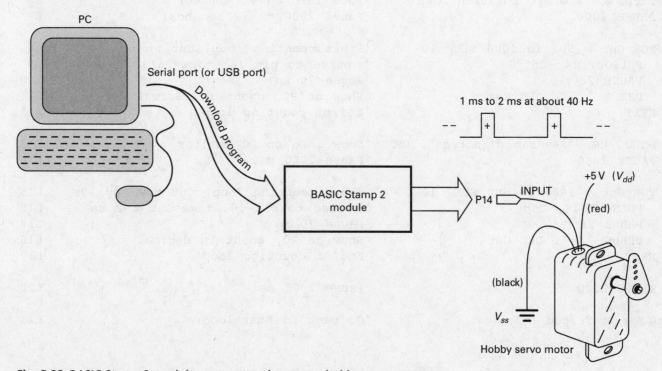

Fig. 5-28 BASIC Stamp 2 module programmed to test a hobby servo motor.

rotate the servo motor. These positive pulses will range from 1 to 2 ms in duration at about 40 Hz. The BS2 module is acting as a PWM (pulse-width modulation) generator.

Programming is accomplished on a MS Windows–based personal computer using the PBASIC editor for the BS2 BASIC Stamp. The program is downloaded using a serial cable (or USB cable) between the PC's serial port and the BASIC Stamp 2 development board. The BS2 module stores the program in EEPROM memory. The PBASIC program is interpreted and executed by the microcontroller on the BS2 module.

The following program will cause the BASIC Stamp 2 module to control the hobby servo motor. Line 2 of the program (**'ServoTestLab5-10**) is the title of the program. The apostrophe symbol (') at the beginning of this line identifies it as a remark statement. Remark statements are not executed by the BS2 IC but are displayed in the PBASIC listing to aid human understanding of the program. Line 3 is a statement used to declare the variable used in the counting loop of the program. The example, line 3 (**Cnt VAR Word 'Declare Cnt as variable, 16 bits**) tells the microcontroller that a variable called **Cnt** will be used in the program and it will be 16 bits long (can represent numbers from 0 to 65535 in decimal). The remark section of line 3 helps clarify the PBASIC statement.

Lines 5 and 6 contain the DEBUG command. In computer jargon, a *bug* is an error in coding that causes the program to produce faulty results. *Debugging* is the art of detecting, locating, and correcting a problem in a program. The PBASIC text editor contains the DEBUG command used for following program operation and debugging.

```
'{$STAMP BS2}                    'For BS2 module                         L1
'ServoTestLab5—10                'Title of program                      L2

Cnt VAR Word                     'Declare Cnt as variable, 16 bits      L3

Startloop:                       'Label for start of continuous loop    L4

DEBUG "   ", CR                  'Show info about program on PC         L5
DEBUG CR, "Start rotation", CR   'Show info on PC monitor               L6
PAUSE 2000                       'Pause 2000 ms (2 seconds)             L7

FOR Cnt = 500 To 1000 STEP 10    'Start counting loop, Cnt up by 10s    L8
  PULSOUT 14, Cnt                '+pulse to pin 14, time=Cnt x 2 us     L9
  PAUSE 20                       'Pause 20 ms                           L10
  DEBUG "   ", DEC Cnt           'Show on PC, count in decimal form     L11
NEXT                             'End of counting loop                  L12

DEBUG CR, "Reverse direction", CR 'Show info on PC monitor              L13
PAUSE 2000                       'Pause 2000 ms                         L14

FOR Cnt = 1000 To 500 STEP 10    'Start counting loop, Cnt down by 10s  L15
  PULSOUT 14, Cnt                '+pulse to pin 14, time=Cnt x 2 us     L16
  PAUSE 20                       'Pause 20 ms                           L17
  DEBUG "   ", DEC Cnt           'Show on PC, count in decimal          L18
NEXT                             'End of counting loop                  L19

PAUSE 2000                       'Pause 2000 ms                         L20

GOTO Startloop                   'Go back to Startloop:                 L21
```

The '**ServoTestLab5-10** program contains a continuous program loop starting with a label (**Startloop:**). Labels are identified by names ending with a colon (:). Labels are places in a program that other lines of code can easily locate and jump to. The end of the loop is the **GOTO Startloop** (line 21) which returns the program to **Startloop:** (line 4).

Two counting loops are used in this program using the FOR-NEXT commands. The first FOR-NEXT loop starts with the **FOR Cnt = 500 TO 1000 STEP 10** (line 8) and ends with the **NEXT** statement (line 12). The first time through the counting loop variable **Cnt** = 500, which is used to generate a positive pulse at port 14. The pulse is generated by the BS2 module using the **PULSOUT 14, Cnt** statement (line 9). The duration of this pulse is 1 ms (2 μs × 500 = 1000 μs = 1 ms). This causes the servo motor to perform full clockwise (CW) rotation. Each time through the first counting loop the **Cnt** variable is increased by 10 until it reaches 1000 (500, 510, 520 . . . 1000), which will cause the servo motor to rotate in rapid steps to its full CCW position. The **PAUSE 20** (line 10) causes the output pulse to stay LOW for 20 ms between positive pulses. The **DEBUG " ", DEC Cnt** code (line 11) will print a few spaces on the PC screen followed by the value of the variable **Cnt** in decimal format. The first loop will generate 51 positive pulses before dropping to the next line of code in line 13.

Line 13, **DEBUG CR, "Reverse direction", CR** will display "Reverse direction" on the PC screen. The first and last **CR** commands stand for carriage returns. In PBASIC, a *carriage return* (**CR** used with **DEBUG** statement) means to move one line down and to the left edge of the PC screen.

The second counting loop starts with line 15 (**FOR Cnt = 1000 TO 500 STEP 10**). The first time through the loop the variable **Cnt** = 1000. The **PULSOUT 14, Cnt** statement (line 16) causes a 2-ms positive pulse output at pin 14 of the BS2 module. This will cause the servo motor to rotate to its full CW position. Each time through the second counting loop the **Cnt** variable will be decreased 10 (1000, 990, 980 . . . 500) with a low of 500. Each time through the loop the decreasing value of the **Cnt** variable will cause the servo motor to rotate in rapid steps CW. The **PAUSE 20** (line 17) allows the output pulse to remain LOW for 20 msec. The **DEBUG " ", DEC Cnt** statement prints to the PC screen. The second loop generates 51 positive pulses before continuing the program in line 20.

Line 20, **PAUSE 2000** allows a 2000-ms (2-second) delay. Line 21, **GOTO Startloop** sends the program back to the beginning of the continuous loop.

For reference, a diagram of Parallax's Board of Education development board is drawn in Fig. 5-29. A few of its features are identified on the drawing, including:

a. Serial input (for downloading PBASIC programs from PC with serial cable)
b. BASIC Stamp 2 module (24-pin DIP)
c. Power connections—either 9-V battery or dc supply with barrel plug
d. ON/OFF switch and power indicator
e. Reset button—for restarting PBASIC programs
f. Voltage regulator
g. Breadboard, solderless
h. Power supply header [V_{dd} (+) and V_{ss}(−)]
i. BS2 input/output (I/O) pins header

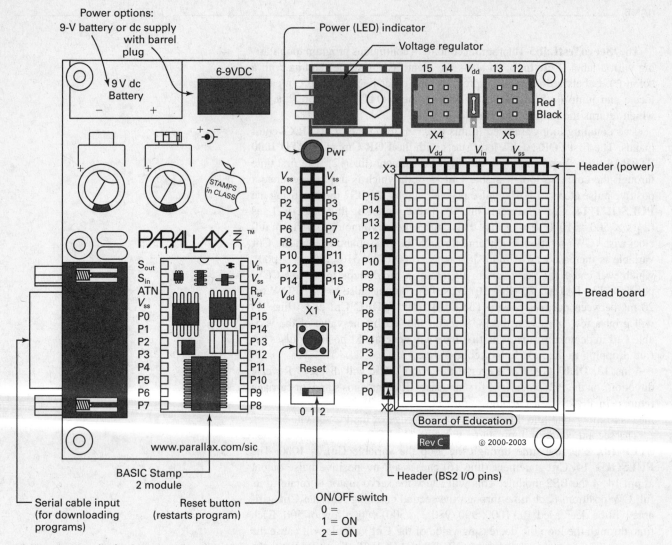

Fig. 5-29 Board of Education development board. (Used by permission of Parallax, Inc.)

Power options:
9-V battery or dc supply with barrel plug

Power (LED) indicator

Voltage regulator

9 V dc Battery

6-9VDC

15 14 V_{dd} 13 12

X4   X5

Red
Black

Header (power)

STAMPS in CLASS

V_{dd} V_{in} V_{ss}

V_{ss} V_{ss}
P0   P1
P2   P3
P4   P5
P6   P7
P8   P9
P10   P11
P12   P13
P14   P15
V_{dd} V_{in}

X1

X3

P15
P14
P13
P12
P11
P10
P9
P8
P7
P6
P5
P4
P3
P2
P1
P0

X2

Bread board

S_{out} V_{in}
S_{in} V_{ss}
ATN R_{st}
V_{ss} V_{dd}
P0   P15
P1   P14
P2   P13
P3   P12
P4   P11
P5   P10
P6   P9
P7   P8

Reset

0 1 2

Board of Education

Rev C   © 2000-2003

Header (BS2 I/O pins)

www.parallax.com/sic

BASIC Stamp 2 module

Serial cable input (for downloading programs)

Reset button (restarts program)

ON/OFF switch
0 =
1 = ON
2 = ON

PROCEDURE

1. Refer to Fig. 5-28. Wire the hobby servo motor to a BS2 module using a development board such as Parallax's BOE (see Fig. 5-29). Check with your instructor on how to connect the servo motor to the development board.
2. Using an MS Windows–based PC, start the BASIC Stamp editor and write the PBASIC program titled 'ServoTestLab5-10.
3. Connect a serial cable (or USB cable) between the serial output port of the PC and the BOE development board.
4. Power ON (BOE board). Download the 'ServoTestLab5-10 program. The program should be running on the BOE.
5. With the serial cable still connected and the servo motor rotating back and forth, observe the debug screen on the PC.
6. Show your instructor your circuit, and be prepared to answer questions about the circuit and the PBASIC program.
7. Power OFF. Take down the circuit, and return all equipment to its proper place.

QUESTIONS

Complete questions 1 to 14.

1. At the heart of the BASIC Stamp 2 module is a programmable device called a _____ (microcontroller, multiplier).

2. The high-level language used to program a BS2 module is _____ (ABLE, PBASIC).

3. Refer to Fig. 5-28. The BS2 module is programmed as a _____ (PWM generator, CR interpolater).

4. Refer to Fig. 5-28. The black wire (GND) of the hobby servo motor is connected to _____ (V_{dd}, V_{ss}) of the power supply.

5. Refer to Fig. 5-28. The hobby servo motor changes angular position based on the pulse width of the positive pulses, which range from about 5 seconds to 10 seconds in duration. (T or F)

6. In computer jargon, _____ (debugging, parsing) is the art of detecting, locating, and correcting a problem in a program.

7. In the program 'ServoTestLab5-10, the _____ (DEBUG " ", CR; FOR Cnt=500 TO 1000 STEP 10) statement is used to start the first counting loop.

8. In the program 'ServoTestLab5-10, the first counting loop is repeated _____ (51, from 500 to 1000) times per cycle through the program.

9. In the program 'ServoTestLab5-10, the variable **Cnt** can hold only one bit of information (either 0 or 1). (T or F)

10. In PBASIC, the statement **PULSOUT 14, 1000** will output a 2 ms TTL level positive pulse from pin 14 of the BS2 module. (T or F)

11. In PBASIC, the statement **PAUSE 2000** causes a pause in program execution for 2000 ms or 2 seconds. (T or F)

12. In PBASIC, a statement such as **Startloop:** is referred to as a _____ (label, remark) and can be found by another program statement.

13. In the program 'ServoTestLab5-10, the statement **DEBUG Dec Cnt** will print on the PC monitor the current _____ (binary, decimal) value contained in variable **Cnt.**

14. Refer to Fig. 5-28 and see the program 'ServoTestLab5-10. In the second counting loop, when **Cnt** = 750, the **PULSOUT 14, Cnt** statement would cause the servo motor to be positioned _____ (fully CCW, in the middle between fully CW and CCW).

1. _____

2. _____

3. _____

4. _____

5. _____

6. _____

7. _____

_____

8. _____

9. _____

10. _____

11. _____

12. _____

13. _____

14. _____

CHAPTER 6

Encoding, Decoding, and Seven-Segment Displays

TEST: ENCODING, DECODING, AND SEVEN-SEGMENT DISPLAYS

Answer the questions in the spaces provided.

1. The decimal number 58 equals _____ in binary.
2. The decimal number 58 equals _____ in 8421 BCD code.
3. The 8421 BCD number 1001 0110 0100 0011 equals _____ in decimal.
4. The decimal number 27 equals _____ in excess-3 code.
5. The excess-3 coded number 1001 0011 equals _____ in decimal.
6. The Gray code _____ (is, is not) a BCD-type code.
7. The Gray code is commonly associated with a shaft encoder used for measuring the angular position of a shaft. (T or F)
8. A 2-bit quadrature code is a type of Gray code that can be used in determing the direction of rotation of a shaft. (T or F)
9. The most popular 7-bit alphanumeric code is _____ (ASCII, NASAB).
10. Refer to Fig. 6-1. During time period t_1, the output indicators read binary _____. [4 bits]
11. Refer to Fig. 6-1. During time period t_2, the output indicators read binary _____. [4 bits]
12. Refer to Fig. 6-1. During time period t_3, the output indicators read binary _____. [4 bits]
13. Refer to Fig. 6-1. During time period t_4, the output indicators read binary _____. [4 bits]
14. What number appears on a seven-segment display when segments b, c, f, and g are lit?
15. Refer to Fig. 6-2. During time period t_1, what decimal number appears on the seven-segment display?
16. Refer to Fig. 6-2. During time period t_2, what decimal number appears on the seven-segment display?
17. Refer to Fig. 6-2. During time period t_3, what decimal number appears on the seven-segment display?
18. Refer to Fig. 6-2. During time period t_4, what segments are lit on the display?
19. The seven-segment LED display shown in Fig. 6-2 is a common-_____ (cathode, anode) type.

1. _____
2. _____
3. _____
4. _____
5. _____
6. _____
7. _____
8. _____
9. _____
10. _____
11. _____
12. _____
13. _____
14. _____
15. _____
16. _____
17. _____
18. _____
19. _____

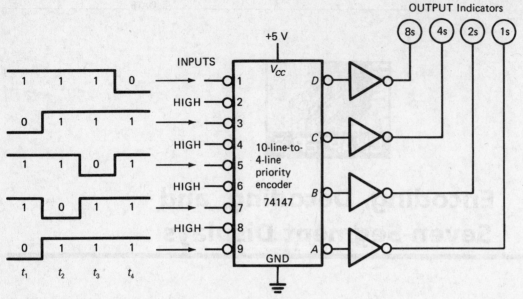

Fig. 6-1 Pulse-train problem, test questions 10 through 13.

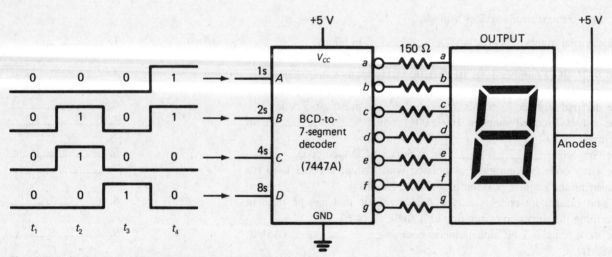

Fig. 6-2 Pulse-train problem, test questions 15 through 19.

20. While the LED display *emits* light, the LCD unit _____ light.

21. Inexpensive monochrome liquid-crystal displays that show black characters on a silvery background are _____ (active-matrix, twisted-nematic field-effect) LCDs.

22. Active-matrix LCDs use _____ (thick-film resistor, thin-film transistor) technology along with red, green, and blue filters in the manufacture of high-quality, thin-panel computer monitors and TV screens.

23. Refer to Fig. 6-3. During time period t_1, what decimal number appears on the seven-segment display?

24. Refer to Fig. 6-3. During time period t_2, what decimal number appears on the seven-segment display?

25. Refer to Fig. 6-3. During time period t_3, what decimal number appears on the seven-segment display?

26. Refer to Fig. 6-3. During time period t_1, what display drive line(s) are in phase with the clock signal?

27. Refer to Fig. 6-3. During time period t_2, what display drive lines are out of phase with the clock signal?

20. _____

21. _____

22. _____

23. _____

24. _____

25. _____

26. _____

27. _____

164

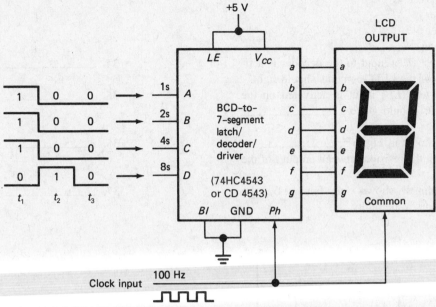

Fig. 6-3 Pulse-train problem, test questions 23 through 27.

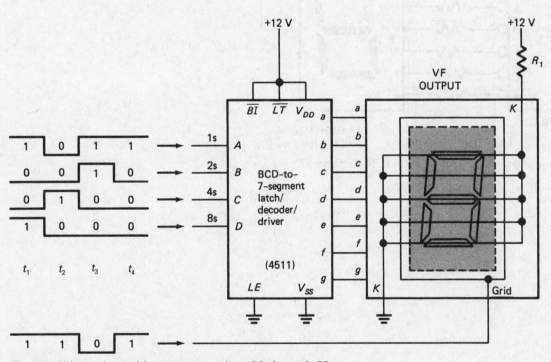

Fig. 6-4 Pulse-train problem, test questions 28 through 32.

28. Refer to Fig. 6-4. The output of this circuit is a decimal formed on the _____ (LCD, LED, VF) display.

28. _____

29. Refer to Fig. 6-4. During time period t_1, what decimal appears on the seven-segment display?

29. _____

30. Refer to Fig. 6-4. During time period t_2, what decimal appears on the seven-segment display?

30. _____

31. Refer to Fig. 6-4. During time period t_3, what decimal appears on the seven-segment display?

31. _____

32. Refer to Fig. 6-4. During time period t_4, what decimal appears on the seven-segment display?

32. _____

33. The 74HC4543 and 4511 ICs contain three functional sections, which are
_____.
   a. Comparator, decoder, and transceiver
   b. Latch, multiplexer, and bus driver
   c. Latch, decoder, and display driver
   d. Parity generator, latch, and counter
34. Refer to Fig. 6-5. When the lamp-test (*LT*) input to the decoder is activated, _____ (all, none) of the LED segments should light.
35. Refer to Fig. 6-5. When the lamp-test (*LT*) input is activated on the
decoder, all of the 7447A IC outputs should go to a _____
logic level.
36. Based on the logic-probe readings shown in Fig. 6-5, _____
(no, some) current is flowing through the resistor between output *f* of the
7447A IC and input *f* of the display.
37. Based on the logic-probe readings shown in Fig. 6-5, the
_____ (display, IC) appears to be faulty.

33. _____

34. _____

35. _____

36. _____

37. _____

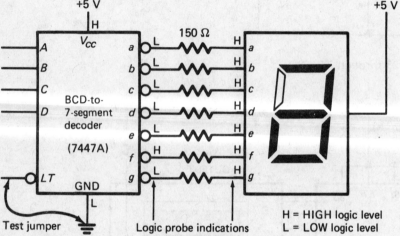

Fig. 6-5 Troubleshooting problem, test questions 34 through 37.

6-1 LAB EXPERIMENT: SEVEN-SEGMENT LED DISPLAYS

OBJECTIVES

1. To test a single LED.
2. To test each segment of a seven-segment LED display.

MATERIALS

Qty.   **Qty.**

1 LED   1 seven-segment LED display,
1 150-Ω, ¼-W resistor   common anode
1 0- to 50-mA dc ammeter   1 5-V dc regulated power supply
   (VOM or DMM)

SYSTEM DIAGRAMS

The schematic diagram in Fig. 6-6 shows a regular LED indicator lamp and a limiting resistor placed across a 5-V dc power supply. A dc ammeter has been placed in the circuit to check the current flow. The limiting resistor is *absolutely necessary*, or the current will be too high and the LED will *burn out*.

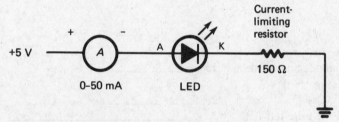

Fig. 6-6 Checking current through an LED.

A commonly used seven-segment LED display is shown in Fig. 6-7. Notice that each of the segments (*a* to *g*) contains a light-emitting diode. This display has the anodes of each LED connected. This display is said to have a *common anode*. The limiting resistor on the right in Fig. 6-7 is used to test each LED segment. The ammeter checks the current flow through each LED.

PROCEDURE

1. Power OFF. Wire the circuit shown in Fig. 6-6.
2. Power ON. Measure the current flow through the LED. Record the results in Table 6-1.
3. Power OFF. Insert the seven-segment LED display into the mounting board. Wire the circuit shown in Fig. 6-7.
4. Power ON. Test each segment of the seven-segment LED display by placing the free end of the 150-Ω limiting resistor to the correct pin. See the instructor for the pin diagram of the LED display. Record the results in Table 6-1.
5. Power OFF. Leave the seven-segment LED display on the mounting board for the next activity.

TABLE 6-1 Testing LED Displays

		Current (mA)	Does it light? (yes or no)
Single LED			
7-segment LED display	Segment *a*		
	Segment *b*		
	Segment *c*		
	Segment *d*		
	Segment *e*		
	Segment *f*		
	Segment *g*		

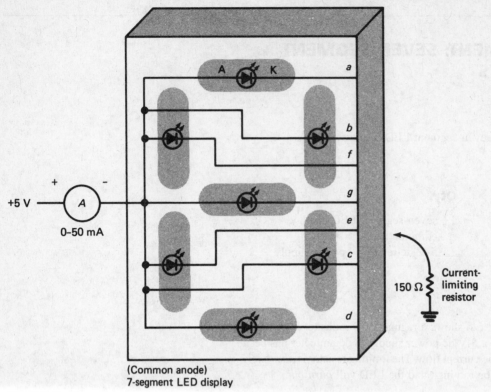

+5 V

A

0–50 mA

150 Ω   Current-limiting resistor

(Common anode)
7-segment LED display

Fig. 6-7 Testing a seven-segment LED common-anode display.

QUESTIONS

Complete questions 1 to 7.

1. Draw the schematic symbol for an LED. Identify the cathode and anode ends of the LED.
2. What is the current flow through a typical LED?
3. Refer to Fig. 6-7. What would happen if the 150-Ω resistor were left out and the seven-segment display were tested? *(Do not try this in the lab.)*
4. Refer to Fig. 6-7. The inputs *a* to *g* for the seven-segment LED display are activated by a _____ (HIGH, LOW).
5. If decimal 8 were displayed on the seven-segment LED display you used in this activity, about how much total current would the display draw?
6. Explain how you identify the cathode lead of a single LED.

7. The seven-segment LED display used in this lab was classified as a _____ (common-anode, common-cathode) type.

1. _____

2. _____

3. _____
   _____

4. _____

5. _____

6. _____
   _____

7. _____

6-2 LAB EXPERIMENT: CODE TRANSLATORS

OBJECTIVES

1. *Option 1—74147 encoder:* Wire and test an encoder-decoder system that converts from decimal to BCD to seven-segment code using the 74147 encoder IC.

2. *Option 2—74148 encoder:* Wire and test an encoder-decoder system that converts from decimal to BCD to seven-segment code using the 74148 encoder IC.

3. *Option 3—Electronics Workbench or Multisim:* To use the circuit simulator to design and test an encoder-decoder system.

MATERIALS

Qty.

1	7404 hex inverter IC
1	74147 or 74148 encoder IC*
1	seven-segment LED display
4	LED indicator lights

Qty.

1	7447 BCD-to-seven-segment decoder IC
1	keypad (0 to 9, N.O. contacts)
7	150-Ω, $\frac{1}{4}$-W resistors
1	5-V dc regulated power supply
1	electronic circuit simulation program— *option 3*

*74LSXXX ICs may be used in place of 74XXX chips.

SYSTEM DIAGRAM

Option 1. An electronic system you will assemble using this option is shown in Fig. 6-8. The 74147 *encoder* translates from decimal to 8421 BCD code. The BCD code appears in BCD form on the LED indicator lights (*D, C, B*

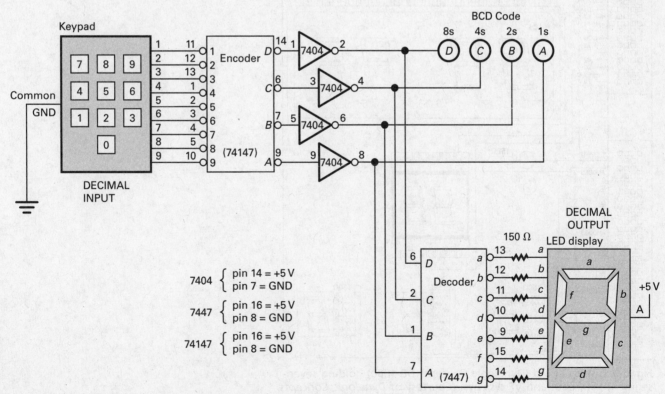

$$7404 \begin{cases} \text{pin } 14 = +5\,\text{V} \\ \text{pin } 7 = \text{GND} \end{cases}$$

$$7447 \begin{cases} \text{pin } 16 = +5\,\text{V} \\ \text{pin } 8 = \text{GND} \end{cases}$$

$$74147 \begin{cases} \text{pin } 16 = +5\,\text{V} \\ \text{pin } 8 = \text{GND} \end{cases}$$

Fig. 6-8 An electronic encoder-decoder system using the 74147 encoder.

and *A*). The 7447 decoder translates from BCD to seven-segment code, causing a decimal number (0–9) to appear on the LED display.

Pay attention to the seven 150-Ω limiting resistors between the decoder and seven-segment LED display. These are *essential* so as not to burn out any of the LEDs. Also notice that both the keypad and seven-segment LED display have only one power connection. The keypad has only a common GND connection. The seven-segment common-anode LED display has only a +5-V (V_{cc}) connection. Notice the small invert bubbles at the output of the 74147 IC. The 7404 inverters change these inverted outputs to regular BCD code.

The interfacing of the keypad to the 74147 TTL encoder in Fig. 6-8 could be improved. Technically, nine pull-up resistors should connect each encoder input to +5 V. In this noncritical design, the pull-up resistors were left out for simplicity. The inputs to the TTL encoder in Fig. 6-8 float HIGH.

The custom commercial display board pictured in Fig. 6-9 makes the mounting and wiring of seven-segment LED, LCD, and VF displays much easier. Notice toward the bottom in Fig. 6-9 that the seven 150-Ω limiting resistors are prewired to the seven-segment LEDs. In this lab you will only use 1 of the 3 seven-segment LED displays near the bottom of the display board. You connect the cathodes (*a* through *g*) and anode of each seven-segment LED display using the solderless breadboard connectors on the left.

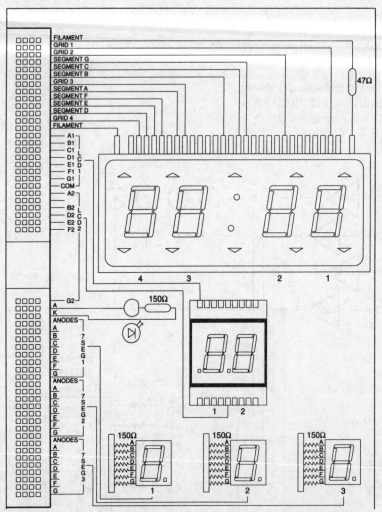

Fig. 6-9 Display board (Dynalogic Concepts DB-1000) holding seven-segment LED, LCD, and VF displays. *(Courtesy of Dynalogic Concepts, 1-800-246-4907.)*

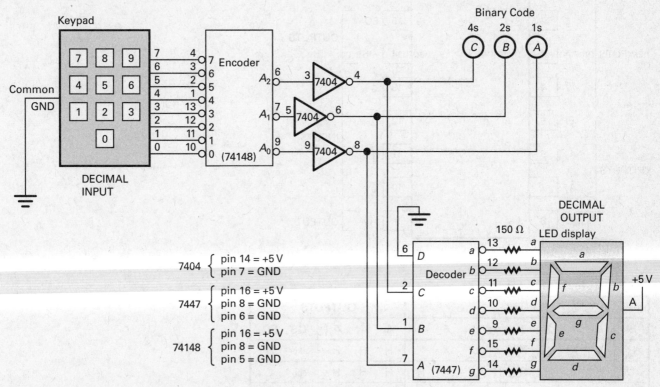

Fig. 6-10 An electronic encoder-decoder system using the 74148 encoder.

Option 2. An electronic system you will assemble using this option is shown in Fig. 6-10. The 74148 *encoder* translates from decimal to 3-bit binary code. The binary code appears on the LED indicator lights (*C, B,* and *A*). The 7447 decoder translates from 3-bit binary to seven-segment code causing a decimal number (0–7) to appear on the LED display.

Pay attention to the seven 150-Ω limiting resistors between the decoder and seven-segment LED display. These are *essential* so as not to burn out any of the LEDs. Also notice that both the keypad and seven-segment LED display have only one power connection. The keypad has only a common GND connection. The seven-segment common-anode LED display has only a +5-V (V_{cc}) connection. Notice the small invert bubbles at the output of the 74148 IC. The 7404 inverters change these inverted outputs to regular binary code.

A pin diagram and truth table for the *74148 encoder IC* are detailed in Fig. 6-11. The manufacturer lists the 74148 as an *8-line-to-3-line priority encoder.* The 74148 encoder features eight active LOW inputs ($\overline{1}$ through $\overline{7}$) and three active-LOW outputs ($\overline{A_0}$ through $\overline{A_2}$). The EI input and E0 and GS outputs will not be used in this lab. The 74148 has a *priority feature,* which means that if two inputs (such as $\overline{1}$ and $\overline{5}$) are both active at the same time, the higher number (5 in this example) will have priority.

The interfacing of the keypad to the 74148 TTL encoder in Fig. 6-10 could be improved. Technically, eight pull-up resistors should connect each encoder input to +5 V. In this noncritical design, the pull-up resistors were left out for simplicity. The inputs to the TTL encoder in Fig. 6-10 float HIGH.

The custom commercial display board pictured in Fig. 6-9 allows easy mounting and wiring of seven-segment LED, LCD, and VF displays. Notice toward the bottom in Fig. 6-9 that the seven 150-Ω limiting resistors are prewired to the seven-segment LEDs. In this lab you will only use 1 of the 3 seven-segment LED displays near the bottom of the display board. You

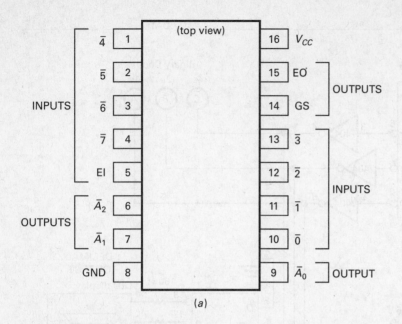

(a)

Truth table for 74148 encoder

EI	0	1	2	3	4	5	6	7	A_2	A_1	A_0	GS	EO
H	X	X	X	X	X	X	X	X	H	H	H	H	H
L	H	H	H	H	H	H	H	H	H	H	H	H	L
L	X	X	X	X	X	X	X	L	L	L	L	L	H
L	X	X	X	X	X	X	L	H	L	L	H	L	H
L	X	X	X	X	X	L	H	H	L	H	L	L	H
L	X	X	X	X	L	H	H	H	L	H	H	L	H
L	X	X	X	L	H	H	H	H	H	L	L	L	H
L	X	X	L	H	H	H	H	H	H	L	H	L	H
L	X	L	H	H	H	H	H	H	H	H	L	L	H
L	L	H	H	H	H	H	H	H	H	H	H	L	H

H = High   L = Low   X = Irrelevant

(b)

Fig. 6-11 8-line-to-3-line priority encoder. (a) Pin diagram (16-pin DIP). (b) Truth table.

connect the cathodes (a through g) and the anode of each seven-segment LED display using the solderless breadboard connectors on the left.

Option 3—Electronics Workbench or Multisim. If assigned by your instructor, use Electronics Workbench or Multisim to design and test an encoder-decoder system something like the one shown in Fig. 6-12.

PROCEDURE

1. Your instructor will tell you which encoder-decoder experiment to perform: (1) *Option 1: 74147 encoder,* (2) *Option 2: 74148 encoder,* or (3) *Option 3: Circuit simulation of encoder-decoder circuit.*
2. *Option 1:* Wire and test the circuit shown in Fig. 6-8. Show your instructor your operating encoder-decoder, and be prepared to answer questions about the circuit.

172

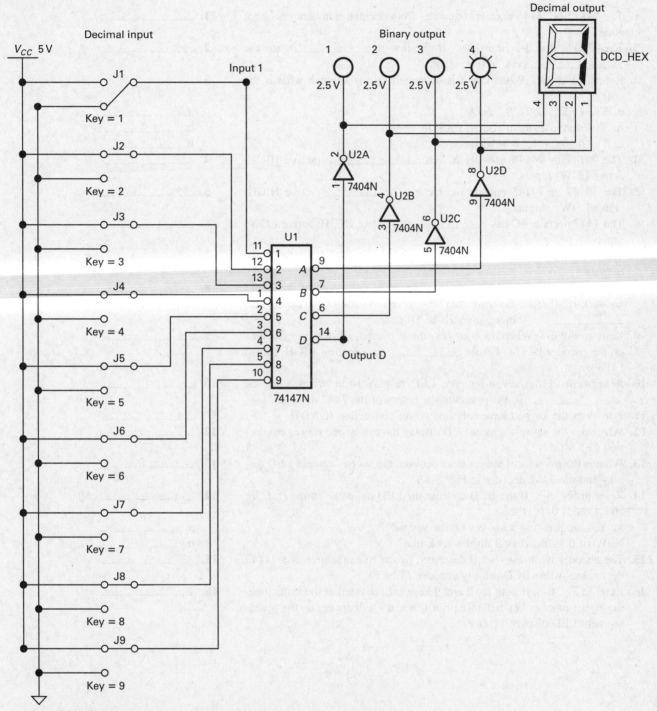

Fig. 6-12 Encoder-decoder system using 74147 encoder and 7447 decoder. (Prepared using Multisim 8.)

3. *Option 2:* Wire and test the circuit shown in Fig. 6-10. Show your instructor your operating encoder-decoder, and be prepared to answer questions about the circuit.

4. *Option 3:* Using Electronics Workbench or Multisim, draw and test the encoder-decoder circuit shown in Fig. 6-12. Show the instructor your operating simulated encoder-decoder, and be prepared to answer questions about the circuit.

5. Power OFF. Take down the circuit, and return all equipment to its proper place.

QUESTIONS

Complete questions 1 to 16.

1. The 74147 or 74148 encoder translates from decimal numbers into what code?

2. The 7447 decoder translates from the _____ to the _____ code at the output.

3. Refer to Fig. 6-8. When key 3 is depressed on the keypad, what is the output?
   a. The LEDs, *D, C, B,* and *A*
   b. The seven-segment code (logical 0s or 1s)
   c. The digital display readout

4. The 74147 or 74148 encoder IC has _____ (active HIGH, active LOW) inputs.

5. The 74147 or 74148 encoder IC has _____ (active HIGH, active LOW) outputs.

6. The 7447 decoder IC has _____ (active HIGH, active LOW) inputs.

7. The 7447 decoder IC has _____ (active HIGH, active LOW) outputs.

8. Refer to Fig. 6-8. The 4-bit code that appears at the output of the 74147 encoder IC (before entering the 7404 inverters) might be called a(n) _____ (binary, inverted BCD) code.

9. Refer to Fig 6-8. When *no input keys are activated*, all of the inputs pins on the encoder IC (74147) are _____ (floating HIGH, forced LOW).

10. A segment of the seven-segment LED display is lit when a logical _____ (0, 1) appears at the output of the 7447 decoder.

11. Why does the keypad have only one power connection (GND)?

12. Why does the seven-segment LED display have only one power connection (+5 V)?

13. What is the purpose of the resistors between the seven-segment LED display and the 7447 decoder in Fig. 6-8?

14. Refer to Fig. 6-8. If the BCD code on the LED indicator lamps *D, C, B,* and *A* reads 0110, then
   a. You are pressing what key on the keypad?
   b. What does the digital display look like?

15. The encoder IC in Fig. 6-8 is described by the manufacturer as a 74147 decimal-to-4-line-BCD priority encoder. (T or F)

16. Refer to Fig. 6-8. If both the 2 and 9 keys are activated at the same time, the higher-number key will take priority and a 9 will appear on the seven-segment LED display. (T or F)

1. _____

2. _____

3.
   a. _____
   b. _____
   b. _____

4. _____

5. _____

6. _____

7. _____

8. _____

9. _____

10. _____

11. _____

12. _____

13. _____

14. _____
   a. _____
   b. _____

15. _____

16. _____

6-3 LAB EXPERIMENT: DRIVING THE LCD DISPLAY

OBJECTIVES

1. To wire and test an LCD driver circuit using a CMOS decoder and 555 timer IC.

2. To observe the in-phase and out-of-phase drive signals present on the inputs to the liquid-crystal display.

MATERIALS

Qty.

1 555 timer IC
1 74HC4543 or CD4543
  BCD-to-seven-segment
  decoder/driver CMOS IC
4 logic switches
1 seven-segment LCD
1 1-kΩ, ¼-W resistor

Qty.

1 10-kΩ, ¼-W resistor
1 1-μF capacitor
1 5-V dc regulated power
  supply
1 dual-trace oscilloscope
  (optional)

SYSTEM DIAGRAMS

A CMOS decoder and astable multivibrator are being used to drive an LCD in Fig. 6-13. The 74HC4543 (or CD4543) IC is used to decode from BCD to seven-segment code. The 555 timer is wired as an astable multivibrator generating

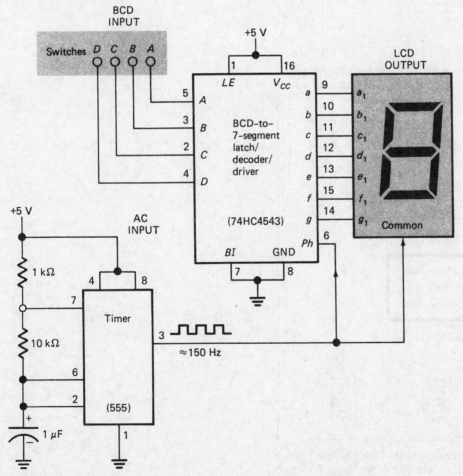

Fig. 6-13 A decoder and clock driving an LCD seven-segment display.

a 150-Hz square-wave signal. The 150-Hz signal is applied to both the common (backplane) of the LCD and the *Ph* (phase) input of the 4543 IC. The active outputs of the 4543 send a 180° out-of-phase signal to the segment, causing it to appear black on a silvery background. Inactive outputs will have an in-phase signal sent to their segment of the LCD.

Liquid-crystal displays are driven by square waves and not dc like LEDs. *Do not apply dc directly to the LCD*, or it can be damaged.

An excellent method of mounting and connecting to the LCD display is the commercial display board by Dynalogic Concepts shown in Fig. 6-14. On the display board in Fig. 6-14, each pin of the LCD is available on the solderless breadboard on the left. If you do not have a commercial display board available, you could mount the LCD unit on two separate solderless mounting boards as shown in Fig. 6-15.

PROCEDURE

1. Insert the 555 IC and 74HC4543 (or CD4543) CMOS IC into the mounting board.

CAUTION CMOS ICs can be damaged by static electricity.

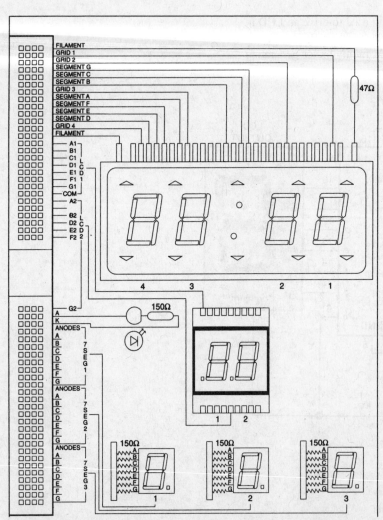

Fig. 6-14 Display board (Dynalogic Concepts DB-1000) holding seven-segment LED, LCD, and VF displays. *(Courtesy of Dynalogic Concepts, 1-800-246-4907.)*

Fig. 6-15 Alternative method of mounting LCD using two separate solderless breadboards.

2. Mount the seven-segment LCD. Be careful because LCDs are fragile— they are made of glass.

   a. *Premounted LCD.* If your LCD is premounted on a PC board, follow the directions of the manufacturer or your instructor. One display board is shown in Fig. 6-14.

   b. *Unmounted LCD.* If your LCD is not mounted, it can be inserted in two separate mounting boards as sketched in Fig. 6-15.

3. Power OFF. Connect power (+5 V and GND) to the 555 and 74HC4543 (or CD4543) ICs.

4. Wire the circuit in Fig. 6-13.

5. Power ON. Operate input switches A, B, C, and D according to the input side of Table 6-2. Record the outputs in the Decimal Output column of Table 6-2.

6. Show your instructor your decoder/LCD driver circuit. Be able to answer questions about the circuit. Ask your instructor if you should do the remaining optional section of the experiment.

7. *OPTIONAL.* Observing phase relationship between the clock and the ON and OFF drive lines to the LCD.

8. Circuit power OFF. Dual-trace oscilloscope power ON. Connect channel 1 of scope to pin 3 (output) of the 555 timer IC. Connect channel 2 of scope to pin 9 (drive line *a*) of the 4543 IC.

9. Power ON. Set input switches to 1000_{BCD}. An 8 should appear on the LCD. Observe and compare the phase relationship of the waveforms. They should be 180° out of phase.

10. Set input switches to 0100_{BCD}. A 4 should appear on the LCD. Observe and compare the phase relationship of the waveforms. They should be in phase.

11. Measure the time duration of one cycle of the clock signal (at pin 3 of the 555 IC) on the scope. Using the measured time t in seconds, calculate the frequency $f = 1/t$. The frequency should be between 50 and 200 Hz.

12. Power OFF. Take down the circuit, and return all equipment to its proper place. The pins of CMOS ICs should be stored in conductive foam.

TABLE 6-2 Readings from LCD

Inputs				Decimal Output
D	C	B	A	
0	0	0	0	
0	0	0	1	
0	0	1	0	
0	0	1	1	
0	1	0	0	
0	1	0	1	
0	1	1	0	
0	1	1	1	
1	0	0	0	
1	0	0	1	

QUESTIONS

Complete questions 1 to 10.

1. The 4543 IC translates from _____ code to _____ code.

2. The liquid-crystal display is driven by a _____ (+5-V dc, 150-Hz ac square-wave) signal.

3. What segments of the LCD appear black when 0110_{BCD} is applied to the circuit in Fig. 6-13?

4. Refer to Fig. 6-13. Which drive lines are 180° out of phase with the clock signal when 0111_{BCD} is applied to the input of the circuit?

5. Only _____ (in-phase, out-of-phase) signals cause segments to appear on the liquid-crystal display.

6. LCDs are driven by _____ (CMOS, TTL) ICs.

7. LCDs are ideal displays to use in a dark environment. (T or F)

8. LCDs are ideal displays for use with solar cells because of their low power consumption. (T or F)

9. The LCD used in this experiment is backlighted to make it more visible under low-light conditions. (T or F)

10. The display used in this experiment was what type of LCD?
    a. Active-matrix color LCD
    b. Active-matrix monochrome LCD
    c. Field-effect monochrome LCD

1. _____ _____

2. _____

3. _____

4. _____

5. _____

6. _____

7. _____

8. _____

9. _____

10. _____

178

6-4 LAB EXPERIMENT: USING CMOS TO DRIVE VF DISPLAYS

OBJECTIVES

1. To test the operation of a vacuum fluorescent display.
2. To wire and test a pulse-counting circuit using a CMOS IC to directly drive a VF display.
3. To observe the operation of a digital circuit operating on two power supplies (+5 V and +12 V dc).

MATERIALS

Qty.   **Qty.**

1   4511 BCD-to-seven-segment   1   keypad (N.O. contacts)
    decoder/driver CMOS IC   1   vacuum fluorescent display
1   74HC08 quad two-input AND   1   47-Ω, $\frac{1}{2}$-W resistor
    gate CMOS IC   5   100-kΩ, $\frac{1}{4}$-W resistors
1   74HC14 hex inverting   1   0.047-μF capacitor
    Schmitt trigger CMOS IC   1   5-V dc regulated power
1   74HC393 dual 4-bit counter   supply
    CMOS IC   1   9-, 12-, or 15-V dc power
1   74C906 hex buffer IC   supply

SYSTEM DIAGRAMS

A single digit of a vacuum fluorescent (VF) display is pictured in Fig. 6-16. Note that two dc power supplies are being used in the testing of this display. The filament, F or heater, is powered by the +5-V supply. The grid and plates are activated by a +12-V dc power supply. You may substitute a 9- or 15-V power supply in place of the +12-V unit in the circuit in Fig. 6-16. The +12-V test lead is used to activate each individual segment (plate) in turn. Note that the negative terminals of the two power supplies must be connected. The 47-Ω resistor serves as a current-limiting resistor for the filament circuit.

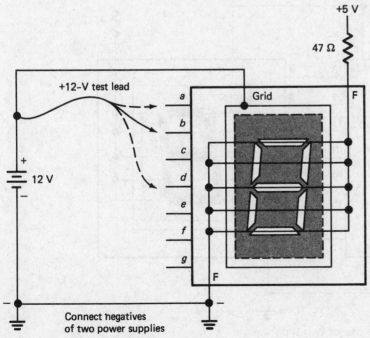

Fig. 6-16 Simple test procedure for a VF display.

Copyright © by McGraw-Hill.

A wiring diagram for a pulse-counting circuit is drawn in Fig. 6-17. The 74HC393 counter should increment one count for each press of the 1 on the keypad. The 74HC14, pull-up resistor, and capacitor form the switch-

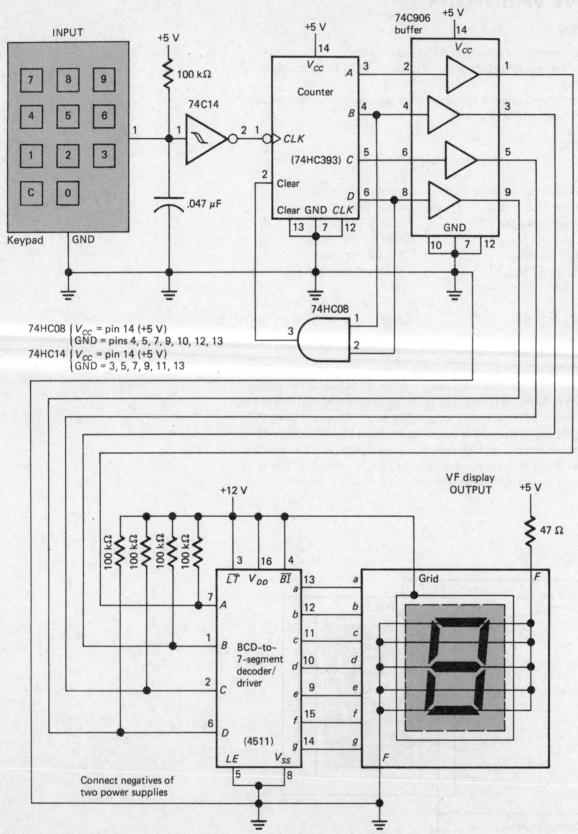

74HC08 $\left\{\begin{array}{l} V_{CC} = \text{pin 14 (+5 V)} \\ GND = \text{pins 4, 5, 7, 9, 10, 12, 13} \end{array}\right.$

74HC14 $\left\{\begin{array}{l} V_{CC} = \text{pin 14 (+5 V)} \\ GND = 3, 5, 7, 9, 11, 13 \end{array}\right.$

Fig. 6-17 Pulse counting with VF display. Note the use of two separate power supplies.

debouncing circuit. The 74HC393 and the AND gate are wired as a decade counter. The counter's BCD output passes through the buffers to the 4511 decoder/driver IC. The 4511 decodes the BCD and drives the VF display.

The most unusual feature of the circuit in Fig. 6-17 is the use of two power supplies. The 74HC393, 74HC14, 74HC08, and 74C906 ICs operate on a +5-V power supply. The VF display and the 4511 decoder/driver operate on a +12-V supply. The 74C906 buffer and four pull-up resistors form the interface between the low-voltage counter and the higher-voltage BCD input to the 4511 decoder IC. The segments and grid of the VF display require the higher 12 V. The filament (also called the cathode or heater) operates on the lower-voltage supply.

The VF display will not operate properly with +5 V on the grids and plates (segments). This is why the high-voltage section is used in this circuit. In practice, the counter, AND gate, and Schmitt trigger inverter could be replaced by equivalent higher-voltage CMOS ICs. This would permit the entire circuit to operate on the higher +12 V. The buffer could also be eliminated if a single 12-V supply were used. This circuit is included as an experiment to demonstrate the interfacing required when dealing with two power supply voltages in the same digital circuit.

PROCEDURE

1. Locate a premounted vacuum fluorescent (VF) display. The DB-1000 display board by Dynalogic Concepts drawn in Fig. 6-18 features solderless connectors at the upper left making connections to internal sections of the four-digit VF display. On the sealed-glass VF display try to identify the filaments (also called cathodes or heaters), plates (segments), and grids.

2. Power OFF. Refer to the circuit in Fig. 6-16. Connect the +5-V power supply and 47-Ω limiting resistor to the filaments (heaters) of the multidigit display. The one set of filaments serves the entire multidigit display.

3. Refer to Fig. 6-16. Connect the +12-V power supply to one of the grids in the multidigit display. Connect the negative terminals of the 5- and 12-V power supplies together as illustrated in Fig. 6-16.

4. Power ON. Test each segment (*a* to *g*) on the VF display. This is done by connecting +12 V to each segment (plate) input. They should each glow as you connect the positive voltage.

5. Power OFF. Disconnect the grid wire, and reconnect to a different grid of the multidigit VF display.

6. Power ON. Test the second digit on the VF display by connecting +12 V to each segment. Show the instructor your test circuit.

7. Power OFF. Insert four ICs (74HC08, 74HC14, 74HC393, and 74C906) into the mounting board. Connect to the +5-V power supply.

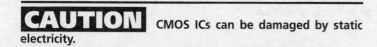

CAUTION CMOS ICs can be damaged by static electricity.

8. Insert the 4511 IC in the mounting board. Connect +12-V power [V_{DD} and V_{SS} (GND)] to the 4511 IC only.

9. Carefully wire the entire circuit shown in Fig. 6-17. Color coding of wires and a neat layout help in complex circuits. Check Appendix A or ask your instructor for connection diagrams to the keypad and VF display.

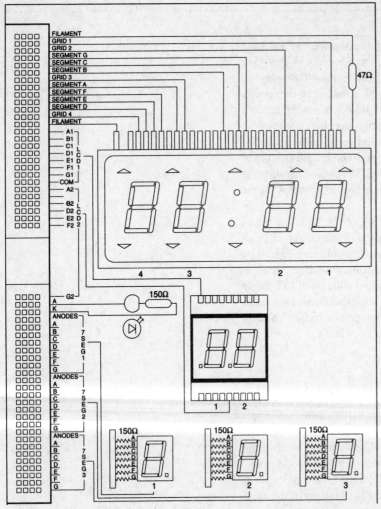

Fig. 6-18 Display board (Dynalogic Concepts DB-1000) holding seven-segment LED, LCD, and VF displays. *(Courtesy of Dynalogic Concepts, 1-800-246-4907.)*

Labels in figure:
FILAMENT
GRID 1
GRID 2
SEGMENT G
SEGMENT C
SEGMENT B
GRID 3
SEGMENT A
SEGMENT F
SEGMENT E
SEGMENT D
GRID 4
FILAMENT

47Ω

A1
B1
C1
D1 L
E1 C
F1 D
G1 1
COM
A2
B2 L
D2 C
E2 D
F2 2

4   3   2   1

G2
A
K
ANODES
A
B
C
D
E
F
G
ANODES
A
B
C
D
E
F
G
ANODES
A
B
C
D
E
F
G

150Ω

7 S E G 1
7 S E G 2
7 S E G 3

1   2

150Ω 150Ω 150Ω

1   2   3

10. Power ON. When you press the 1 on the keypad, the decimal count on the display should increment by one. The 74HC393 IC is wired as a decade counter.

11. Have your instructor check your circuit. Be prepared to answer questions about the pulse-counting circuit (Fig. 6-17).

12. *OPTIONAL.* Turning on all of the displays.

13. Power OFF. Connect +12 V to all of the grids for seven-segment digits on your VF display.

14. Power ON. Repeatedly press the 1 on the keypad, and observe the VF display. All digits should be displaying a number.

15. Power OFF. Take down the circuit, and return all equipment to its proper place. The pins of CMOS ICs should be stored in conductive foam.

QUESTIONS

Complete questions 1 to 15.

1. The VF display emits a _____ color when lit.
2. The cathodes of VF displays are also called _____ or _____.
3. Refer to Fig. 6-16. What are the results of touching the +12-V test lead to the plate input *d* of the VF display?

1. _____
2. _____ _____
3. _____

4. Refer to Fig. 6-17. The 74HC393 and 74HC08 form a _____ counter.

4. _____

5. Refer to Fig. 6-17. What components form the interface between the low-voltage counter and the high-voltage decoder/driver?

5. _____

6. Refer to Fig. 6-17. What components form the switch-debouncing circuit?

6. _____
_____

7. Refer to Fig. 6-17. If 7 lights, which segment inputs to the VF display have +12 V applied to them?

7. _____

8. On a 4000 series IC, the V_{SS} pin is connected to the _____ (negative, positive) of the power supply.

8. _____

9. On a 4000 series IC, the V_{DD} pin is connected to the _____ (negative, positive) of the power supply.

9. _____

10. Refer to Fig. 6-19. The section of the VF display labeled A is called the _____.

10. _____

    a. Grid
    b. Filament, cathode, or heater
    c. Plate
    d. Collector

11. Refer to Fig. 6-19. The section of the VF display labeled B is called the _____.

11. _____

    a. Grid
    b. Filament, cathode, or heater
    c. Plate
    d. Collector

12. Refer to Fig. 6-19. The sections of the VF display labeled C are called the _____.

12. _____

    a. Grids
    b. Filaments, cathodes, or heaters
    c. Plates

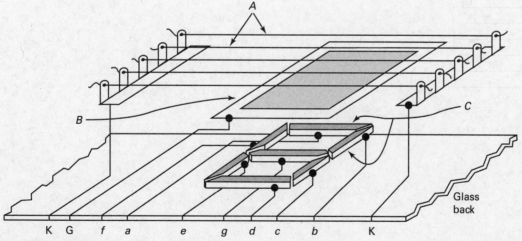

Fig. 6-19 VF display.

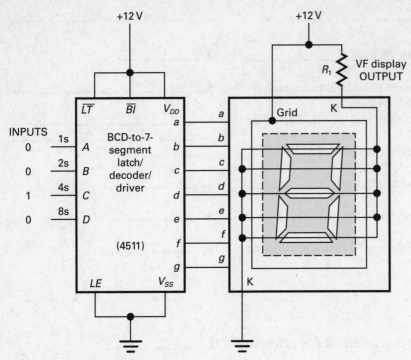

+12 V

+12 V

INPUTS

1s   0   A

2s   0   B

4s   1   C

8s   0   D

$\overline{LT}$ $\overline{BI}$ V_{DD}

BCD-to-7-segment latch/decoder/driver

(4511)

LE V_{SS}

a, b, c, d, e, f, g

R_1

VF display
OUTPUT

Grid   K

K

Fig. 6-20 Decoded/VF display.

13. Refer to Fig. 6-20. The +12-V power supply is being used because the _____ (CMOS, TTL) 4511 IC and the VF display work well at that voltage.

14. Refer to Fig. 6-20. What is the decimal reading on the VF display if the input is 0100_{BCD}?

15. Refer to Fig. 6-20. With the input to the 4511 IC at 0100_{BCD}, the VF display grid activated with +12 V, and the filaments or cathodes (K) heated, what are the voltages at each of the seven segment inputs to the VF display?

segment a = _____

segment b = _____

segment c = _____

segment d = _____

segment e = _____

segment f = _____

segment g = _____

13. _____

14. _____

6-5 TROUBLESHOOTING PROBLEM: DECODER/DISPLAY CIRCUIT

OBJECTIVES

1. To wire and operate a normal decoder/LED display circuit.
2. To troubleshoot a decoder/LED display circuit.
3. To determine which display segment is faulty and whether it is open or partially shorted.

MATERIALS

Qty.		Qty.	
1	7447 BCD-to-seven-segment decoder TTL IC	7	150-Ω, $\frac{1}{4}$-W resistors
1	*faulty* seven-segment LED display (get from instructor)	1	5-V dc regulated power supply
		1	logic probe
1	*good* seven-segment LED display	1	voltmeter (VOM or DMM)

SYSTEM DIAGRAM

You will troubleshoot the decoder/LED display circuit shown in Fig. 6-21. Your instructor will furnish a *faulty* (or perhaps good) seven-segment LED display. Your task will be to troubleshoot the circuit. You will determine which, if any, segments are bad and the nature of the fault (open or shorted segment).

In troubleshooting, it is important to know the characteristics of a *normal* circuit. You will first construct the normal decoder/LED display circuit to verify the test data shown in Fig. 6-21. Normally, all LED display segments would be ON when the LT (lamp test) input to the 7447A IC is activated. Notice that both logic probe (Ls and Hs) and approximate dc voltages are recorded on the schematic in Fig. 6-21.

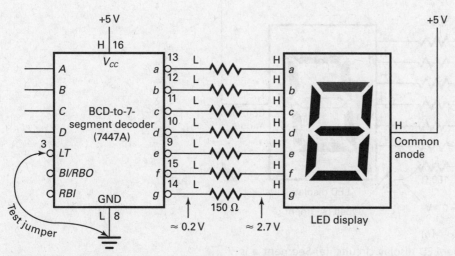

Fig. 6-21 Normal test data for decoder/LED display circuit.

Faulty decoder/LED display circuits are sketched in Fig. 6-22. Segment *a* is not lighting in Fig. 6-22(*a*). The logic probe readings are satisfactory. The dc voltage reading at input *a* to the seven-segment LED display is too high. It is at 4 to 5 V as compared to the normal of about 2.6 to 3.0 V. This would be the effect of a *shorted* or *partially shorted* segment *a* of the LED display.

Segment *g* is not lighting on the display in Fig. 6-22(*b*). When taking logic probe readings, all are correct except input *g* to the display is LOW instead of HIGH. The dc voltage at input *g* to the display is also very low at near 0 V, while it should be about 2.8 V. This would be the effect of an *open* in segment *g* of the seven-segment LED display.

Replacing the display unit is the solution to the problem in either faulty circuit detailed in Fig. 6-22. The pin diagram for the replacement LED must be the same as the faulty unit.

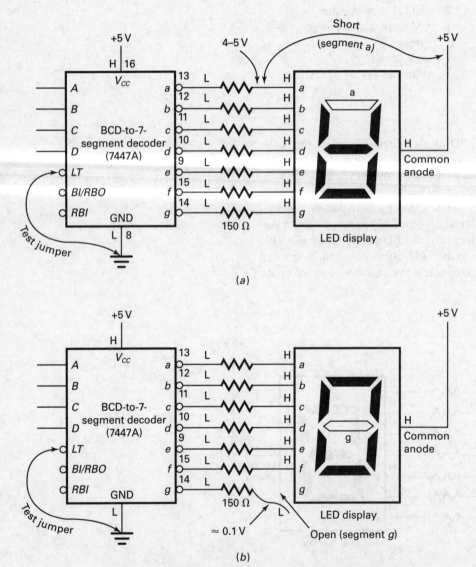

(a)

(b)

Fig. 6-22 Test data for faulty decoder/LED display circuits. (*a*) Segment *a* is shorted. (*b*) Segment *g* is open.

PROCEDURE

1. Power OFF. Wire the normal circuit shown in Fig. 6-21 using a good seven-segment LED display. Connect a temporary test jumper wire from the *LT* (lamp-test) input of the 7447 IC to GND.
2. Power ON. Observe the display. All segments should now be lit.
3. Use a logic probe and voltmeter (VOM or DMM) to verify the reading for the normal circuit sketched in Fig. 6-21.
4. Refer to Fig 6-22(*a*). Place a temporary jumper wire to short out segment *a* of the LED display. Use a logic probe and voltmeter to take the test reading for this faulty circuit. Record your results. Remove the shorting wire, and return the circuit to normal operation.
5. Refer to Fig 6-22(*b*). Disconnect the resistor from input *g* of the LED. Use a logic probe and voltmeter to take the test reading for this faulty circuit. Record your results.
6. Power OFF. Report your results to your instructor, and be prepared to answer selected questions about the circuit and results. *Pick up a faulty seven-segment LED from your instructor.*
7. Power OFF. Wire the circuit shown in Fig. 6-21 using the faulty seven-segment LED.
8. Power ON. Observe the display. Use a logic probe and dc voltmeter to troubleshoot the circuit.
9. Power OFF. Report your results to your instructor. *Return the faulty seven-segment LED to your instructor.*
10. Power OFF. Take down the circuit, and return all equipment to its proper place.

6-6 MULTISIM EXPERIMENT: DECODING 8-BIT BINARY TO HEXADECIMAL

OBJECTIVES

1. To construct and test a 8-bit binary counter with decoding to drive a two-digit hexadecimal display.

2. To use seven-segment displays to show all 16 hexadecimal characters.

MATERIALS

Electronics Workbench or Multisim software on a computer system

SYSTEM DIAGRAM

It is common in digital systems to transfer data in byte (8-bit) groups. A block diagram of a simple 8-bit binary to two-digit hexadecimal is sketched in Fig. 6-23. The binary input is generated by switches in Fig. 6-23. In this example, it is assumed that the hexadecimal output is shown on 2 seven-segment LED displays. The decoder block in Fig. 6-23 converts the 8-bit binary to two-digit hexadecimal code. The driver block suggests that the outputs are compatible (both voltage and current), matching the LED outputs. Earlier you have used limiting resistors to do this job.

If the input from the decoder/driver circuit in Fig. 6-23 were binary 11000011, then the output would read C3 in hexadecimal. This system does not have any memory characteristics. This means that when any bit on the binary input is changed, the hex displays will change immediately. In Fig. 6-23, what would be the decoded hex output if the input were binary 11111111? The output would read FF in hexadecimal.

A circuit simulator version (Multisim) of an 8-bit counter sending byte-wide counts to a two-digit hexadecimal decoder/display is detailed in Fig. 6-24. The three inputs to the 8-bit counter are *reset* (H = reset display to 00 or L = allow to count), *enable* (H = allow to count or L = pause count), and *CLK* (each positive pulse increments the count by 1). The CLK input is triggered by the digital clock (astable MV).

The output of the counter is 8-bit binary. The displays contain both the blocks shown in Fig. 6-23 (decoder and driver blocks). Assume that the outputs shown in Fig. 6-24 are seven-segment LED displays.

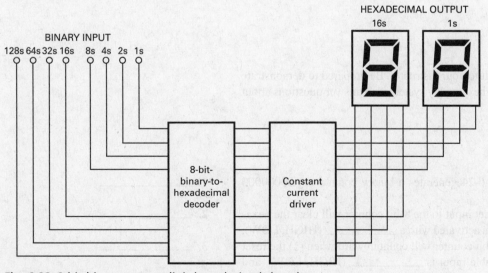

Fig. 6-23 8-bit-binary-to-two-digit-hexadecimal decoder.

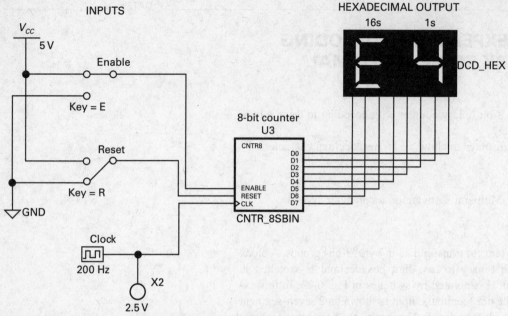

Fig. 6-24 Circuit simulator version of 8-bit counter decoded and driving a hexadecimal display (prepared using Multisim).

PROCEDURE

1. Construct the binary counter/decoder/hex display system shown in Fig. 6-24 using electronic circuit simulation software. Experiment with the digital clock frequency setting (try from 50 to 300 Hz). The simulation runs slower than real time, depending on the computer system.
2. Operate the counter/decoder/hex display system. *Enable* is an active HIGH input. *Reset* is an active HIGH input. The clock will run continuously.
3. Observe the action of the enable and reset inputs.
4. Observe the format the hexadecimal digits (especially for formation of A, B, C, D, E, and F) on the seven-segment display. Record this below.

5. Show your simulated circuit to your instructor. Be prepared to demonstrate the binary counter/decoder/hex display system, and answer questions about its operation.

QUESTIONS

Complete questions 1 to 13.

1. The 8-bit counter in Fig. 6-24 generates a binary count from 00000000 to 11111111. (T or F)

1. _____

2. Refer to Fig. 6-24. The reset input to the 8-bit counter will clear the hexadecimal display to 00 when activated with a _____ (HIGH, LOW).

2. _____

3. Refer to Fig. 6-24. The 8-bit counter will count upward when (1) the reset input is LOW, (2) the enable input is _____ (HIGH, LOW), and (3) clock pulses reach the CLK input.

3. _____

190

4. Refer to Fig. 6-24. What is the binary count if the hex display reads E4?

5. Refer to Fig. 6-24. If the output of the counter is binary 10101111, the hexadecimal output will read _____ (two hex digits).

6. Refer to Fig. 6-24. If the output of the counter is binary 00101100, the hexadecimal output will read _____ (two hex digits).

7. Refer to Fig. 6-24. The count can range from 00000000 to 11111111 in binary. This range equals 00 in hex to _____ (two hex digits) or 0 to _____ (in decimal).

8. Refer to Fig. 6-24. What will happen to the hexadecimal output if the count was C2 in hex and the reset input goes from LOW to HIGH?

9. Refer to Fig. 6-24. Imagine the hexadecimal output is sequencing upward F0, Fl, F2, and F3 when the enable input goes from HIGH to LOW. What is the count showing on the hexadecimal display?

10. The decoder in Fig. 6-23 does not have a memory characteristic, but disabling the enable input to the counter in Fig. 6-24 leaves the last count on the display which means that U3 (counter) has digital memory. (T or F)

11. Refer to Fig. 6-24. It is understood that the **hexadecimal output** block at the upper right contains (1) seven-segment displays, (2) constant current drivers, and (3) _____ (decoders, multiplexers).

12. An 8-bit binary group is commonly called a _____ (byte, crawl).

13. Draw the seven-segment format used by the decoder in Fig. 6-24 for the hexadecimal characters for A, B, C, D, E, and F.

4. _____

5. _____

6. _____

7. _____ _____

8. _____
   _____

9. _____

10. _____

11. _____

12. _____

CHAPTER 7

Flip-Flops

TEST: FLIP-FLOPS

Answer the questions in the spaces provided.

1. If the basic building block for combinational logic circuits is the logic
   gate, the basic building block for sequential logic circuits is the
   a. Flip-flop
   b. Multiplexer
   c. Sequencer
   d. Timer

2. The flip-flop shown in Fig. 7-1 is a(n) _____ type.
   a. Asynchronous
   b. Synchronous

3. Refer to Fig. 7-1. Output Q of the flip-flop is at logical _____
   (0, 1) during pulse t_1.

4. Refer to Fig. 7-1. Output Q of the flip-flop is at logical _____
   (0, 1) during pulse t_2.

5. Refer to Fig. 7-1. Output Q of the flip-flop is at logical _____
   (0, 1) during pulse t_3.

6. Refer to Fig. 7-1. Output Q of the flip-flop is at logical _____
   (0, 1) during pulse t_4.

7. Refer to Fig. 7-1. The R-S flip-flop is in the *set* mode of operation
   during pulse(s) _____.

8. Refer to Fig. 7-1. The R-S flip-flop is in the *reset* mode of operation
   during pulse(s) _____.

9. Refer to Fig. 7-1. The R-S flip-flop is in the *hold* mode of operation
   during pulse(s) _____.

1. _____

2. _____

3. _____

4. _____

5. _____

6. _____

7. _____

8. _____

9. _____

```
  0 | 1 | 1 | 1  →  ○ R      Q ├─ ?

  t₁  t₂  t₃  t₄          FF

  1 | 1 | 0 | 1  →  ○ S      Q̄ ├─
```

Fig. 7-1 Pulse-train problem.

10. Refer to Fig. 7-2. Output Q of the clocked R-S flip-flop will be at a
    logical _____ (0, 1) after pulse t_1.

10. _____

11. Refer to Fig. 7-2. Output Q of the clocked R-S flip-flop will be at a logical _____ (0, 1) after pulse t_2.

11. _____

12. Refer to Fig. 7-2. Output Q of the clocked R-S flip-flop will be at a logical _____ (0, 1) after pulse t_3.

12. _____

13. Refer to Fig. 7-2. Output Q of the clocked R-S flip-flop will be at a logical _____ (0, 1) after pulse t_4.

13. _____

14. Refer to Fig. 7-2. The clocked R-S flip-flop is in the *hold* mode of operation during pulse(s) _____.

14. _____

15. Refer to Fig. 7-2. The complementary output $\overline{Q}$ of the clocked R-S flip-flop is at a logical _____ (0, 1) after pulse t_4.

15. _____

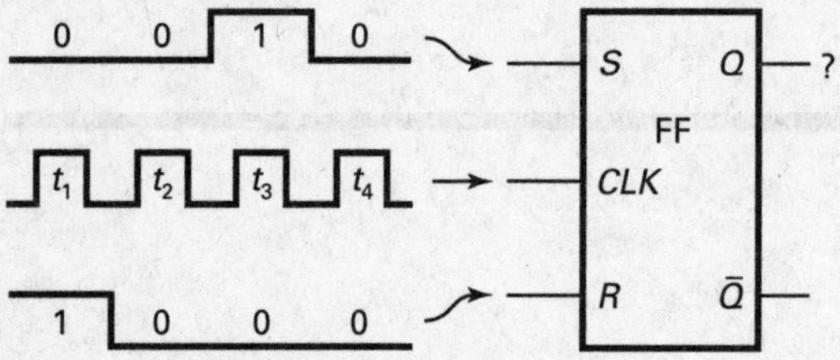

Fig. 7-2 Pulse-train problem.

16. Refer to Fig. 7-3. This flip-flop is sometimes referred to as a
   a. Delay flip-flop
   b. Demon flip-flop

16. _____

17. Refer to Fig. 7-3. Output Q of the D flip-flop will be at a logical _____ (0, 1) after pulse t_1.

17. _____

18. Refer to Fig. 7-3. Output Q of the D flip-flop will be at a logical _____ (0, 1) after pulse t_2.

18. _____

19. Refer to Fig. 7-3. Output Q of the D flip-flop will be at a logical _____ (0, 1) after pulse t_3.

19. _____

20. Refer to Fig. 7-3. Output Q of the D flip-flop will be at a logical _____ (0, 1) after pulse t_4.

20. _____

21. Refer to Fig. 7-3. Output Q of the D flip-flop will be at a logical _____ (0, 1) after pulse t_5.

21. _____

22. Refer to Fig. 7-3. The asynchronous inputs to this D flip-flop are the _____ inputs.
   a. *CLK* and Q
   b. *CLR* and *PS*
   c. *D* and *CLK*

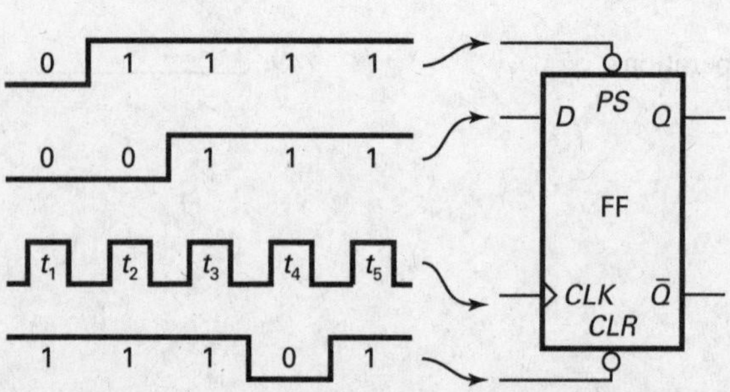

22. _____

Fig. 7-3 Pulse-train problem.

23. Refer to Fig. 7-4. Output Q of the J-K flip-flop will be at a logical _____ (0, 1) after pulse t_1.

23. _____

24. Refer to Fig. 7-4. Output Q of the J-K flip-flop will be at a logical _____ (0, 1) after pulse t_2.

24. _____

25. Refer to Fig. 7-4. Output Q of the J-K flip-flop will be at a logical _____ (0, 1) after pulse t_3.

25. _____

194

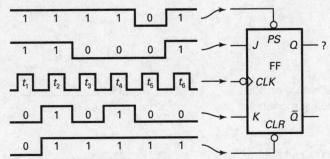

Fig. 7-4 Pulse-train problem.

26. Refer to Fig. 7-4. Output Q of the J-K flip-flop will be at a logical _____ (0, 1) after pulse t_4.

26. _____

27. Refer to Fig. 7-4. Output Q of the J-K flip-flop will be at a logical _____ (0, 1) after pulse t_5.

27. _____

28. Refer to Fig. 7-4. Output Q of the J-K flip-flop will be at a logical _____ (0, 1) after pulse t_6.

28. _____

29. Refer to Fig. 7-4. The J-K flip-flop is in what mode of operation during pulse t_1?
    a. Asynchronous clear
    b. Asynchronous preset
    c. Hold
    d. Set
    e. Toggle

29. _____

30. Refer to Fig. 7-4. The J-K flip-flop is in what mode of operation during pulse t_2?
    a. Asynchronous clear
    b. Asynchronous preset
    c. Hold
    d. Set
    e. Toggle

30. _____

31. Refer to Fig. 7-4. The J-K flip-flop is in what mode of operation during pulse t_3?
    a. Asynchronous clear
    b. Asynchronous preset
    c. Hold
    d. Set
    e. Toggle

31. _____

32. Refer to Fig. 7-4. The J-K flip-flop is in what mode of operation during pulse t_4?
    a. Asynchronous preset
    b. Hold
    c. Reset
    d. Set
    e. Toggle

32. _____

33. Refer to Fig. 7-4. The J-K flip-flop is in what mode of operation during pulse t_5?
    a. Asynchronous preset
    b. Hold
    c. Reset
    d. Set
    e. Toggle

33. _____

34. Refer to Fig. 7-4. The J-K flip-flop is in what mode of operation during pulse t_6?
  a. Asynchronous preset
  b. Hold
  c. Reset
  d. Set
  e. Toggle

35. Refer to Fig. 7-4. The outputs of this flip-flop change states on the _____ edge of the clock pulse according to the logic symbol.
  a. Negative-going (falling)
  b. Positive-going (rising)

36. When a D flip-flop is temporarily used to store a bit of data and hold it until needed, it is commonly called a _____.

37. The normal output Q of a flip-flop is at a logical _____ (0, 1) when the unit is *reset*.

38. A Schmitt trigger IC is commonly used for
  a. Counting
  b. Multiplexing
  c. Wave shaping

39. Refer to Fig. 7-5. The embedded part labeled A in the Hall-effect switch is a(n) _____ (op-amp, Schmitt trigger) providing the "snap-action" digital output.

40. Refer to Fig. 7-5. The 74LS112 J-K flip-flops are both in the _____ (hold, toggle) mode of operation.

41. Refer to Fig. 7-5. When the S pole of the magnet approaches the Hall-effect sensor, the output transistor of the 3141 IC turns_____ (off, on) and the input to FF1 goes from HIGH to LOW toggling the flip-flop.

42. Refer to Fig. 7-5. The Hall-effect switch is triggering a 2-bit _____ (counter, shift-register) circuit.

34. _____

35. _____

36. _____

37. _____

38. _____

39. _____

40. _____

41. _____

42. _____

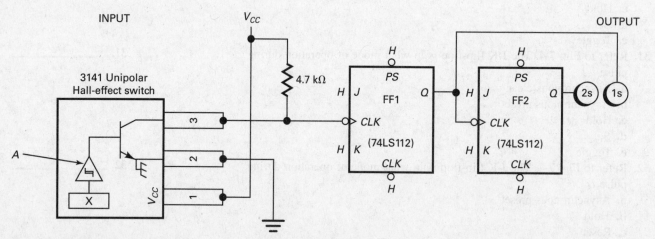

Fig. 7-5 Unipolar Hall-effect switch used to trigger a flip-flop circuit.

7-1 LAB EXPERIMENT: R-S FLIP-FLOPS

OBJECTIVES

1. To wire and observe the operation of an R-S flip-flop.
2. To wire and observe the operation of a clocked R-S flip-flop.
3. *OPTIONAL:* Electronics Workbench or Multisim. To use the EWB circuit simulator software to draw and test the operation of an R-S flip-flop with active HIGH inputs.

MATERIALS

Qty.

1   7400 two-input NAND gate IC
1   clock (single pulses)
1   5-V dc regulated power supply

Qty.

3   logic switches
2   LED indicator-light assemblies
1   electronic circuit simulation software (optional)

SYSTEM DIAGRAM

Figure 7-6 shows the flip-flops you will construct in this experiment.

PROCEDURE

1. Draw a logic diagram for the R-S flip-flop shown in Fig. 7-6(*a*). Use 2 two-input NAND gates.
2. Insert the 7400 IC into the mounting board.
3. Power OFF. Connect power to the 7400 IC (V_{CC} and GND).
4. Construct the circuit you drew in step 1. Wire the input switches *R* and *S*, the 7400 IC, and the two LED output indicator-light assemblies.
5. Power ON. Operate input switches *R* and *S* as shown in the truth table in Table 7-1. Observe and record the results in the *Q* and $\overline{Q}$ columns.

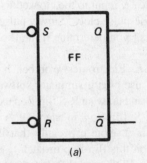

(a)

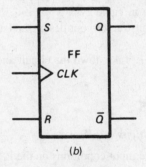

(b)

TABLE 7-1 Truth Table for R-S Flip-Flop

INPUTS		OUTPUTS		
S	**R**	**Q**	**$\overline{Q}$**	**Name of condition**
0	0			Prohibited
0	1			
1	0			
1	1			
				Hold, reset, or set

6. In the right column of Table 7-1, write the name of the condition of the outputs. Use the term "hold," "set," or "reset."
7. Draw a logic diagram for the clocked R-S flip-flop shown in Fig. 7-6(*b*). Use 4 two-input NAND gates.
8. Power OFF. Construct the circuit you drew in step 7. Wire the input switches *R* and *S*, the single-pulse clock input *CLK*, the 7400 IC, and the two LED output indicator-light assemblies.
9. Power ON. Operate the circuit and record the results of the experiment in Table 7-2.

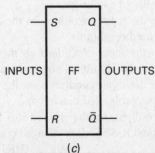

Fig. 7-6 Logic symbols. (*a*) R-S flip-flop (implement with two NAND gates). (*b*) Clocked R-S flip-flop (implement with four NAND gates). (*c*) R-S flip-flop (implement with two NOR gates).

TABLE 7-2 Truth Table for Clocked R-S Flip-Flop

INPUTS			OUTPUTS				
Clock	Data		Before clock pulse		After clock pulse		Name of condition
CLK	S	R	Q	$\bar{Q}$	Q	$\bar{Q}$	
⊓	0	0	0	1			
⊓	0	1	0	1			
⊓	1	0	0	1			
⊓	1	1	0	1			Prohibited
⊓	0	0	1	0			
⊓	0	1	1	0			
⊓	1	0	1	0			
⊓	1	1	1	0			Prohibited
							Hold, reset, or set

10. In the right column of Table 7-2 write the name of the condition of the outputs.

11. Move the *CLK* input to the clocked R-S flip-flop to a logic switch instead of the single-pulse clock. Slowly pulse the flip-flop (start at 0, then 1, and then 0). Can you determine if the flip-flop is positive- or negative-edge-triggered?

12. *OPTIONAL:* Electronics Workbench or Multisim. If assigned by your instructor, use circuit simulator software to:

   a. Design and draw an R-S flip-flop using two NOR gates (like the NAND gate R-S flip-flop in step 1). A block logic symbol is shown in Fig. 7-6(c). Note especially that the data inputs *S* and *R* are active HIGH on this flip-flop or latch.

   b. Test the NOR-gate flip-flop, generating a truth table something like the one in Table 7-1.

13. Show the instructor your design, and answer questions on your NOR-gate latch.

14. Power OFF. Take down the circuit and return all equipment to its proper place.

QUESTIONS

Complete questions 1 to 6.

1. Describe the input conditions on the R-S flip-flop [Fig. 7-6(a)] for the set, reset, and hold conditions.

2. Describe the input conditions on the clocked R-S flip-flop for the set, reset, and hold conditions.

3. Describe the input conditions on the R-S flip-flop [Fig. 7-6(a)] that are considered the prohibited state or condition.

4. Describe the input conditions on the clocked R-S flip-flop that are considered the prohibited condition.

5. What two inputs on the clocked R-S flip-flop are considered data inputs?

6. The clocked R-S flip-flop changed output states when the clock pulse went from _____ (HIGH to LOW, LOW to HIGH).

1. _____

2. _____

3. _____

4. _____

5. _____

6. _____

198

7-2 LAB EXPERIMENT: D FLIP-FLOPS

OBJECTIVES

1. To operate and test D flip-flops using a 7474 IC.
2. *OPTIONAL:* Electronics Workbench or Multisim. To use EWB circuit simulator software to wire and test a CMOS D flip-flop (4013 IC).

MATERIALS

Qty.

Qty.

1   7474 D-type flip-flop IC
1   clock (single pulses)
1   5-V dc regulated power
    supply

4   logic switches
2   LED indicator-light
    assemblies
1   electronic circuit simulation
    software (optional)

SYSTEM DIAGRAM

You will wire and operate D-type flip-flops in this experiment. Figure 7-7 shows the wiring of the 7474 TTL IC you will use. Notice that the inputs are grouped as synchronous and asynchronous inputs.

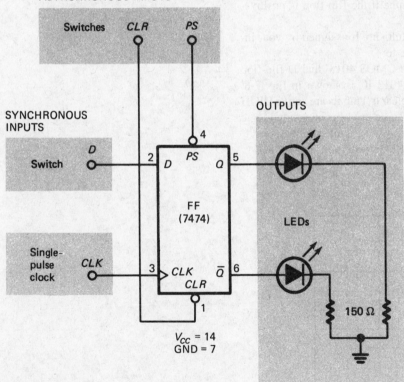

Fig. 7-7 Wiring a D flip-flop.

PROCEDURE

1. Insert the 7474 IC into the mounting board.
2. Power OFF. Connect power to the 7474 IC (V_{CC} and GND).
3. Refer to Fig. 7-7. Wire the input switches D, CLR, and PS; the single-pulse clock input CLK; the 7474 IC; and the two LED output indicator-light assemblies.

4. Power ON. Operate the asynchronous inputs *CLR* and *PS*, and record the results in Table 7-3.
5. In the right-hand column of Table 7-3 write the condition of the outputs. Your choices are listed.
6. Disable the asynchronous inputs (*PS* and *CLR* to 1).
7. Power ON. Operate the synchronous inputs *D* and *CLK* of the 7474 IC according to the truth table in Table 7-4. Observe and record the results in Table 7-4.

TABLE 7-3 Truth Table for 7474 D Flip-Flop (Asynchronous Inputs)

INPUTS		OUTPUTS		
Clear	Preset	Q	$\bar{Q}$	Name of condition
0	0			Prohibited
0	1			
1	0			
1	1			
				Clear Q to 0 Preset Q to 1 Disable asynchro- -nous inputs

TABLE 7-4 Truth Table for D Flip-Flop

INPUTS		OUTPUTS			
Clock	Data	Before clock pulse		After clock pulse	
CLK	D	Q	$\bar{Q}$	Q	$\bar{Q}$
⊓	0	0	1		
⊓	0	1	0		
⊓	1	0	1		
⊓	1	1	0		

Test with PS = 1 and CLR = 1

8. Move the *CLK* input to a switch, and slowly pulse the D flip-flop (start at 0, then 1, and then 0). Can you determine if the flip-flop is positive- or negative-edge-triggered?
9. *OPTIONAL:* Electronics Workbench or Multisim. If assigned by your instructor, use the circuit simulator software to:
   a. Wire and test one D flip-flop from the CMOS 4013 dual D flip-flop IC. Data sheet information about the 4013 IC is shown in Fig. 7-8. Note that the asynchronous set (*S*) and clear (*R*) inputs are active HIGH on the 4013. For simplicity, switches can be used for inputs and logic probes for output indicators as you wire your circuit on EWB.
   b. Test the D flip-flop. Observe and record the results in Table 7-5.

TABLE 7-5 Truth Table for 4013 D Flip-Flop

Mode of operation	INPUTS				OUTPUTS	
	Asynchronous		Synchronous			
	S	R	CLK	D	Q	$\bar{Q}$
Asynchronous reset	0	1	x	x		
Asynchronous set	1	0	x	x		
Prohibited	1	1	x	x	1	1
Set	0	0	↑	1		
Reset	0	0	↑	0		
	set	reset	clock	data		

0 = LOW
1 = HIGH
↑ = LOW-to-HIGH transition of clock pulse
x = Irrelevant

10. Show the instructor your wiring and test results on the 4013 D flip-flop, and answer questions about your test circuit and the IC.
11. Power OFF. Take down the circuit, and return all equipment to its proper place.

BLOCK DIAGRAM

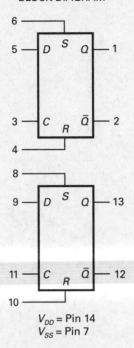

V_{DD} = Pin 14
V_{SS} = Pin 7

TRUTH TABLE

INPUTS				OUTPUTS		
CLOCK†	DATA	RESET	SET	Q	$\bar{Q}$	
⟋	0	0	0	0	1	
⟋	1	0	0	1	0	
⟍	x	0	0	Q	$\bar{Q}$	No Change
x	x	1	0	0	1	
x	x	0	1	1	0	
x	x	1	1	1	1	

x = Don't Care
† = Level Change

Fig. 7-8 Block diagram and truth table for 4013 dual D flip-flop.

QUESTIONS

Complete questions 1 to 9.

1. List the synchronous inputs of the D flip-flop.
2. List the asynchronous inputs of the D flip-flop (7474) that you used.
3. List the outputs of the D flip-flop.
4. Which output column in Table 7-4 is exactly the same as the input D column on the D flip-flop?
5. A logical _____ at *PS* will preset the Q output of the 7474 D flip-flop to a logical 1 or High.
6. The synchronous inputs of the 7474 D flip-flop only operate when the *PS* and *CLR* inputs are _____ (disabled, enabled) with a logical _____.
7. The 7474 D flip-flop is a _____ (negative-, positive-) edge-triggered flip-flop.

1. _____ _____
2. _____ _____
3. _____ _____
4. _____
5. _____
6. _____
7. _____

8. Explain why the D flip-flop is often referred to as the *delay* flip-flop.

9. Complete the *mode of operation* section of the truth table using Table 7-6 for the 7474 D flip-flop. (Use answers such as synchronous set, synchronous reset, asynchronous set, and asynchronous reset.)

8. _____

9. _____

TABLE 7-6 Truth Table for 7474 D Flip-Flop

Mode of operation	INPUTS				OUTPUTS	
	Asynchronous		Synchronous			
	PS	CLR	CLK	D	Q	$\overline{Q}$
	0	1	x	x	1	0
	1	0	x	x	0	1
Prohibited	0	0	x	x	1	1
	1	1	↑	1	1	0
	1	1	↑	0	0	1

0 = LOW
1 = HIGH
x = Irrelevant
↑ = LOW-to-HIGH transition of clock pulse

7-3 LAB EXPERIMENT: J-K FLIP-FLOPS

OBJECTIVE

1. To operate and test J-K flip-flops.
2. *OPTIONAL:* Electronics Workbench or Multisim. To use circuit simulator software to wire and test a CMOS J-K flip-flop (4027 IC).

MATERIALS

Qty.

1 74LS112 J-K flip-flop IC
1 clock (single pulses)
1 5-V dc regulated power supply

Qty.

5 logic switches
2 LED indicator-light assemblies
1 electronic circuit simulation software (optional)

SYSTEM DIAGRAM

Figure 7-9 shows a wiring diagram for the J-K flip-flop you will wire and test in this experiment. Notice the grouping of the asynchronous and synchronous inputs.

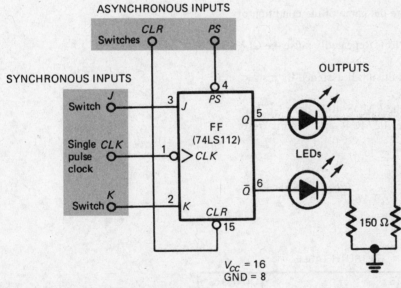

Fig. 7-9 Testing a 74LS112 J-K flip-flop.

PROCEDURE

1. Insert the 74LS112 IC into the mounting board.
2. Power OFF. Connect power to the 74LS112 IC.
3. Refer to Fig. 7-9. Wire the asynchronous inputs *PS* and *CLR* to two switches. Wire the synchronous inputs *J* and *K* to switches and the *CLK* input to a single-pulse clock. Wire outputs *Q* and $\overline{Q}$ to two LED indicator-light assemblies.
4. Power ON. Operate the asynchronous inputs *PS* and *CLR* and record the results in Table 7-7.
5. In the right-hand column of Table 7-7 write the condition of the output. Choices are listed.
6. Disable the asynchronous inputs (*PS* and *CLR* to 1).

TABLE 7-7 Truth Table for 74LS112 J-K Flip-Flop (Asynchronous Inputs)

INPUTS		OUTPUTS		
Clear	Preset	Q	$\overline{Q}$	Name of condition
0	0			Prohibited
0	1			
1	0			
1	1			
				Clear Q to 0 Disable Preset Q to 1

TABLE 7-8 Truth Table for J-K Flip-Flop (Synchronous Inputs)

INPUTS			OUTPUTS				
Clock	Data		Before clock pulse		After clock pulse		Name of condition
CLK	J	K	Q	$\overline{Q}$	Q	$\overline{Q}$	
↑	0	0	0	1			
↑	0	1	0	1			
↑	1	0	0	1			
↑	1	1	0	1			
↑	0	0	1	0			
↑	0	1	1	0			
↑	1	0	1	0			
↑	1	1	1	0			
PS and CLR = 1							Hold, reset, set or toggle

0 = LOW
1 = HIGH
↑ = LOW-to-HIGH transition of the clock pulse

7. Power ON. Operate the synchronous inputs J, K, and CLK of the 74LS112 IC according to the truth table. Observe and record the results in columns Q and $\overline{Q}$, Table 7-8.

8. In the right-hand column of Table 7-8 write the name of the condition of the output. Your choices are listed.

9. Power ON. All inputs (J, K, CLR, and PS) to 1. Repeatedly pulse the CLK input. The flip-flop is now *toggling*.

10. *OPTIONAL:* Electronics Workbench or Multisim. If assigned by instructor, use the circuit simulator software to:
   a. Wire and test one J-K flip-flop from the CMOS 4027 dual J-K flip-flop IC. Data sheet information about the 4027 IC is shown in Fig. 7-10.

BLOCK DIAGRAM

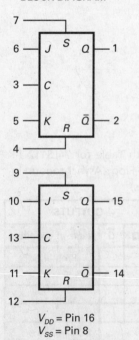

TRUTH TABLE

INPUTS						OUTPUTS*	
C†	J	K	S	R	Q_n‡	Q_{n+1}	$\overline{Q_{n+1}}$
⤒	1	x	0	0	0	1	0
⤒	x	0	0	0	1	1	0
⤒	0	x	0	0	0	0	1
⤒	x	1	0	0	1	0	1
⤒	1	1	0	0	Qo	$\overline{Qo}$	Qo
⤓	x	x	0	0	x	Q_n	$\overline{Q_n}$ No Change
x	x	x	1	0	x	1	0
x	x	x	0	1	x	0	1
x	x	x	1	1	x	1	1

V_{DD} = Pin 16
V_{SS} = Pin 8

x = Don't Care   ‡ = Present State
† = Level Change   * = Next State

Fig. 7-10 Block diagram and truth table for 4027 (Motorola MC14027B) dual J-K flip-flop.

204

Note that the asynchronous set (S) and clear (R) inputs are active HIGH on the 4027. For simplicity, switches can be used for inputs and logic probes for output indicators as you wire your circuit.

b. Test the J-K flip-flop. Observe and record the results in Table 7-9.

TABLE 7-9 Truth Table for 4027 J-K Flip-Flop

| Mode of operation | INPUTS | | | | | OUTPUTS | |
| | Asynchronous | | Synchronous | | | | |
	PS	CLR	CLK	J	K	Q	$\overline{Q}$
Asynchronous set	0	1	x	x	x		
Asynchronous reset	1	0	x	x	x		
Prohibited	0	0	x	x	x	1	1
Hold	1	1	↑	0	0	No change	
Reset	1	1	↑	0	1		
Set	1	1	↑	1	0		
Toggle	1	1	↑	1	1		
	preset	clear	clock	data	data		

0 = LOW
1 = HIGH
x = Irrelevant
↑ = LOW-to-HIGH transition of clock pulse

11. Show the instructor your wiring and test results on the 4027 J-K flip-flop and answer questions about your test circuit and the IC.

12. Power OFF. Take down the circuit, and return all equipment to its proper place.

QUESTIONS

Complete questions 1 to 9.

1. Draw a logic symbol for a J-K flip-flop. Label the inputs *J, K, CLK, PS,* and *CLR* and the outputs *Q* and $\overline{Q}$.

2. List the synchronous inputs of the J-K flip-flop.

2. _____

_____

3. List the asynchronous inputs of the J-K flip-flop that you used.

3. _____

_____

4. The 74LS112 J-K flip-flop uses what type of triggering?

4. _____

5. The truth table for the asynchronous inputs is the same as for what other flip-flop we have used?

5. _____

6. The invert bubbles at the *PS* and *CLR* inputs of the 74LS112 J-K flip-flop mean that a logical _____ will enable or activate these inputs.

6. _____

7. What is the meaning of the inverter bubble on the clock input of the 74LS112 J-K flip-flop?

7. _____

8. Describe the toggle mode of operation of this flip-flop.

8. _____

9. Complete the *mode of operation* section of the truth table using Table 7-10 for the 74LS112 J-K flip-flop. (Use answers such as asynchronous set, asynchronous reset, toggle, hold, synchronous set, and synchronous reset.)

9.

TABLE 7-10 Truth Table for 74LS112 J-K Flip-Flop

Mode of operation	INPUTS					OUTPUTS	
	Asynchronous		Synchronous				
	PS	CLR	CLK	J	K	Q	$\overline{Q}$
	0	1	x	x	x	1	0
	1	0	x	x	x	0	1
Prohibited	0	0	x	x	x	1	1
	1	1	↓	0	0	No change	
	1	1	↓	0	1	0	1
	1	1	↓	1	0	1	0
	1	1	↓	1	1	Opposite state	

0 = LOW
1 = HIGH
x = Irrelevant
↓ = HIGH-to-LOW transition of clock pulse

7-4 LAB EXPERIMENT: USING A LATCH

OBJECTIVES

1. *Option 1—74147 10-line-to-4-bit encoder:* To wire and test an encoder-decoder system that latches data at the inputs to a 7447 decoder on each keystroke. To observe the action of the latch-enable circuitry.
2. *Option 2—74148 8-line-to-3-bit encoder:* To wire and test an encoder-decoder system that latches data at the inputs to the 7447 decoder on each keystroke. To observe the action of the latch-enable circuitry.

MATERIALS

Qty.

1   7410 three-input NAND gate IC
1   7420 four-input NAND gate IC
1   7432 two-input OR gate IC
1   7447 BCD-to-seven-segment decoder IC
1   7475 4-bit transparent latch IC
1   74121 one-shot multivibrator IC
1   74147 10-line-to-4-line encoder IC or
1   74148 8-line-to-3-line encoder IC
1   common-anode seven-segment LED display

Qty.

1   keypad (0 to 9, N.O. contacts)
8   LED indicator-light assemblies
7   150-Ω, ¼-W resistors
1   10-kΩ, ¼-W resistor
1   250-pF capacitor
1   5-V dc regulated power supply
1   logic probe

SYSTEM DIAGRAMS

Your instructor will announce which option you will follow (*option 1,* using 74147, or *option 2,* using 74148 encoder).

Option 1: 74147 Encoder

The complex encoder-decoder system you will wire and operate is shown in simplified block diagram form in Fig. 7-11. The encoder translates 1 of 10 inputs from the keypad to inverted BCD form. This is displayed on four LEDs at the left. The *latch-enable circuitry* generates a *short pulse* for each keystroke. This short pulse, called the *latch-enable pulse,* causes the transfer of the encoder output to the complementary $\overline{Q}$ outputs of the 4-bit latch. This latched normal BCD data can be read on the four LEDs at the top right. The 4-bit latch holds BCD data at the decoder inputs until another switch on the keypad is pressed. The decoder translates from BCD to seven-segment code. This code lights the proper segments on the seven-segment display.

A wiring diagram for the encoder (74147 IC)–decoder system is drawn in Fig. 7-12. The 7410, 7420, 7432, and 74121 ICs at the lower left make up the latch-enable circuitry. This circuitry produces a short positive pulse routed to the 7475 latch's enable inputs (pins 4 and 13). A positive pulse at pins 4 and 13 of the 7475 latch IC will "load" and then "latch" the data from the keypad and 74147 encoder.

The interfacing of the keypad to the 74147 encoder in Fig. 7-12 could be improved. Technically, nine pull-up resistors should connect each encoder input to +5 V. In this noncritical design, the pull-up resistors were left out for

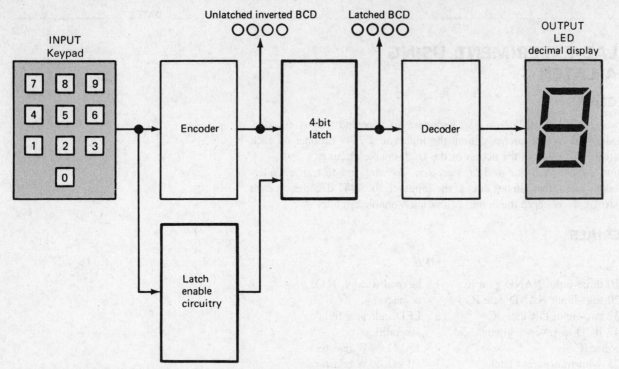

Fig. 7-11 Simplified block diagram of latched encoder-decoder system.

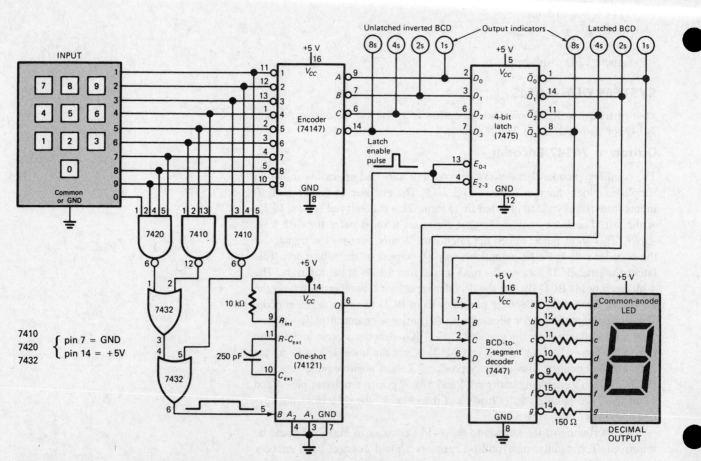

Fig. 7-12 Option 1: 74147 IC. Wiring diagram for latched encoder-decoder system using TTL ICs and LED display.

simplicity. The reason the circuit works without the pull-up resistors is because inputs to the TTL encoder in Fig. 7-12 float HIGH.

Option 2: 74148 Encoder

The complex encoder-decoder system you will wire and operate is shown in simplified block diagram form in Fig. 7-11. The encoder translates one of eight inputs from the keypad to inverted 3-bit binary form. This is displayed on three LEDs at the left. The *latch-enable circuitry* generates a *short pulse* for each keystroke. This short pulse, called the *latch-enable pulse,* causes the transfer of the encoder output to the complementary $\overline{Q}$ outputs of the 3-bit latch. This latched normal 3-bit binary data can be read on the three LEDs at the top right. The latch holds 3-bit binary data at the decoder inputs. The decoder translates from 3-bit binary to seven-segment code. This code lights the proper segments on the seven-segment display.

A wiring diagram for the encoder (74148 IC)–decoder system is drawn in Fig. 7-13. The 7420, 7432, and 74121 ICs at the lower left make up the latch-enable circuitry. This circuitry produces a short positive pulse routed to the 7475 latch's enable inputs (pins 4 and 13). A positive pulse at pins 4 and 13 of the 7475 latch IC will "load" and then "latch" the data from the keypad and 74148 encoder.

The interfacing of the keypad to the 74148 encoder in Fig. 7-13 could be improved. Technically, eight pull-up resistors should connect each encoder input to +5 V. In this noncritical design, the pull-up resistors were left out for simplicity. The reason the circuit works without the pull-up resistors is because inputs to the TTL encoder in Fig. 7-13 float HIGH.

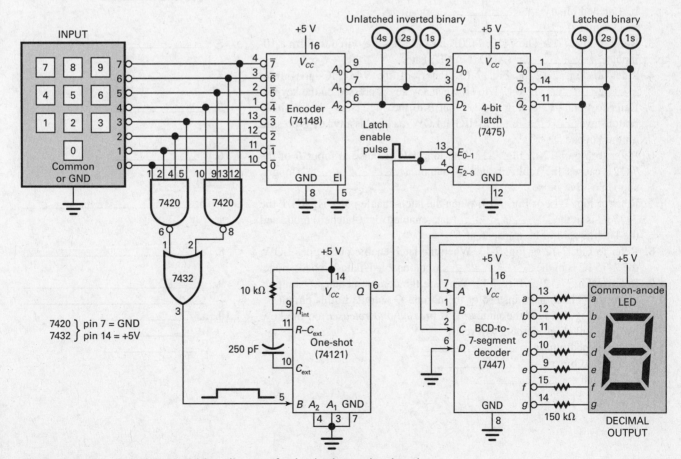

Fig. 7-13 Option 2: 74148 IC. Wiring diagram for latched encoder-decoder system using TTL ICs and LED display.

PROCEDURE

1. Your instructor will tell you which encoder-decoder experiment to perform: (1) *Option 1: 74147 encoder* or (2) *Option 2: 74148 encoder.*
2. *Option 1:* Wire and test the circuit shown in Fig. 7-12. Show your instructor your operating encoder-decoder, and be prepared to answer questions about the circuit.
3. *Option 2:* Wire and test the circuit shown in Fig. 7-13. Show your instructor your operating encoder-decoder, and be prepared to answer questions about the circuit.
4. Verify that a very short pulse is present at the output of the 74121 (pin 6 when key is pressed) one-shot multivibrator IC. Use a logic probe that has pulse-stretching or pulse-memory capabilities.
5. Answer the following questions 1 through 10 below before you take down your circuit.
6. Power OFF. Take down the circuit, and return all equipment to its proper place.

QUESTIONS

Complete questions 1 to 10.

1. Refer to Fig. 7-12 or Fig. 7-13. If the number 7 is pressed and *not released* on the keypad, the outputs will read:
   a. Unlatched binary = _____
   b. Latched binary = _____
   c. Decimal = _____

   1. a. _____
   b. _____
   c. _____

2. Refer to Fig. 7-12 or Fig. 7-13. If the number 7 is *pressed and released* on the keypad, the outputs will read:
   a. Unlatched inverted binary = _____
   b. Latched binary = _____
   c. Decimal = _____

   2. a. _____
   b. _____
   c. _____

3. Refer to Fig. 7-12. The 7410, 7420, and 7432 ICs are wired to form a 10-input _____ (AND, NAND) gate.

   3. _____

4. Refer to Fig. 7-12 or Fig. 7-13. Pin 6 of the 7432 IC goes from _____ (H to L, L to H) when any key is pressed on the keypad.

   4. _____

5. Refer to Fig. 7-12 or Fig. 7-13. Input B to the 74121 one-shot multivibrator stays _____ (HIGH, LOW) as long as any key is pressed on the keypad.

   5. _____

6. Refer to Fig. 7-12 or Fig. 7-13. The long HIGH pulse at input B of the 74121 causes the multivibrator to output a _____ (longer, short) positive pulse.

   6. _____

7. Refer to Fig. 7-12 or Fig. 7-13. When the latch-enable pulse is HIGH, the 7475 IC is in the _____ (data-enabled, data-latched) mode and the latch is said to be transparent.

   7. _____

8. Refer to Fig. 7-12 or Fig. 7-13. When the latch-enable pulse goes LOW, the 7475 IC is in the _____ (data-enabled, data-latched) mode.

   8. _____

9. Refer to Fig. 7-12 or Fig. 7-13. Why are the complementary $\overline{Q}$ outputs of the 7475 latch used instead of the normal Q outputs in this circuit?

   9. _____
   _____

10. Refer to Fig. 7-13. If the number 0 is *pressed and released* on the keypad, the following outputs will read:
    a. Unlatched binary = _____
    b. Latched binary = _____
    c. Decimal = _____

    10. a. _____
    b. _____
    c. _____

7-5 LAB EXPERIMENT: THE SCHMITT TRIGGER

OBJECTIVES

1. To use a Schmitt trigger IC for wave shaping.
2. To compare the wave-shaping action of a regular TTL IC with a Schmitt trigger IC.

MATERIALS

Qty.

1 7404 inverter TTL IC
1 7414 Schmitt trigger inverter TTL IC
1 5-V dc regulated power supply

Qty.

1 function generator (2 to 4 V p-p ac)
1 oscilloscope

SYSTEM DIAGRAM

In this simple experiment, you will compare the wave-shaping capabilities of a Schmitt trigger IC with a regular TTL IC. Figure 7-14 shows a sine wave being fed into two inverters. You will observe the shape of the waveforms at the inverter outputs on an oscilloscope.

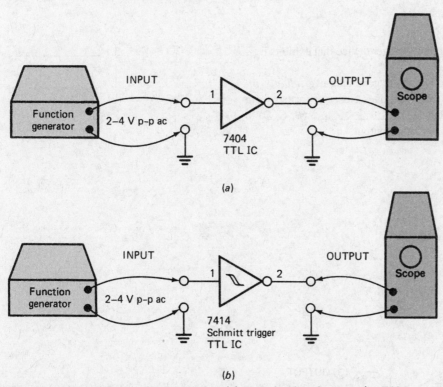

(a)

(b)

Fig. 7-14 Wave-shaping characteristics. (a) Test circuit using regular 7404 TTL IC. (b) Test circuit using Schmitt trigger 7414 TTL IC.

PROCEDURE

1. Insert the 7404 and 7414 ICs into the mounting board.
2. Power OFF. Connect power (V_{CC} and GND) to both ICs.

3. Wire *both* circuits in Fig. 7-14. You will test one after the other.
4. Power ON. Set the function generator to sine wave. Set the frequency from 50 to 200 Hz. Adjust the function generator voltage to 2 to 4 V p-p ac as measured on the calibrated oscilloscope.

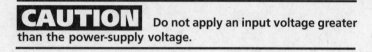

CAUTION Do not apply an input voltage greater than the power-supply voltage.

5. Observe the output waveforms for the circuits in Fig. 7-14(*a*) and (*b*). Which IC seems to do the best job of converting the sine wave into a sharp square wave?

_____

6. Show your instructor your best wave-shaping circuit in operation.
7. *OPTIONAL.* Change the function generator to triangular wave. Adjust the frequency from 50 to 200 Hz. Adjust the function generator voltage to 2 to 4 V p-p as measured on a calibrated oscilloscope.
8. Observe the output waveforms for the circuits in Fig. 7-14(*a*) and (*b*).
9. Power OFF. Take down the circuits, and return all equipment to its proper place.

QUESTIONS

Complete questions 1 to 4.

1. The 7414 is a _____ (regular, Schmitt trigger) TTL IC.   1. _____
2. The Schmitt trigger feature of an IC is used in   2. _____
   a. Multiplexing
   b. Latching
   c. Signal conditioning
3. Draw the symbol associated with a Schmitt trigger device that exhibits hysteresis.   3. _____
4. The Hall-effect sensor labeled *A* in Fig. 7-15 generates a *gradually increasing or decreasing voltage* as a magnet moves closer or farther from the sensor. A Schmitt trigger circuit labeled _____ (*B, C, D*) in Fig. 7-15 will cause a "snap action" or digital output.   4. _____

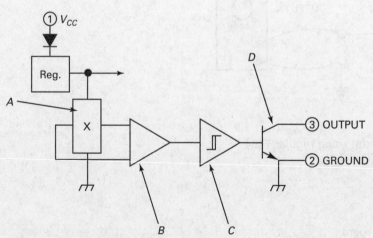

Fig. 7-15 Functional block diagram of a Hall-effect switch IC.

212

7-6 LAB EXPERIMENT: A LATCHED ENCODER-DECODER USING CMOS/LCD

OBJECTIVES

1. To implement a latched encoder-decoder system using CMOS ICs and a liquid-crystal display.
2. To demonstrate the priority feature of the encoder IC.
3. To observe the circuit's action when the latch-enable pulse is disabled.

MATERIALS

Qty.		Qty.	
2	555 timer ICs	8	LED indicator-light assemblies
1	74HC08 quad two-input AND gate CMOS IC	1	1-kΩ, ¼-W resistor
1	74HC14 hex inverter CMOS IC	2	10-kΩ, ¼-W resistors
		10	100-kΩ, ¼-W resistors
1	74HC147 10-to-4-line priority encoder CMOS IC	2	0.01-μF capacitors
		1	1-μF capacitor
1	74HC4543 BCD-to-seven-segment latch-decoder-driver CMOS IC	1	5-V dc regulated power supply
1	seven-segment liquid-crystal display (LCD)	1	keypad (0 to 9, N.O. contacts)

SYSTEM DIAGRAMS

The complex encoder-decoder system you will wire and operate is shown in simplified block form in Fig. 7-16. This is the CMOS/LCD version of the latched encoder-decoder in Fig. 7-12.

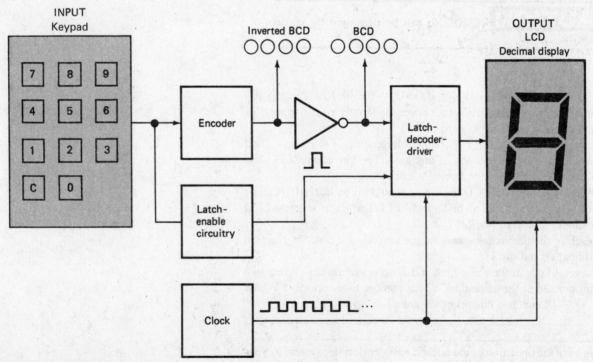

Fig. 7-16 Simplified block diagram of latched encoder-decoder system using CMOS ICs and LCD.

The encoder in Fig. 7-16 translates the 10 inputs from the keypad to inverted BCD form. This is displayed on four LEDs at the top left. Inversion occurs, and the data are displayed in true BCD form on the four LEDs at the top right. The latch-enable circuitry generates a short pulse for each keystroke which *latches* the BCD in the latch-decoder-driver. The decoder translates from BCD to seven-segment code. The clock is a free-running multivibrator which generates a 150-Hz square-wave signal. The clock signal is sent to both the LCD and the driver chip. The driver IC and clock signal activate the proper LCD segments. The latch, decoder, and driver are shown in a single block because they are all contained in a single CMOS IC.

The wiring diagram for the complex CMOS/LCD encoder-decoder system is drawn in Fig. 7-17. The 74HC147 IC encodes the keyboard input to inverted BCD. The four 74HC14 inverters complement the data to true BCD form. This can be observed on the BCD display. The 74HC08 and single 74HC14 inverter form a five-input NAND gate which emits a HIGH any time a key is pressed. The one-shot multivibrator (MV) emits a single very short positive pulse. This latch-enable pulse latches the BCD data in the 74HC4543 IC. Next, the 74HC4543 IC decodes and drives the liquid-crystal display. The clock furnishes the required low-frequency square-wave signal used to drive the LCD decimal display.

Notice in Fig. 7-16 that the 555 timer ICs are wired either as one-shot or as free-running MVs. To save gates, the five-input NAND gate is not connected directly to the keyboard inputs as in Fig. 7-12 but is still able to detect any key closure.

PROCEDURE

1. Insert the 555s, 74HC08, 74HC14, 74HC147, and 74HC4543 ICs into the mounting board. Locate the keypad near the 74HC147 IC. Locate the liquid-crystal display near the 74HC4543 IC. The LCD may be mounted on a commercial display board such as the DB-1000 by Dynalogic Concepts.

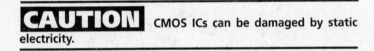

CAUTION CMOS ICs can be damaged by static electricity.

2. Power OFF. Connect power (V_{CC} and GND) to the six ICs. The 555 ICs have unusual power connections. Connect GND only to the ground or common connection on the keypad.
3. Carefully wire the circuit in Fig. 7-17. Color coding of wires is a great help in wiring complex circuits of this kind. The "inverted BCD" and "BCD" outputs are LED indicator-light assemblies.
4. Power ON. The inverted BCD indicators should all be lit (HHHH). The true BCD indicators should all be OFF (LLLL). The seven-segment LCD output should read zero (0).
5. Press each of the decimal numbers on the keypad, and note the reaction of the liquid-crystal display.
6. Press several keys at the same time and observe the results. When two keys are pressed at the same time, which number takes priority? Which IC in Fig. 7-17 contains this priority feature?

_____

_____

7. *Demonstrate* the operation of the latched encoder-decoder system to your instructor. Be prepared to answer questions about the system. Have your instructor OK your circuit.

214

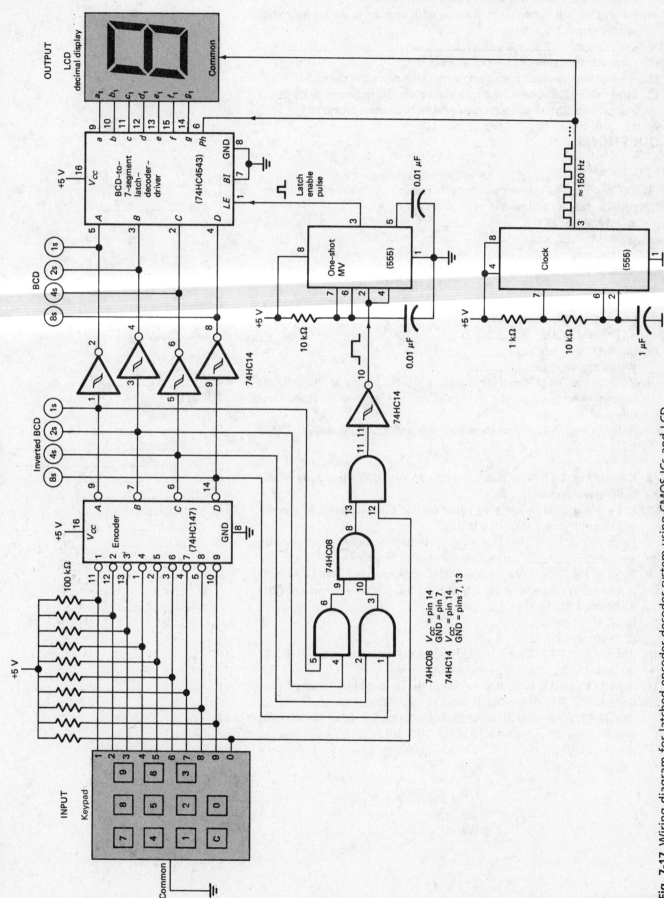

Fig. 7-17 Wiring diagram for latched encoder-decoder system using CMOS ICs and LCD.

8. Power OFF. *Disable* the latch-enable circuitry by placing a jumper wire from pin 1(*LE* = latch enable) of the 74HC4543 IC to +5 V.

9. Power ON. Try the circuit by pressing each number on the keypad. What are the results?

_____

10. Clear the fault (remove jumper wire).

11. Answer the questions below before you take down the circuit.

12. Power OFF. Take down the circuit and return all equipment to its proper place. Store CMOS ICs with their pins in conductive foam.

QUESTIONS

Complete questions 1 to 12.

1. Refer to Fig. 7-17. If the number 6 is pressed and released on the keypad, the outputs will read:
   a. Inverted BCD = _____
   b. BCD = _____
   c. Decimal = _____

2. Refer to Fig. 7-17. The 74HC08 and single 74HC14 inverter are wired to form a five-input _____ (NAND, OR) gate.

3. Refer to Fig. 7-17. If both the numbers 1 and 8 were pressed on the keypad, the outputs would read:
   a. Inverted BCD = _____
   b. BCD = _____
   c. Decimal = _____

4. Refer to Fig. 7-17. Which IC contains a priority feature so that only the larger number is output if two keypad numbers are pressed at the same time?

5. Refer to Fig. 7-17. What three jobs are performed by the 74HC4543 IC?

6. Refer to Fig. 7-17. What is the job of the 555 timer IC wired as an astable multivibrator (clock)?

7. Refer to Fig. 7-17. What is the job of the 555 timer IC wired as a monostable multivibrator (one-shot MV)?

8. Refer to Fig. 7-17. The drive signals entering the liquid-crystal display are _____ (pulsing dc, steady dc) in nature.

9. Refer to Fig. 7-17. If the latch enable (*LE*) input to the 74HC4543 IC is tied to +5 V, what will be the outputs if the 5 key is *pressed and released*?
   a. Inverted BCD = _____
   b. BCD = _____
   c. Decimal = _____

10. Refer to Fig. 7-17. The ten 100-kΩ resistors at the inputs to the 74HC147 are called _____ resistors.

11. Refer to Fig. 7-17. What ICs are part of the latch-enable circuitry?

12. Refer to Fig. 7-17. Whenever an input key is pressed, a _____ (HIGH, LOW) appears at one of the inputs to the five-input NAND gate, which causes the output (pin 10 of the 74HC14) to go _____ (HIGH, LOW).

1. a. _____
   b. _____
   c. _____

2. _____

3. a. _____
   b. _____
   c. _____

4. _____

5. _____
   _____
   _____
   _____

6. _____
   _____
   _____

7. _____

8. _____

9. a. _____
   b. _____
   c. _____

10. _____

11. _____
    _____

12. _____ _____

CHAPTER 8

Counters

TEST: COUNTERS

Answer the questions in the spaces provided.

1. Each J-K flip-flop in Fig. 8-1 is in the _____ mode of operation because both J and K inputs are HIGH.
   a. Hold
   b. Reset
   c. Set
   d. Toggle

2. The counter shown in Fig. 8-1 is best described as a _____ type counter.
   a. Decade-
   b. Ripple-
   c. Self-stopping-
   d. Synchronous-

3. Refer to Fig. 8-1. The binary output of the counter after pulse t_1 will be _____. [3 bits]

4. Refer to Fig. 8-1. The binary output of the counter after pulse t_2 will be _____. [3 bits]

5. Refer to Fig. 8-1. The binary output of the counter after pulse t_3 will be _____. [3 bits]

6. Refer to Fig. 8-1. The binary output of the counter after pulse t_4 will be _____. [3 bits]

7. Refer to Fig. 8-1. The binary output of the counter after pulse t_5 will be _____. [3 bits]

8. The counter shown in Fig. 8-1 can be called a mod-_____ counter. [Answer with a decimal number.]

1. _____

2. _____

3. _____

4. _____

5. _____

6. _____

7. _____

8. _____

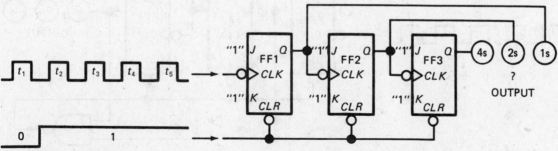

Fig. 8-1 Pulse-train problem.

9. The counter shown in Fig. 8-2 can be described as a _____ counter.

   a. Ripple down
   b. Ripple up
   c. Self-stopping
   d. Synchronous

9. _____

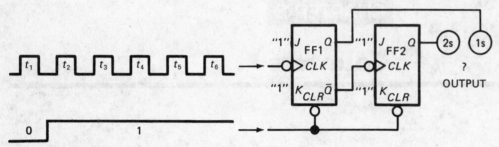

Fig. 8-2 Pulse-train problem.

10. Refer to Fig. 8-2. The binary output of the counter after pulse t_1 will be _____. [2 bits]

11. Refer to Fig. 8-2. The binary output of the counter after pulse t_2 will be _____. [2 bits]

12. Refer to Fig. 8-2. The binary output of the counter after pulse t_3 will be _____. [2 bits]

13. Refer to Fig. 8-2. The binary output of the counter after pulse t_4 will be _____. [2 bits]

14. Refer to Fig. 8-2. The binary output of the counter after pulse t_5 will be _____. [2 bits]

15. Refer to Fig. 8-2. The binary output of the counter after pulse t_6 will be _____. [2 bits]

16. The counter shown in Fig. 8-2 can be called a mod- _____ counter. [Answer with a decimal number.]

17. Refer to Fig. 8-3. The binary output of the counter after pulse t_1 will be _____. [3 bits]

18. Refer to Fig. 8-3. The binary output of the counter after pulse t_2 will be _____. [3 bits]

19. Refer to Fig. 8-3. The binary output of the counter after pulse t_3 will be _____. [3 bits]

10. _____

11. _____

12. _____

13. _____

14. _____

15. _____

16. _____

17. _____

18. _____

19. _____

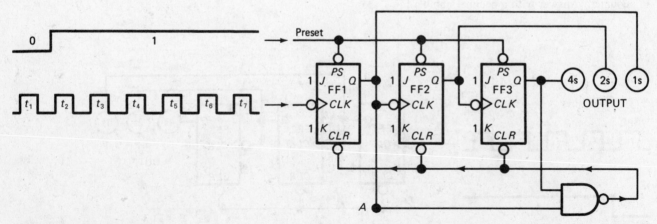

Fig. 8-3 Pulse-train problem.

218

20. Refer to Fig. 8-3. The binary output of the counter after pulse t_4 will be _____. [3 bits]

21. Refer to Fig. 8-3. The binary output of the counter after pulse t_5 will be _____. [3 bits]

22. Refer to Fig. 8-3. The binary output of the counter after pulse t_6 will be _____. [3 bits]

23. Refer to Fig. 8-3. The binary output of the counter after pulse t_7 will be _____. [3 bits]

24. The counter shown in Fig. 8-3 can be called a mod-_____ counter. [Answer with a decimal number.]

25. Refer to Fig. 8-3. If the input frequency at the *CLK* of FF1 is 100 Hz, the output frequency at point *A* will be _____ Hz.

26. Counters are commonly used in digital clocks and frequency counters to
   a. OR pulses
   b. Divide frequency
   c. Multiply frequency
   d. Subtract pulses

27. The 74192 TTL IC being used in Fig. 8-4 is best described as a _____ counter.
   a. 4-bit up
   b. Ripple up
   c. Ripple down
   d. Synchronous up/down

28. Refer to Fig. 8-4. The binary output from the counter during time period t_1 is _____. [4 bits]

29. Refer to Fig. 8-4. The binary output from the counter during time period t_2 is _____. [4 bits]

30. Refer to Fig. 8-4. The binary output from the counter during time period t_3 is _____. [4 bits]

31. Refer to Fig. 8-4. The binary output from the counter during time period t_4 is _____. [4 bits]

32. Refer to Fig. 8-4. The binary output from the counter during time period t_5 is _____. [4 bits]

33. Refer to Fig. 8-4. The binary output from the counter during time period t_6 is _____. [4 bits]

20. _____
21. _____
22. _____
23. _____

24. _____
25. _____
26. _____

27. _____

28. _____
29. _____
30. _____
31. _____
32. _____
33. _____

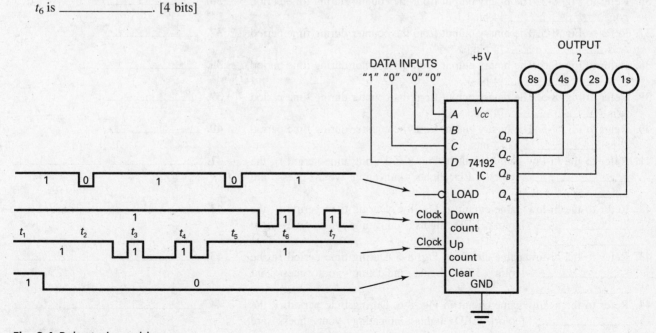

Fig. 8-4 Pulse-train problem.

34. Refer to Fig. 8-4. The binary output from the counter during time period t_7 is _____. [4 bits]

34. _____

35. Refer to Fig. 8-5. The binary output from the counter during time period t_1 is _____. [4 bits]

35. _____

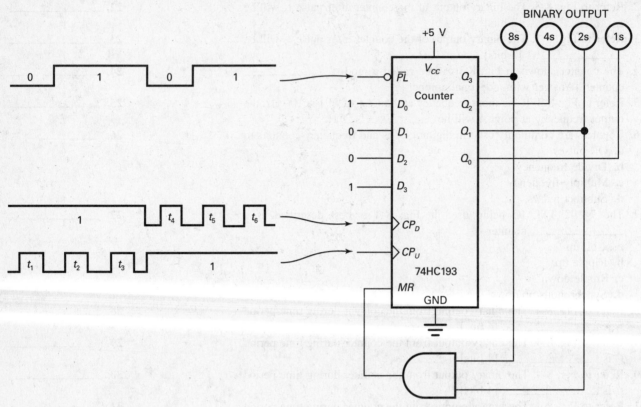

Fig. 8-5 Pulse-train problem.

36. Refer to Fig. 8-5. The binary output from the counter during time period t_2 is _____. [4 bits]

36. _____

37. Refer to Fig. 8-5. The binary output from the counter during time period t_3 is _____. [4 bits]

37. _____

38. Refer to Fig. 8-5. The binary output from the counter during time period t_4 is _____. [4 bits]

38. _____

39. Refer to Fig. 8-5. The binary output from the counter during time period t_5 is _____. [4 bits]

39. _____

40. Refer to Fig. 8-5. The binary output from the counter during time period t_6 is _____. [4 bits]

40. _____

41. Refer to the hi-low game circuit in Fig. 8-6. During time period t_1, the _____ [color] LED lights indicating your guess are _____.

41. _____ _____

42. Refer to the hi-low game circuit in Fig. 8-6. During time period t_2, the _____ [color] LED lights indicating your guess are _____.

42. _____ _____

43. Refer to the hi-low game circuit in Fig. 8-6. During time period t_3, the _____ [color] LED lights indicating your guess are _____.

43. _____ _____

44. Refer to the hi-low game circuit in Fig. 8-6. During time period t_4, the _____ [color] LED lights indicating your guess are _____.

44. _____ _____

220

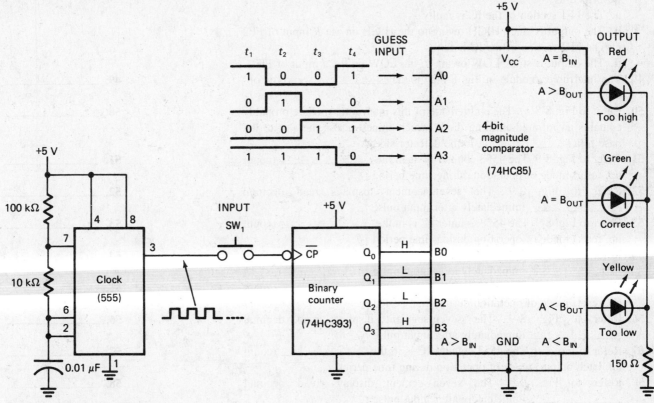

Fig. 8-6 Pulse-train problem.

45. Refer to the hi-low game circuit in Fig. 8-6. The 555 IC is used as a clock or _____ (astable, bistable) MV in this circuit.

46. Refer to Fig. 8-7. During troubleshooting, an instrument called a _____ could be used to insert pulses at the clock input (point *A*) of the 2-bit counter.

47. Refer to Fig. 8-7. For proper counter operation, both flip-flops in the 2-bit counter should be in the _____ mode.

45. _____

46. _____

47. _____

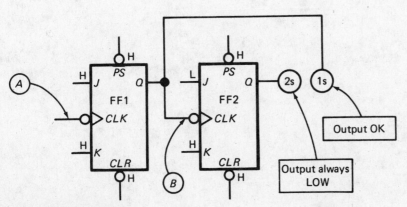

H = HIGH on logic probe
L = LOW on logic probe

Fig. 8-7 Troubleshooting problem.

221

48. Refer to Fig. 8-7. Assuming you insert pulses at point *B* while troubleshooting and FF2 still does not toggle, the problem could best be described as
  a. The FF1 section of the IC is faulty
  b. The output is stuck HIGH owing to the HIGH on the *K* input of FF2
  c. The FF2 section of the IC is faulty
  d. The output is stuck LOW owing to the LOW on the *J* input of FF2

49. The interrupter module in Fig. 8-8(*a*) is the _____ -type optical encoder.

50. Refer to Fig. 8-8(*b*). The emitter side of this interrupter module probably contains an infrared-emitting diode which shines across the slot to the base of a _____ in the detector side.

51. Refer to Fig. 8-9. The 4553 counter IC is in the _____ (count up, reset) mode of operation during time period t_1.

52. Refer to Fig. 8-9. The seven-segment displays read decimal _____ immediately after input pulse t_1.

53. Refer to Fig. 8-9. The 4553 counter IC is in the _____ (count up, reset) mode of operation during time period t_2.

54. Refer to Fig. 8-9. The seven-segment displays read decimal _____ immediately after input pulse t_2.

55. Refer to Fig. 8-9. The 4553 counter IC is in the _____ (count up, reset) mode of operation during time period t_5.

56. Refer to Fig. 8-9. The seven-segment displays read decimal _____ immediately after input pulse t_5.

57. Refer to Fig. 8-9. The 4553 counter IC is in the _____ (count up, latch output) mode of operation during time period t_6.

58. Refer to Fig. 8-9. The seven-segment displays read decimal _____ immediately after input pulse t_6.

59. Refer to Fig. 8-9. The seven-segment LED displays are turned on one at a time and are lit in rapid succession by the 4553 IC, three PNP transistors, and 4543 decoder/driver IC. Lighting the displays in this manner is called _____ (display multiplexing, display sequencing).

48. _____

49. _____

50. _____

51. _____

52. _____

53. _____

54. _____

55. _____

56. _____

57. _____

58. _____

59. _____

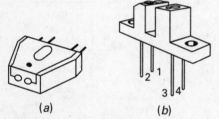

(*a*)   (*b*)

Fig. 8-8 Optical sensors.

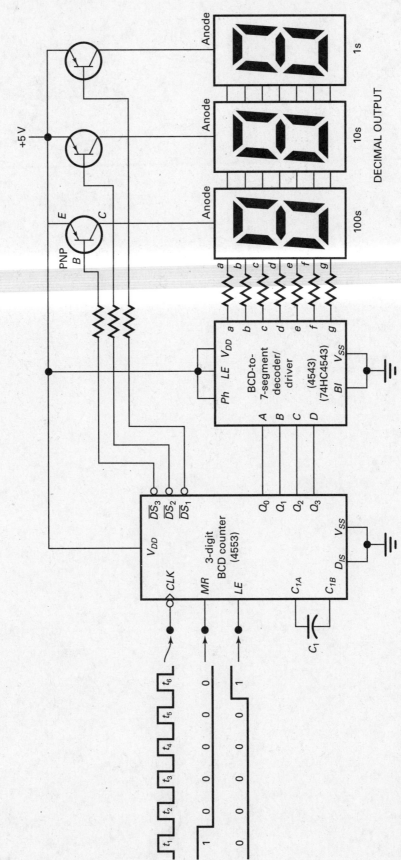

Fig. 8-9 BCD counter circuit.

8-1 LAB EXPERIMENT: RIPPLE COUNTERS

OBJECTIVES

1. To draw a logic diagram for, wire, and test a 4-bit ripple up counter using J-K flip-flops.
2. To draw a logic diagram for, wire, and test a 4-bit ripple down counter using J-K flip-flops.
3. To design, wire, and test a modulo-10 ripple counter with digital readout.

MATERIALS

Qty.   **Qty.**

1   7400 two-input NAND gate IC   1   7447 BCD-to-seven-segment
2   7476 dual J-K flip-flop ICs   decoder/driver IC
1   clock (single pulses)   1   logic switch
1   seven-segment LED display   4   LED indicator-light
1   5-V dc regulated power   assemblies
    supply   7   150-Ω, ¼-W resistors

SYSTEM DIAGRAM

Figure 8-10 diagrams the three circuits you will partially design, construct, and test in this experiment. You must design and draw in detail the counters. You will use flip-flops when designing these counters. The counter in Fig. 8-10(c) will also need a NAND gate.

PROCEDURE

1. Draw a logic diagram of the 4-bit ripple up counter shown in Fig. 8-10(a). Use four J-K flip-flops and four LED output indicator-light assemblies.

2. Insert two 7476 ICs into the mounting board.
3. Power OFF. Connect power (V_{CC} and GND) to each 7476 IC.
4. Wire the circuit you drew in step 1. Connect the *CLK* input to a single-pulse clock. Wire the 7476 ICs and the four LED output indicator-light assemblies.
5. Power ON. Operate the 4-bit counter. Observe and record the results in the 4-Bit Ripple Up Counter column in Table 8-1.
6. Draw a logic diagram of the 4-bit ripple down counter shown in Fig. 8-10(b). Use the same parts as in the last counter. Notice the addition of a *PS* input for setting the output of the counter to 1111.

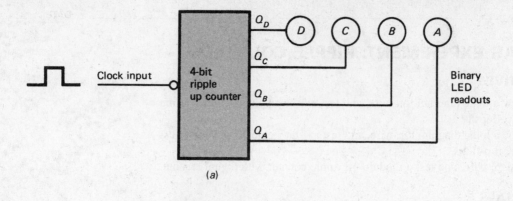

(a)

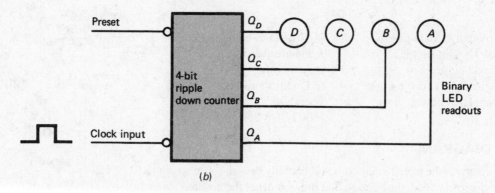

(b)

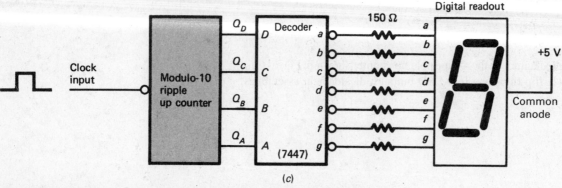

(c)

Fig. 8-10 (a) Block diagram of a 4-bit ripple up counter (*student must draw logic diagram*). (b) Block diagram of a 4-bit ripple down counter (*student must draw logic diagram*). (c) Block diagram of a modulo-10 ripple up counter with LED digital readout (*student must draw logic diagram for decade counter only*).

7. Power OFF. Rewire the 7476 J-K flip-flops to get the down counter you drew in step 6.

8. Power ON. Preset (*PS* to 0 and then back to 1) the outputs to 1111.

9. Operate the down counter by pulsing the *CLK* input. Observe and record the results in the 4-Bit Ripple Down Counter column, Table 8-1.

10. Draw a logic diagram of a modulo-10 ripple up counter shown in the block in Fig. 8-10(*c*). Use four J-K flip-flops (7476 ICs) and one 2-input NAND gate (7400 IC).

TABLE 8-1 Truth Table for Three Counters

INPUT	OUTPUTS								
Pulse number	4-bit ripple up counter				4-bit ripple down counter				Decade counter with digital readout
	D	C	B	A	D	C	B	A	
0	0	0	0	0	1	1	1	1	*0*
1									
2									
3									
4									
5									
6									
7									
8									
9									
10									
11									
12									
13									
14									
15									
16									
17									

11. Attach the 7447 BCD-to-seven-segment decoder, the seven-segment LED display, and the seven limiting resistors to your decade counter. You now should have the system shown in Fig. 8-10(*c*).

12. Insert the 7447 IC and the seven-segment LED display near one another on the mounting board. Use a commercial display board such as the DB-1000 by Dynalogic Concepts, if available.

13. Power OFF. Connect power (V_{CC} and GND) to the 7447 IC. Connect +5 V (V_{CC}) only to the seven-segment LED display.

14. Construct the counter. Wire the *CLK* input, the two 7476 ICs, and the 7400 IC.

15. Wire the counter outputs (Q_D, Q_C, Q_B, and Q_A) to the inputs of the 7447 decoder (*D*, *C*, *B*, and *A*).

16. *Carefully* connect the seven 150-Ω limiting resistors between the 7447 decoder and the seven-segment LED display.

17. Power ON. Operate the modulo-10 counter by pulsing the input. Observe and record the results in the Decade Counter with Digital Readout column, Table 8-1.

18. Power OFF. Leave the 7447 decoder and seven-segment display on the mounting board for the next experiment.

QUESTIONS

Complete questions 1 to 7.

1. Draw the logic diagram for a 2-bit ripple up counter. Use two J-K flip-flops. Label the input *CLK;* label the output indicators *B* and *A*.

2. List the counting sequence of the 2-bit counter you drew in question 1.

   2. _____

3. Draw a 3-bit ripple down counter that will count from 111 to 000 (7 to 0 in decimal). Use three J-K flip-flops. Label the input *CLK;* include and label a *PS* input. Label the output indicators *C, B,* and *A*.

4. List the counting sequence of the 3-bit down counter you drew in question 3.

   4. _____

5. Draw a modulo-6 ripple up counter. Use three J-K flip-flops and one 2-input NAND gate. You must use the *CLR* inputs of the J-K flip-flops. Label the clock input as *CLK;* label the output indicators *C, B,* and *A*.

6. List the counting sequence of the mod-6 counter you drew in question 5.

   6. _____

7. The modulo-10 block in Fig. 8-10(*c*) is more commonly known as a _____ (decade, cascaded) counter.

   7. _____

228

8-2 LAB EXPERIMENT: TTL IC RIPPLE COUNTERS

OBJECTIVES

1. To design, draw, construct, and test a 4-bit counter using the 7493 IC.
2. To add a 7447 decoder and seven-segment LED display to the 4-bit counter for a digital readout.
3. To design, draw, construct, and test a modulo-10 counter with binary and digital readouts.

MATERIALS

Qty. **Qty.**

1 7447 BCD-to-seven-segment decoder/driver IC
2 logic switches
4 LED indicator-light assemblies
7 150-Ω, ¼-W resistors

1 7493 4-bit binary counter IC
1 clock (single pulses)
1 seven-segment LED display
1 5-V dc regulated power supply

SYSTEM DIAGRAMS

Figure 8-11 diagrams three circuits you will partially design, construct, and test in this experiment. You must design and draw the counters in the blocks. The counters are based upon the 7493 IC counter. Manufacturers' data sheets and your textbook will help in your design work.

PROCEDURE

1. Draw a wiring diagram of the 4-bit counter shown in Fig. 8-11(a). Use a 7493 IC and four LED output indicator-light assemblies.

2. Insert the 7493 IC into the mounting board.
3. Power OFF. Connect power (V_{CC} and GND) to the 7493 IC.
4. Wire the circuit you drew in step 1. Connect input A to a single-pulse clock. Wire the 7493 IC and the four LED output indicator-light assemblies.
5. Power ON. Operate the 4-bit counter. Observe and record the results in the 4-Bit Counter, Binary Readout column, Table 8-2.
6. Power OFF. The digital readout will now be added to the 4-bit counter you have already constructed.

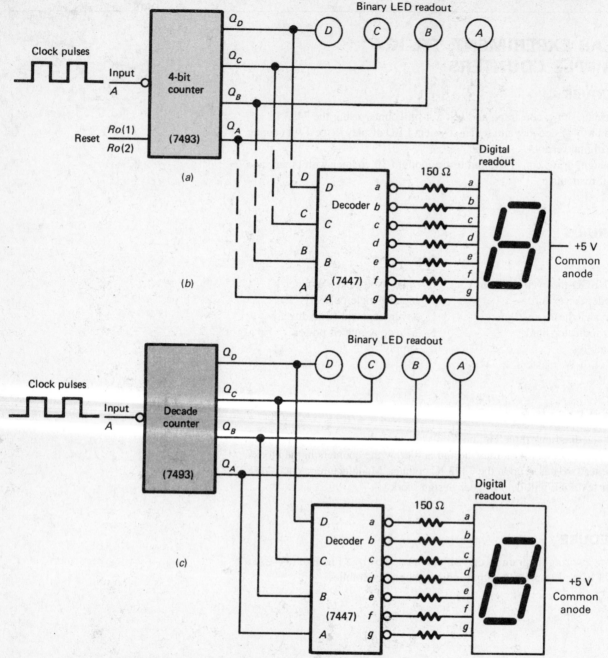

Fig. 8-11 (a) Block diagram of a 4-bit counter (*student must draw logic diagram*). (b) Add-on decoder/driver and display circuit. (c) Diagram of a decade counter with both binary and LED digital readouts (*student must draw logic diagram for decade counter only*).

7. Insert a 7447 IC and a seven-segment LED display near one another on the mounting board. Use a commercial display board such as the DB-1000 by Dynalogic Concepts, if available.

8. Connect power (V_{CC} and GND) to the 7447 IC. Connect +5 V (V_{CC}) only to the seven-segment LED display.

9. Wire the counter outputs, the decoder, and the digital display as shown in Fig. 8-11(a) and (b).

10. Use extra *caution* to make sure the seven 150-Ω limiting resistors between the 7447 decoder and the seven-segment display are properly connected.

11. Power ON. Operate the counter. Observe and record the results in the 4-Bit Counter, Digital Readout column, Table 8-2.

12. Power OFF. Leave all the ICs and displays wired on the mounting board.

13. Draw a wiring diagram of the modulo-10 counter (decade counter) shown in Fig. 8-11(*c*). Use a 7493 IC.

14. Power OFF. Wire the circuit you drew in step 13. Connect input *A* to a single-pulse clock. Wire the 7493 IC. Wire the four binary LED indicator-light assemblies. Wire the counter to the decoder. Wire the decoder to the seven-segment display with seven 150-Ω resistors. You should now have wired the system shown in Fig. 8-11(*c*).

15. Power ON. Operate the decade counter. Observe and record the results in the Decade Counter column, Table 8-2.

16. Power OFF. Remove the counter. Leave the 7447 and the seven-segment display on the mounting board for the next experiment.

TABLE 8-2   Truth Table for Two Counters

INPUT	OUTPUTS									
	4-bit counter					Decade counter				
Pulse number	Binary readout				Digital readout	Binary readout				Digital readout
	D	C	B	A		D	C	B	A	
0	0	0	0	0	0	0	0	0	0	0
1										
2										
3										
4										
5										
6										
7										
8										
9										
10										
11										
12										
13										
14										
15										
16										
17										

QUESTIONS

Complete questions 1 to 7.

1. Draw a logic diagram of a modulo-9 counter using a 7493 IC. Label the input clock *CLK;* label the binary output indicators *D, C, B,* and *A.*

2. List the counting sequence of the modulo-9 counter you drew in question 1.

    2. _____

3. The 7493 IC is a _____ (ripple-, synchronous-) type counter.

    3. _____

4. The 7493 counter will count _____ (up, down, both up and down).

    4. _____

5. What is the purpose of the 7447 IC in the circuit in Fig. 8-11(*c*)?

    5. _____

6. Draw a logic diagram of a modulo-6 counter using a 7493 IC. Label the clock input as input *B;* label the binary outputs *C, B,* and *A.*

7. List the counting sequence of the modulo-6 counter you drew in question 6.

    7. _____

232

8-3 LAB EXPERIMENT: TTL IC
SYNCHRONOUS UP/DOWN COUNTERS

OBJECTIVES

1. To wire and test a decade up counter using a 74192 IC.
2. To wire and test a decade down counter using a 74192 IC.

MATERIALS

Qty.   **Qty.**

1   7447 BCD-to-seven-segment   1   74192 synchronous decade
    decoder/driver IC   up/down counter IC
1   logic switch   1   clock (single pulses)
4   LED indicator-light   1   seven-segment LED display
    assemblies   1   5-V dc regulated power
7   150-Ω, ¼-W resistors   supply

SYSTEM DIAGRAM

Figure 8-12 is a wiring diagram for a decade up counter and a decade down counter. You will be wiring both these counters using a 74192 IC counter. Notice that the counter in Fig. 8-12 has both binary and digital readouts. To count up, the clock pulses are connected to the count up input (pin 5) of the 74192 IC. To count down, the clock is connected to the count down input (pin 4) of the 74192 IC.

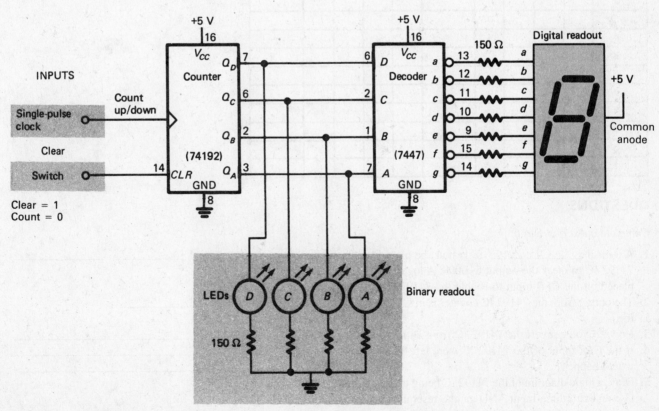

Fig. 8-12 Wiring diagram for a decade up/down counter with both binary and LED digital readouts.

PROCEDURE

1. Insert the 74192 IC into the mounting board.
2. Power OFF. Connect power (V_{CC} and GND) to the 74192 IC.
3. The 7447 IC and the seven-segment display already should be on the mounting board and connected to power. Use a commercial display board such as the DB-1000 by Dynalogic Concepts, if available.
4. Wire the counter as shown in Fig. 8-12. Connect the count up input (pin 5) to the single-pulse clock. Connect the CLR input to a switch. Finish wiring the 74192 IC, the 7447 IC, the seven limiting resistors and seven-segment display, and the LED output indicator-light assemblies.
5. Move the CLR input to 0 so the 74192 will count.
6. Power ON. Operate the decade up counter. Observe and record the results in the Decade Up Counter column, Table 8-3.
7. Power OFF. Change the clock input to the count down input (pin 4) of the 74192 IC. You now have converted the counter to a decade down counter.
8. Power ON. Operate the decade down counter. Observe and record the results in the Decade Down Counter column, Table 8-3.
9. Power OFF. Leave the parts connected on the mounting board for the next experiment.

TABLE 8-3 Truth Table for Two Counters

INPUT	OUTPUTS									
	Decade up counter					Decade down counter				
Pulse number	Binary readout				Digital readout	Binary readout				Digital readout
	D	C	B	A		D	C	B	A	
0										
1										
2										
3										
4										
5										
6										
7										
8										
9										
10										
11										

QUESTIONS

Complete questions 1 to 6.

1. A logical _____ (0, 1) must be placed on the CLR input of the 74192 IC to clear the output to 0000. A logical _____ (0,1) is placed on the CLR input to enable the 74192 IC to count.

1. _____   _____

2. The output from the 74192 IC counter is in _____ (BCD, octal) form.

2. _____

3. How do you convert the 74192 IC from an up to a down counter?

3. _____

4. If the CLR input of the 74192 IC were left floating (not connected), what would happen?

4. _____
_____

5. Draw a logic diagram of the 74192 IC being used as a modulo-7 up counter. Use an extra three-input AND gate to reset the counter to 0000. The AND gate feeds back into the CLR on the 74192 IC.

6. List the counting sequence of the modulo-7 up counter you drew in question 5.

6. _____

234

8-4 LAB EXPERIMENT: CASCADING COUNTERS

OBJECTIVES

1. To wire and test a 0-to-99 digital TTL up counter.
2. To rewire the digital up counter to form a 99-to-0 digital down counter.

MATERIALS

Qty.

2   7447 BCD-to-seven-segment decoder/driver ICs
1   logic switch
2   seven-segment LED displays
1   5-V dc regulated power supply

Qty.

2   74192 synchronous decade up/down counter ICs
1   clock (free-running)
14 150-Ω, ¼-W resistors

SYSTEM DIAGRAMS

Figure 8-13 is a wiring diagram for a 0-to-99 digital up counter. We say that the 74192 ICs are *cascaded* (in series).

Figure 8-14 is a system that counts down in decimals from 99 to 0. The count down and borrow portions in Fig. 8-14 show the *only changes* in the up counter to make a down counter.

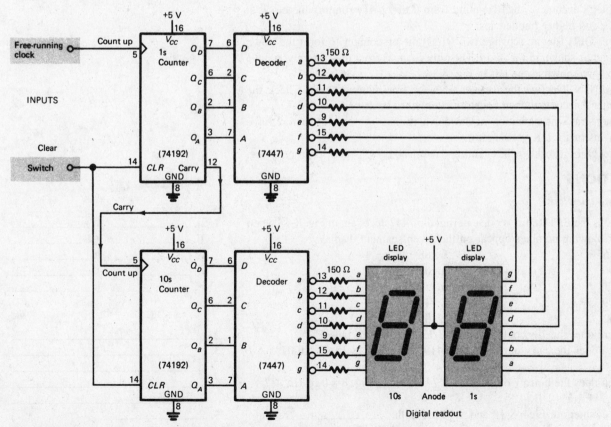

Fig. 8-13 Wiring diagram for a 0-to-99 digital up counter with LED displays.

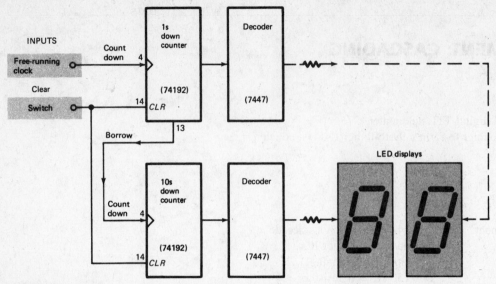

Fig. 8-14 Block diagram of a digital down counter (99 to 0) with seven-segment LED displays.

PROCEDURE

1. Insert all four ICs and LED displays into the mounting board. Place displays with the 10s digit on left and the 1s digit on right. Use a commercial display board such as the DB-1000 by Dynalogic Concepts, if available.
2. Power OFF. Connect power (V_{CC} and GND) to all four ICs. Connect +5 V (V_{CC}) only to the LED displays.
3. Wire the circuit in Fig. 8-13. Use a free-running clock for the count up input. Use an input switch for the *CLR* input. Finish wiring the circuit.
4. Place the *CLR* switch to the 0 position so the counters may count.
5. Power ON. Operate the counter with the free-running clock. Check the counter's accuracy when counting from 0 to 99. Try running the clock at lower and higher frequencies.
6. Power OFF. Rewire (change two wires) the up counter to form the down counter as shown in Fig. 8-14. The only changes from the system you have wired are shown at the left in Fig. 8-14.
7. Power ON. Operate the counter with the free-running clock. Check the counter's accuracy when counting from 99 to 0.
8. Demonstrate the operation of the down counter to your instructor. Have your instructor OK your circuit.
9. Power OFF. Take down the circuit, and return all equipment to its proper place.

QUESTIONS

Complete questions 1 to 5.

1. What is the BCD (8421) code entering the 7447 decoders in Fig. 8-13 when the following numbers appear on the seven-segment displays?
   a. 75   **f.** 88
   b. 00   **g.** 02
   c. 31   **h.** 50
   d. 99   **i.** 44
   e. 62   **j.** 05

   1. a. _____ f. _____
      b. _____ g. _____
      c. _____ h. _____
      d. _____ i. _____
      e. _____ j. _____

2. Where does the 10s counter in Fig. 8-13 get its input pulse?

   2. _____

3. When does the carry output of the 74192 IC in Fig. 8-13 go HIGH?

   3. _____
      _____

4. When does the borrow output of the 74192 IC in Fig. 8-14 go HIGH?

   4. _____
      _____

5. The counters in Figs. 8-13 and 8-14 are both _____ (recirculating, self-stopping)-type counters.

   5. _____

8-5 LAB EXPERIMENT: USING COUNTERS FOR FREQUENCY DIVISION

OBJECTIVES

1. To use a 74192 TTL IC as a divide-by-10 counter.
2. To wire and test a 4-bit counter used as a frequency divider.
3. *OPTIONAL:* Electronics Workbench or Multisim. To use the circuit simu-lator software to wire a 3-bit binary ripple counter, to observe frequency division, and to calculate frequency.

MATERIALS

Qty.   **Qty.**

2   7476 J-K flip-flop ICs   1   400-Hz square-wave signal
1   74192 synchronous decade   source (trainer clock, audio
     counter IC   generator, or function
1   5-V dc regulated power   generator)
     supply   1   oscilloscope (optional)
1   frequency counter   1   electronic circuit simulation
                                            software (optional)

SYSTEM DIAGRAMS

The 74192 decade counter IC in Fig. 8-15 is being used as a divide-by-10 unit. Note that only the Q_D output is being used. A 4-bit counter, wired using 7476 J-K flip-flops, is used to divide frequency in Fig. 8-16. You are to determine the frequency at points $B, C, D,$ and E in Fig. 8-16, using a frequency counter.

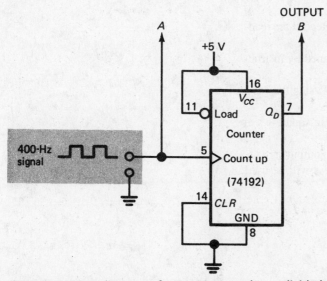

Fig. 8-15 Wiring diagram of a 74192 IC used as a divide-by-10 counter.

PROCEDURE

1. Insert the 74192 IC into the mounting board.
2. Power OFF. Connect power (V_{CC} and GND) to the 74192 IC.
3. Complete the wiring of the circuit in Fig. 8-15.
4. Connect a 400-Hz TTL-level signal to pin 5 of the 74192C. Check with your instructor about which signal source to use in this experiment. If you *do not* use the trainer clock, adjust the generator for a square wave with a voltage of about 2 to 3 V p-p as measured by a calibrated oscilloscope.

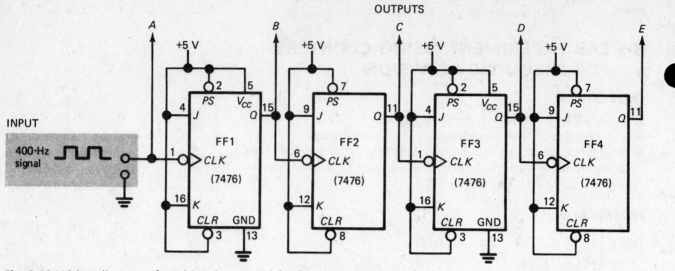

Fig. 8-16 Wiring diagram of a 4-bit counter used for frequency division.

5. Power ON. Using a frequency counter, measure the frequency at point *A* to GND and carefully adjust for 400 Hz.
6. With a frequency counter, measure the frequency at point *B* in Fig. 8-15. Record this frequency in the decade counter row under output *B*, Table 8-4.
7. Power OFF. Insert two 7476 ICs in the mounting board.
8. Connect power (V_{CC} and GND) to both ICs.
9. Complete the wiring of the 4-bit counter shown in Fig. 8-16.
10. Connect a 400-Hz TTL-level signal to pin 1 of FF1 of the circuit shown in Fig. 8-16.
11. Power ON. Measure the frequency at point *A* to GND, and carefully adjust to 400 Hz.
12. Measure the output frequencies at points *B*, *C*, *D*, and *E*. Record the readings in Table 8-4.
13. *OPTIONAL:* Electronics Workbench or Multisim. If assigned by instructor, use EWB to:
   a. Wire the 3-bit binary ripple counter circuit in Fig. 8-17 using the 7493 ripple counter IC.
   b. Use EWB's function generator as a source of a 1-kHz square-wave signal.
   c. Use EWB's dual-trace oscilloscope to compare input to output signals. Note the time period (in seconds) for each output; these will be used to calculate the frequency.
   d. Calculate the frequency at each output ($f = 1/t$).

TABLE 8-4 Frequency Measurements for Two Counters

	Measured frequency				
	INPUT	OUTPUT			
	A	*B*	*C*	*D*	*E*
Decade counter Fig. 8-15	400				
4-bit counter Fig. 8-16	400				
	Hz	Hz			

238

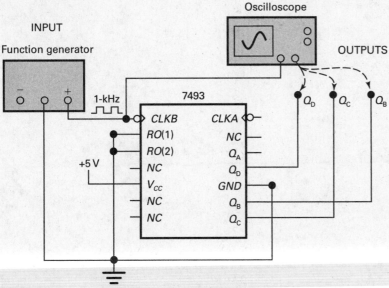

Fig. 8-17 Optional Electronics Workbench or Multisim problem. Measuring output frequencies from a divide-by-*X* counter using an oscilloscope.

14. Show instructor your EWB counter circuit, and explain how you measured time period and calculated frequency for each output signal.

15. Power OFF. Take down the circuit, and return all equipment to its proper place.

QUESTIONS

Complete questions 1 to 8.

1. The 74192 IC in Fig. 8-15 was used as a _____ -by- _____ counter.

1. _____ _____

2. Output *B* of the 4-bit unit in Fig. 8-16 is a _____ -by- _____ counter.

2. _____ _____

3. Output *C* of the 4-bit unit in Fig. 8-16 is a _____ -by- _____ counter.

3. _____ _____

4. Output *D* of the 4-bit unit in Fig. 8-16 is a _____ -by- _____ counter.

4. _____ _____

5. Output *E* of the 4-bit unit in Fig. 8-16 is a _____ -by- _____ counter.

5. _____ _____

6. Output Q_B from the 7493 counter in Fig. 8-17 would emit a frequency of _____ Hz.

6. _____

7. Output Q_C from the 7493 counter in Fig. 8-17 would emit a frequency of _____ Hz.

7. _____

8. Output Q_D from the 7493 counter in Fig. 8-17 would emit a frequency of _____ Hz.

8. _____

8-6 LAB EXPERIMENT: CMOS IC COUNTERS

OBJECTIVES

1. To design, draw, construct, and test a 4-bit CMOS counter using the 74HC393 IC.
2. To design, draw, construct, and test a CMOS decade counter using the 74HC393 IC.
3. To monitor the output of the decade counter using a liquid-crystal display.
4. *OPTIONAL*: Electronics Workbench or Multisim. To use electronic circuit simulation software to construct and operate a two-digit hexadecimal counter using a 74HC393 binary counter IC.

MATERIALS

Qty.		Qty.	
1	555 timer IC	4	LED indicator-light
1	74HC08 two-input AND gate CMOS IC		assemblies
		1	seven-segment LCD
1	74HC393 4-bit ripple counter CMOS IC	1	1-kΩ, ¼-W resistor
		1	10-kΩ, ¼-W resistor
1	74HC4543 BCD-to-seven-segment latch/decoder/ driver CMOS IC	1	1-μF capacitor
		1	5-V dc regulated power supply
1	clock (free-running)	1	electronic circuit simulation software (optional)

SYSTEM DIAGRAMS

Figures 8-18 and 8-19 diagram two circuits you will partially design, construct, and test in this experiment. You must design and draw the counters in the blocks. The counters are based on the 74HC393 CMOS dual 4-bit binary ripple counter IC. The decade counter design in Fig. 8-19 will also require a two-input AND gate. Manufacturers' data sheets and your textbook will help in your design work.

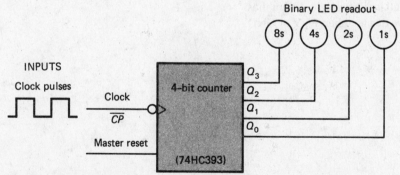

Fig. 8-18 Block diagram of a 4-bit CMOS counter (*student must draw logic diagram*).

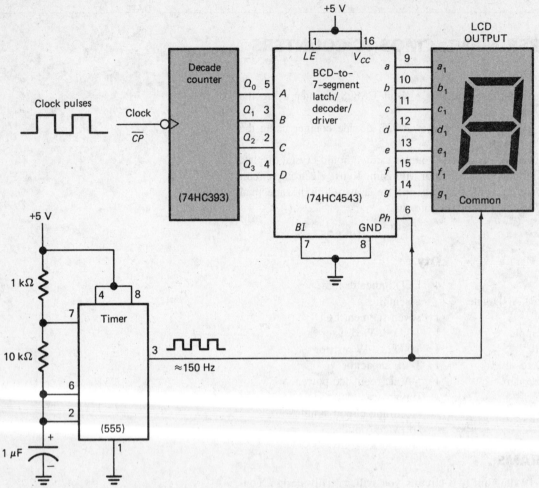

Fig. 8-19 Diagram for a CMOS decade counter circuit with LCD digital readout (*student must draw logic diagram for the decade counter only*).

PROCEDURE

1. Draw a wiring diagram of the 4-bit counter shown in Fig. 8-18. Use a 74HC393 CMOS IC and four LED output indicator-light assemblies. Remember that *unused CMOS inputs* must be tied to either +5 V or GND.

2. Insert the 74HC393 IC into the mounting board.

CAUTION CMOS ICs can be damaged by static electricity.

3. Power OFF. Connect power (V_{CC} and GND) to the 74HC393 IC.
4. Wire the circuit you drew in step 1.
5. Power ON. Connect the clock input to a free-running clock. Operate the 4-bit counter. Observe the results. Disconnect the clock input.

6. Power OFF. Redesign the 4-bit counter to form a decade counter as in Fig. 8-19. You will need to add a two-input AND gate (such as the 74HC08 IC) to the 74HC393 counter IC. Remember: Unused CMOS inputs go to +5 V or GND.

7. Add the 74HC08, 74HC4543, and 555 ICs to the mounting board. Connect power (V_{CC} and GND) to all the ICs.

8. Wire the counter circuit you designed in step 6. Wiring detail for the 555 timer, 4543 decoder/driver IC, and liquid-crystal display is shown in Fig. 8-19. Check with your instructor if you have not used an LCD before for mounting details and pin diagrams.

9. Power ON. Connect the clock input to a free-running clock. Operate your LCD decade counter circuit.

10. Show your instructor your CMOS counter designs, and demonstrate the operation of the LCD decade counter circuit.

11. Disconnect the clock input.

12. Power OFF. Take down the circuit, and return all equipment to its proper place. The pins of CMOS ICs should be stored in conductive foam.

13. *OPTIONAL:* Electronics Workbench or Multisim. If assigned by your instructor, use Electronics Workbench software to:

   a. Construct and operate the two-digit hexadecimal counter shown in Fig. 8-20 using the two 4-bit counters from the 74HC393 IC.

   b. Show your instructor your two-digit hexadecimal counter, and be able to answer selected questions on the circuit.

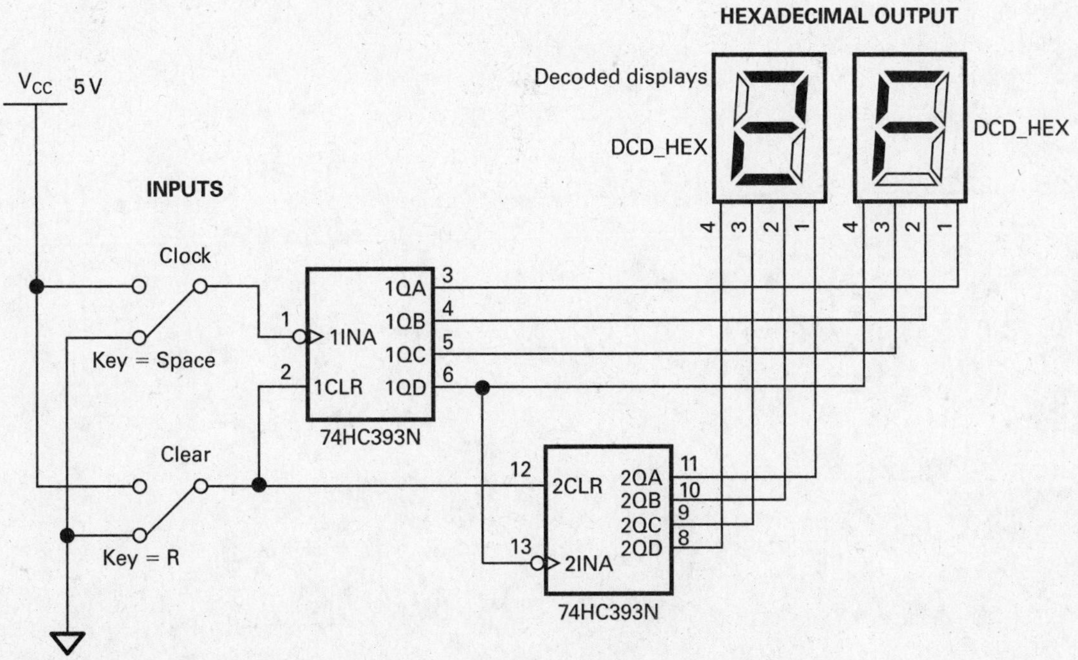

Fig. 8-20 Two-digit hexadecimal counter circuit (prepared using Multisim 8).

QUESTIONS

Complete questions 1 to 7.

1. The 74HC393 is a _____ -bit ripple counter _____ (CMOS, TTL) IC.

2. The master reset pin to the 74HC393 is an _____ (active HIGH, active LOW) input.

3. The 74HC393 IC contains _____ (one, two, four) complete 4-bit up counters.

4. List the decimal counting sequence for the counter shown in Fig. 8-19.

5. Draw just the decade counter block shown in Fig. 8-19 using the 74HC08 and 74HC393 ICs.

6. Refer to Fig. 8-19. What is the purpose of the 74HC4543 IC in this circuit?

7. Refer to Fig. 8-19. What is the purpose of the 555 timer IC in this circuit?

1. _____ _____

2. _____

3. _____

4. _____
   _____
   _____
   _____

6. _____
   _____

7. _____

8-7 LAB EXPERIMENT: AN OPTICAL ENCODER DRIVING A COUNTER

OBJECTIVES

1. To wire and test an optical encoder.
2. To wire and test a counter system driven by an optical encoder.
3. *OPTIONAL:* To test a CdS photoresistive cell as an optical encoder.

MATERIALS

Qty.

1 7414 hex Schmitt trigger inverter IC
1 7447 BCD-to-seven-segment decoder/driver IC
1 74192 synchronous decade up/down counter IC
1 H21A1 opto-coupled interrupter module
1 seven-segment LED display, common anode
8 150-Ω resistors
1 10-kΩ resistor
1 logic probe
1 5-V regulated power supply
• *OPTIONAL:* CdS photocell and 1-kΩ potentiometer

SYSTEM DIAGRAM

A slot-type opto-coupled interrupter module is being used in Fig. 8-21(*a*) to detect the holes in a shaft encoder disk. When the interrupter module is wired as shown in Fig. 8-21(*b*), it will become an optical encoder. When the slot is open, the logic level at the output is LOW. When a solid material (such as a card) drops into the slot blocking the infrared light from reaching the photo-transistor, the logic level at the output is pulled HIGH by the 10-kΩ pull-up resistor. The L-to-H and H-to-L logic shifts can drive digital circuits such as counters.

The optical encoder is the input device shown driving a counter system in Fig. 8-22. The 74192 IC counter will trigger on a L-to-H signal, which means a H-to-L signal on the left side of the 7414 inverter. A H-to-L signal is generated at output pin 3 of the H21A1 module just when the infrared light beam goes from interrupted to not interrupted. The 7414 Schmitt trigger inverter is

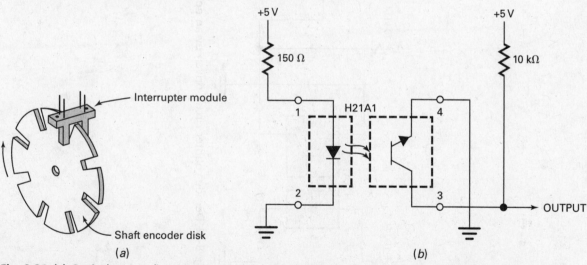

Fig. 8-21 (*a*) Optical sensor (interrupter module) detecting the slots in a shaft encoder disk. (*b*) Wiring diagram of an optical encoder circuit using the H21A1 opto-coupled interrupter module.

245

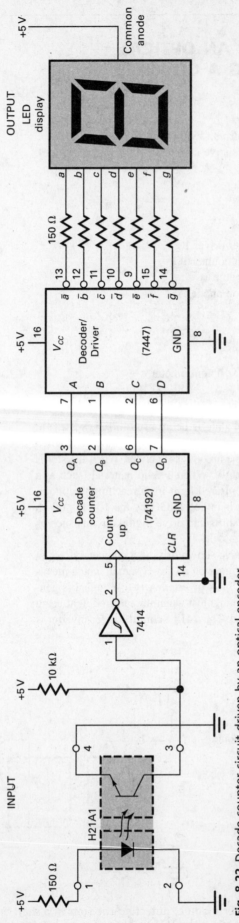

Fig. 8-22 Decade counter circuit driven by an optical encoder.

used for waveshaping to make sure the signal to the counter has a fast rise time. The 74192 IC is the decade counter which is wired as a decade up counter and also functions as a temporary memory to hold the current count. The 7447 IC decodes the BCD input to a seven-segment code and also drives the LED display. The LED is a common-anode-type seven-segment display.

PROCEDURE

1. Insert the opto-coupled interrupter module in a mounting board, and wire the optical encoder circuit in Fig. 8-21(*b*).
2. Power ON. Repeatedly drop a piece of cardboard into the slot of the interrupter module and remove it. Monitor the output of the optical encoder with a logic probe. Observe and record your results. _____

_____

3. Power OFF. Insert three ICs and the LED display into the mounting board (see Fig. 8-22). A commercial display board such as the DB-1000 from Dynalogic Concepts should be used, if available.
4. Power OFF. Connect power (V_{CC} and GND) to all three ICs. Connect +5 V (V_{CC}) only to the common anode of the LED display.
5. Wire the circuit in Fig. 8-22.
6. Power ON. Operate the counter by dropping a piece of cardboard into the slot of the interrupter module and removing it. Observe *when* the counter increments. When does the counter increase its count? _____

_____

7. Show your instructor your circuit, and be prepared to answer questions about its operation.
8. *OPTIONAL:* Using a CdS photoresistive cell as optical encoder.
9. Replace the optical encoder circuit (H21A1 and resistors) with a much less sophisticated optical sensor, a *cadmium sulfide (CdS) photoresistive cell.* This photoresistive cell will respond to white light. It can detect increases and decreases in white light striking its surface. A CdS photoresistive cell has *high resistance when it is dark* and *low resistance when white light strikes its surface.* This increase/decrease in resistance can be converted into L-to-H and H-to-L signals which can trigger the counter. The 7414 Schmitt trigger is required when using the CdS cell because the signal from the photocell has a very slow rise/fall time. A simple experimental optical encoder circuit using the CdS cell is drawn in Fig. 8-23(*a*).

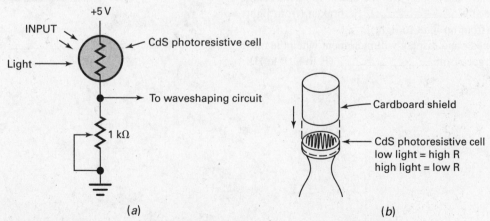

(*a*) (*b*)

Fig. 8-23 Optional lab. Using a simple CdS photocell as an experimental optical encoder. (*a*) Optical encoder circuit using CdS photoresistive cell as light sensor. (*b*) Constructing a shield for CdS photocell to keep out surrounding light.

10. Replace the old interrupter module with the experimental CdS photo-resistive cell optical encoder circuit [Fig. 8-23(a)] in the counter circuit.

11. Power ON. Adjust the 1-kΩ potentiometer to the middle of its range. Try the circuit by alternately shining a bright light at the surface of the CdS photocell and darkening the surface of the device. To get the counter to respond properly, you may have to adjust the resistance of the potentiometer. Because the surrounding (ambient) light affects the CdS photocell, you may have to build a simple shade around the unit as sketched in Fig. 8-23(b). *Do not expect the experimental CdS photocell optical encoder to work as well as the interrupter module.*

12. Show your instructor your optional circuit using the CdS photoresistive cell as an input transducer. Be prepared to answer questions about the circuit.

13. Power OFF. Take down the circuit, and return all equipment to its proper place.

QUESTIONS

Complete questions 1 to 10.

1. Refer to Fig. 8-22. What main device performs the task of the optical encoder?

1. _____

2. Refer to Fig. 8-22. The primary purpose of the 7414 IC in this circuit is for _____ (differentiating, waveshaping).

2. _____

3. Refer to Fig. 8-22. The 74192 IC has two jobs in this circuit: to count upward on each input pulse and as a temporary _____ (magnetic encoder, memory device).

3. _____

4. Refer to Fig. 8-22. The 7447 IC has two jobs in this circuit: to decode BCD code to seven-segment code and to act as a display _____ (driver, multiplexer).

4. _____

5. Refer to Fig. 8-22. The 7447 IC has active _____ (HIGH, LOW) outputs. When an LED display segment is activated (lights), the 7447 IC is said to be _____ (sinking the current, sourcing the current).

5. _____

6. The H21A1 is a _____ (reflective-type, slot-type) interrupter module.

6. _____

7. The H21A1 interrupter module uses _____ (infrared, white) light to couple the emitter side to the detector side of the unit which is not greatly affected by surrounding light conditions.

7. _____

8. Refer to Fig. 8-22. The counter and display will increment when the cardboard is _____ (moved into slot, removed from slot).

8. _____

9. Refer to Fig. 8-22. The counter and display will increment when the beam of infrared light across the slot _____ [is broken (from light to no light), begins again (from no light to light)].

9. _____

10. Refer to Fig. 8-22. The counter and display will increment when the output from the 7414 inverter goes from _____ (H to L, L to H).

10. _____

248

8-8 LAB EXPERIMENT: HI-LOW GUESSING GAME

OBJECTIVE

To construct and play an electronic binary hi-low guessing game.

MATERIALS

Qty.

1	555 timer IC
1	74HC32 quad two-input OR gate CMOS IC
1	74HC85 4-bit magnitude comparator CMOS IC
1	74HC393 4-bit ripple counter CMOS IC
5	logic switches
1	red LED

Qty.

1	green LED
1	yellow LED
1	150-Ω, ¼-W resistor
1	10-kΩ, ¼-W resistor
1	100-kΩ, ¼-W resistor
1	0.01-μF capacitor
1	5-V dc regulated power supply

SYSTEM DIAGRAM

The wiring diagram for a simple electronic game is detailed in Fig. 8-24. The clock (555 timer IC) generates a 600-Hz square-wave signal which is sent to the OR gate. When the control input of the OR gate (pin 1) goes LOW, the clock signal passes through to the $\overline{CP}$ input of the 74HC393 IC, causing it to count. Moving the control input (pin 1) of the OR gate back to HIGH causes the gate to block the clock signal. The counter stops at some random binary number from 0000 to 1111. This is the "unknown number" the player must try to guess.

Next the player tries to guess the unknown number. If the guess is correct, the $A = B$ output of the 4-bit magnitude comparator IC is activated and the green LED lights. If the guess is too low, the $A < B$ output goes HIGH, lighting the yellow LED. If the guess is too high, the $A > B$ output of the 74HC85 magnitude comparator IC is activated and the red LED lights. The object of the game is to guess the unknown binary number in the fewest tries.

PROCEDURE

1. Insert the four ICs into the mounting board.

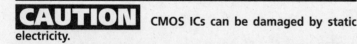

CAUTION CMOS ICs can be damaged by static electricity.

2. Power OFF. Connect power (V_{CC} and GND) to the four ICs. Note that several "extra" pins are connected to V_{CC} and GND on most of the ICs. This is because all *unused inputs* of CMOS ICs must be connected to either V_{CC} or GND.
3. Wire the hi-low guessing game circuit in Fig. 8-24. Use five logic switches for the inputs. Colored LEDs are very useful outputs. However, regular LED indicator-light assemblies may be used.
4. Power ON. Try the hi-low guessing game by first generating a random binary number. Next try guessing the unknown number.
5. Show your instructor your hi-low binary guessing game. Be prepared to answer questions about the circuit's operation.
6. Power OFF. Take down the circuit, and return all equipment to its proper place. The pins of CMOS ICs should be stored in conductive foam.

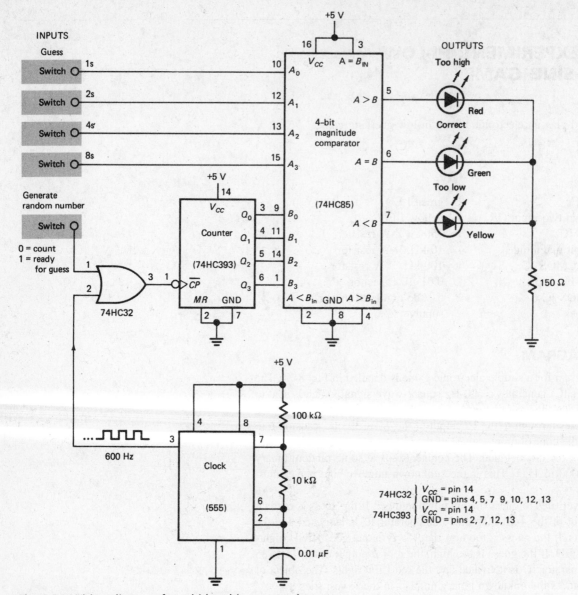

Fig. 8-24 Wiring diagram for a hi-low binary guessing game.

QUESTIONS

Complete questions 1 to 4.

1. Refer to Fig. 8-24. What is the purpose of the 74HC393 IC in this circuit?

   1. _____

2. Refer to Fig. 8-24. What is the purpose of the 74HC85 IC in this circuit?

   2. _____

3. Refer to Fig. 8-24. What is the purpose of the 74HC32 IC in this circuit?

   3. _____

4. Refer to Fig. 8-24. What is the purpose of the 555 IC in this circuit?

   4. _____

8-9 LAB EXPERIMENT: AN EXPERIMENTAL TACHOMETER

OBJECTIVES

1. To wire and test an experimental tachometer using Hall-effect switch input.
2. To observe the effect of latched outputs.
3. To observe the operation of display multiplexing.

MATERIALS

Qty.   **Qty.**

1 74HC04 inverter IC   3 10-kΩ resistors
1 555 timer IC   1 33-kΩ resistor
1 4543 BCD-to-seven-segment   1 150-kΩ resistor
   decoder/driver IC   1 500-kΩ potentiometer
1 4553 three-digit BCD counter IC   2 0.01-μF disc capacitors
1 3141 unipolar Hall-effect switch   1 1-μF electrolytic capacitor
1 magnet (mounted on wood/   1 10-μF electrolytic capacitor
   plastic wheel for attachment   1 clock (single negative pulse)
   to variable-speed power drill) 1 LED indicator light assembly
3 2N3906 PNP transistors   1 5-V dc regulated power
3 seven-segment common-anode   supply
   LED displays (use Dynalogic   1 logic probe
   Concepts DB-1000 display   1 oscilloscope (optional
   board, if possible   for calibration of 6-second
7 150-Ω resistors (or use   count-up pulse)
   DB-1000 display board)

SYSTEM DIAGRAM

An instrument that will measure the speed of angular rotation of a shaft has been designed using only subsystems you may have already studied. The experimental tachometer (see Fig. 8-25) features a Hall-effect switch input, a three-digit counter IC, and multiplexed seven-segment LED display outputs. The 4553 IC contains several functional sections besides the three-digit BCD counters. These "extra" sections of the 4553 IC include waveshaping at the *CLK* input, a *latch-enable (LE)* input for freezing the three-digit BCD count at any time, and a display multiplexing section. Although this tachometer is not practical, it does demonstrate how real-world inputs using a Hall-effect sensor can be translated by simple digital electronic subsystems into an instrument that will output the speed of rotation of a shaft in revolutions per minute (RPMs).

A complete wiring diagram for the experimental tachometer is detailed in Fig. 8-25. It will be useful to mount the magnet on a wood or plastic disc with a shaft through the middle so a variable-speed portable drill can be used as a drive. Notice in Fig. 8-25 that the south pole of the magnet should face outward to trigger the 3141 unipolar Hall-effect switch. This arrangement will produce one pulse per revolution of the shaft, which is sent to the 4553 counter IC via the *CLK* input.

The negative "trigger pulse" input shown at the lower left in Fig. 8-25 starts action by (1) clearing the BCD counters, and (2) triggering the one-shot MV, which generates a 6-second "count-up" pulse. The BCD counters will count the number of pulses from the Hall-effect switch in 6 seconds and then stop the counting. This is like counting the RPMs by 10s (10, 20, . . . 990, 1000, 1010, etc.). When the count-up pulse goes HIGH after the 6-second count-up

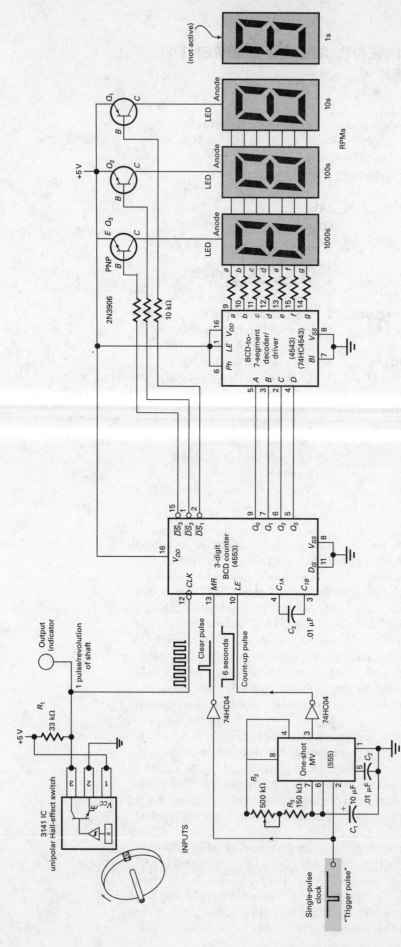

Fig. 8-25 An experimental tachometer using counters, latched outputs, and multiplexed displays.

period, the *LE* input is activated and the current three-digit BCD count is frozen at the inputs to the output section of the 4553 IC. It is interesting to note that the BCD counters in the 4553 IC continue to count upward even after the *LE* input is activated, but this advancing count does not show on the seven-segment displays.

The output section of the 4553 IC contains a display multiplexer. The task of the multiplexer is to sequentially light each seven-segment LED display, in turn, showing the output count. As an example, if the latched count after 6 seconds were 135, the 10s display would first be lit by turning on Q_1, activating the anode of the LED display. At the same time the display multiplexer section of the 4553 IC would output 0101_{BCD} to the 4543 decoder IC which drives the proper segments to form decimal 5. The next instant, Q_1 would be turned off and Q_2 turned on. The 100s display would be lit with renewed segment data of 0011_{BCD} decoded and sent to the displays. A decimal 3 would appear on the 100s LED display. Finally, Q_2 would be turned off and Q_3 turned on. The 1000s display would be lit with renewed segment data of 0001_{BCD} decoded and sent to the displays. A decimal 1 would appear on the 1000s LED display. This sequence would be repeated 50 to 100 times per second, showing an output of 1350 RPM.

The sequence of lighting each seven-segment display one at a time in turn is repeated 50 to 100 times per second. The frequency of the multiplexing is set by the value of the external capacitor C_3 shown in Fig. 8-25. The display multiplexing can be slowed down enough to observe the "multiplexing action" by increasing the value of capacitor C_3 to 0.1 or 1.0 μF.

Display multiplexing is used to save components and simplify circuitry. In this case, it would save 2 decoders and 14 resistors. Display multiplexing becomes more important as the number of digits increases. Display multiplexing is very common.

PROCEDURE

1. Insert the four ICs, three PNP transistors, and one Hall-effect switch into the mounting board.

 **CAUTION** CMOS ICs can be damaged by static electricity.

2. Power OFF. Wire the entire circuit. Color coding of wires is useful on complex circuits. A commercial display board such as the DB-1000 by Dynalogic Concepts is recommended when implementing this circuit.
3. Power OFF. Use the trainer clock (negative single pulse) for the "trigger pulse" input. Mount the magnet in a wood/plastic disc with a shaft for chucking in a variable-speed electric drill.
4. Power ON. Rotate the magnet at lower speeds near the 3141 Hall-effect switch, and observe the indicator light near the *CLK* input to the 4553 counter IC. Record your observations.

5. Power ON. Calibrate the duration of the "count-up pulse" by adjusting the value of potentiometer R_2. A logic probe can be used at the *LE* input (pin 10) of the 4553 IC to determine how long this count-up pulse stays LOW. It should stay LOW for 6 seconds. A more accurate calibration could be done with an oscilloscope.

6. Power ON. While rotating the magnet near the Hall-effect switch, generate an input trigger pulse to (1) clear the counter, and (2) generate the 6-second count-up pulse. After the 6-second count-up time the display will read the speed of the rotating magnet in RPMs. You must add a zero in the 1s place to get RPM. The "not active" 1s display shown in Fig. 8-25 means adding a zero as the LSD in the output of the tachometer.

7. Power OFF. Slow down the frequency of the multiplexing by replacing C_3 with a 1-μF capacitor. Record your observations.

_____

8. Show your instructor the operation of your experimental tachometer. Be prepared to answer selected questions about this circuit.

9. Power OFF. Take down the circuit, and return all equipment to its proper place. Carefully remove and properly store the CMOS ICs in conductive foam.

QUESTIONS

Complete questions 1 to 12.

1. Refer to Fig. 8-25. The *CLK* input to the 4553 three-digit counter IC is generated by a _____ (Hall-effect switch and magnet, trainer clock).

2. Refer to Fig. 8-25. Which two important inputs to the 4553 IC are generated by the input "trigger pulse"?

3. Refer to Fig. 8-25. The 6-second count-up pulse which controls the *LE* input to the 4553 IC is generated by the 555 timer IC wired as a _____ (free-running MV, one-shot MV).

4. Refer to Fig. 8-25. The short positive pulse that triggers the *MR* input to the 4553 IC causes the BCD counters to immediately _____ (reset to 0000 0000 0000$_{BCD}$, be placed in the hold mode of operation).

5. Refer to Fig. 8-25. External capacitor C_3 controls the _____ (duration of the count-up pulse entering the 4553 IC, frequency of the display multiplexing).

6. Refer to Fig. 8-25. When the *LE* input goes _____ (HIGH, LOW) after the 6-second count-up pulse, the current data in the BCD counters are latched (or frozen) at the inputs to the display multiplexer inside the 4553 IC.

7. Refer to Fig. 8-25. Display multiplexing means to turn on all displays at one time and then turning them all off for a short time to save energy, and then repeating this sequence. (T or F)

8. Refer to Fig. 8-25. If the BCD counters in the 4553 IC hold the count of 216 as the *LE* input goes HIGH after the 6-second count-up pulse, the speed of rotation of the input wheel is _____ RPM.

9. Refer to Fig. 8-25. With the display multiplexing working properly, only *one of the three* PNP digit driver transistors will be turned on at a time. (T or F)

10. Refer to Fig. 8-25. What discrete component is used to adjust the time duration of the count-up pulse?

11. Refer to Fig. 8-25. If the three PNP transistors are referred to as digit drivers, then the _____ (4543, 4553) IC would be the segment decoder/driver for the displays.

12. Refer to Fig. 8-25. The seven-segment LED displays are of the _____ (common-anode, common-cathode) type.

1. _____

2. _____
   _____

3. _____

4. _____
   _____

5. _____
   _____

6. _____

7. _____

8. _____

9. _____

10. _____

11. _____

12. _____

8-10 DESIGN PROBLEM: A 0-TO-99 COUNTER WITH LCD DECIMAL OUTPUT

OBJECTIVES

1. To design, draw, construct, and test a 0-to-99 counter.
2. To implement your design of the counter circuit using CMOS ICs and a liquid-crystal display.

MATERIALS

Qty.		Qty.	
1	555 timer IC	8	LED indicator-light assemblies
1	74HC08 quad two-input AND gate CMOS IC	1	Two-digit seven-segment liquid-crystal display
1	74HC393 dual 4-bit ripple counter CMOS IC	1	1-kΩ, ¼-W resistor
2	74HC4543 BCD-to-seven-segment latch/decoder/driver CMOS IC	1	10-kΩ, ¼-W resistor
		1	1-μF capacitor
1	clock (free-running)	1	5-V dc regulated power supply

SYSTEM DIAGRAM

The block diagram in Fig. 8-26 illustrates the functional sections of a 0-to-99 counter. The counter will be implemented with 74HC00 series CMOS ICs and a two-digit liquid-crystal display.

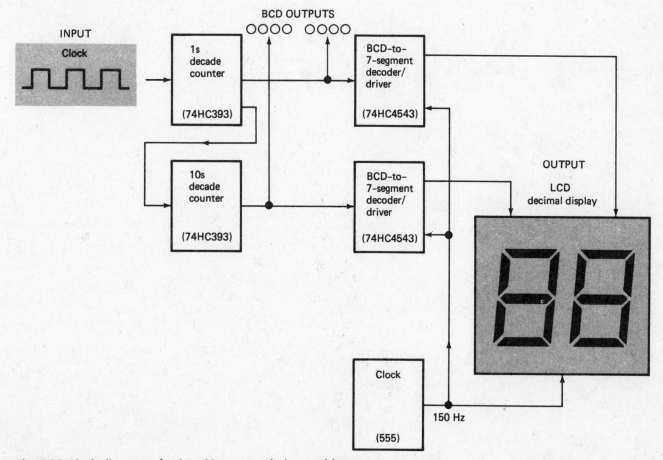

Fig. 8-26 Block diagram of a 0-to-99 counter design problem.

The input to the counter circuit in Fig. 8-26 is a clock signal. It enters the clock input of the 1s decade counter. The Q_d output from the 1s counter is sent to the clock input of the 10s counter (counters are cascaded). The 74HC393 4-bit counters must be converted to decade counters by adding 2 two-input AND gates. The outputs of the counters are displayed in BCD on eight LED indicator-light assemblies. The counter outputs are also decoded by two 74HC4543 BCD-to-seven-segment decoder/driver ICs. A 555 timer IC is wired as an astable multivibrator forming the display driver clock. The clock IC furnishes the 150-Hz square-wave signal needed to drive the liquid-crystal displays.

PROCEDURE

1. Design the system shown in Fig. 8-26. Manufacturers' data sheets and your textbook will be helpful. The circuit in Fig. 8-19 will also be useful.
2. On a separate sheet of paper, draw a wiring diagram of your counter system. The materials list contains suggested parts. Pin numbers are available in Appendix A.
3. Have your instructor approve your wiring diagram.
4. Wire and operate the CMOS/LCD counter system.

CAUTION CMOS ICs can be damaged by static electricity.

5. Have your instructor approve your operating system.

CHAPTER 9

Shift Registers

TEST: SHIFT REGISTERS

Answer the questions in the spaces provided.

1. The unit shown in Fig. 9-1 is a _____ load shift register.
   a. Broadside
   b. Multibit
   c. Parallel
   d. Serial

2. The shift register shown in Fig. 9-1 is best described as a 4-bit _____ -type unit.
   a. Shift-left nonrecirculating
   b. Shift-left recirculating
   c. Shift-right nonrecirculating
   d. Shift-right recirculating

3. Refer to Fig. 9-1. The logic states appearing at the output indicators after pulse t_1 are _____. [4 bits, with bit A on the left and bit D on the right]

4. Refer to Fig. 9-1. The logic states appearing at the output indicators after pulse t_2 are _____. [4 bits, with bit A on the left and bit D on the right]

5. Refer to Fig. 9-1. The logic states appearing at the output indicators after pulse t_3 are _____. [4 bits, with bit A on the left and bit D on the right]

1. _____

2. _____

3. _____

4. _____

5. _____

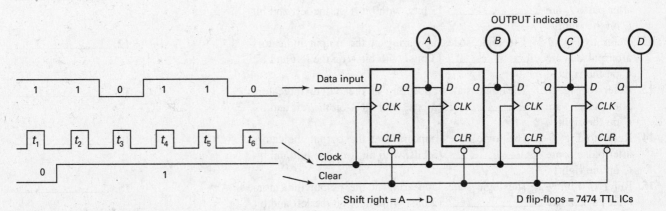

Fig. 9-1 Pulse-train problem.

6. Refer to Fig. 9-1. The logic states appearing at the output indicators after pulse t_4 are _____. [4 bits, with bit A on the left and bit D on the right]

7. Refer to Fig. 9-1. The logic states appearing at the output indicators after pulse t_5 are _____. [4 bits, with bit A on the left and bit D on the right]

8. Refer to Fig. 9-1. The logic states appearing at the output indicators after pulse t_6 are _____. [4 bits, with bit A on the left and bit D on the right]

9. The shift register shown in Fig. 9-2 is best described as a 3-bit _____ -type unit.
  a. Shift-left nonrecirculating
  b. Shift-left recirculating
  c. Shift-right nonrecirculating
  d. Shift-right recirculating

6. _____

7. _____

8. _____

9. _____

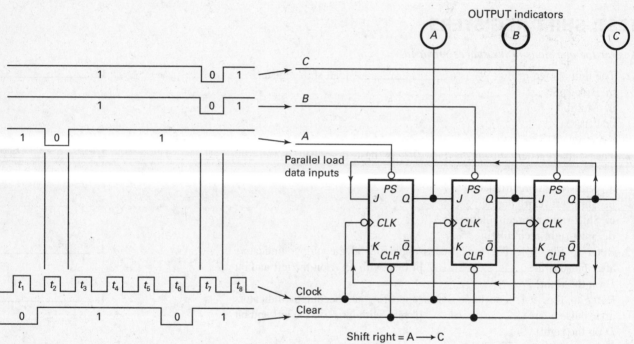

Fig. 9-2 Pulse-train problem.

10. Refer to Fig. 9-2. The logic states appearing at the output indicators after pulse t_1 are _____. [3 bits, with bit A on the left and bit C on the right]

11. Refer to Fig. 9-2. The logic states appearing at the output indicators after pulse t_2 are _____. [3 bits, with bit A on the left and bit C on the right]

12. Refer to Fig. 9-2. The logic states appearing at the output indicators after pulse t_3 are _____. [3 bits, with bit A on the left and bit C on the right]

13. Refer to Fig. 9-2. The logic states appearing at the output indicators after pulse t_4 are _____. [3 bits, with bit A on the left and bit C on the right]

14. Refer to Fig. 9-2. The logic states appearing at the output indicators after pulse t_5 are _____. [3 bits, with bit A on the left and bit C on the right]

15. Refer to Fig. 9-2. The logic states appearing at the output indicators after pulse t_6 are _____. [3 bits, with bit A on the left and bit C on the right]

10. _____

11. _____

12. _____

13. _____

14. _____

15. _____

258

16. Refer to Fig. 9-2. The logic states appearing at the output indicators after pulse t_7 are _____. [3 bits, with bit A on the left and bit C on the right]

16. _____

17. Refer to Fig. 9-2. The logic states appearing at the output indicators after pulse t_8 are _____. [3 bits, with bit A on the left and bit C on the right]

17. _____

18. The unit shown in Fig. 9-2 is a(n) _____ load shift register.
  a. Auto
  b. Parallel
  c. Serial
  d. Synchronous

18. _____

19. Refer to Fig. 9-3. The shift-left operation is accomplished with the 74194 IC when S_0 is_____(HIGH, LOW), S_1 is HIGH, and the clock pulse goes from LOW to HIGH.

19. _____

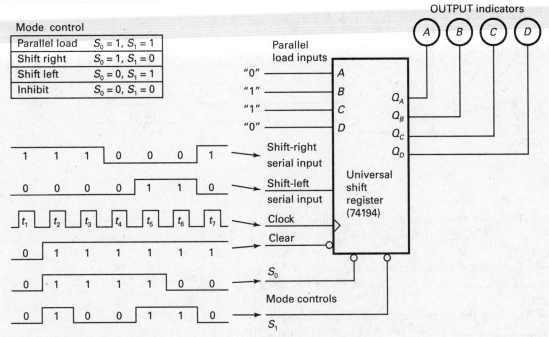

Fig. 9-3 Pulse-train problem.

20. Refer to Fig. 9-3. The logic states appearing at the output indicators after pulse t_1 are _____. [4 bits, with bit A on the left and bit D on the right]

20. _____

21. Refer to Fig. 9-3. The logic states appearing at the output indicators after pulse t_2 are _____. [4 bits, with bit A on the left and bit D on the right]

21. _____

22. Refer to Fig. 9-3. The logic states appearing at the output indicators after pulse t_3 are _____. [4 bits, with bit A on the left and bit D on the right]

22. _____

23. Refer to Fig. 9-3. The logic states appearing at the output indicators after pulse t_4 are _____. [4 bits, with bit A on the left and bit D on the right]

23. _____

24. Refer to Fig. 9-3. The logic states appearing at the output indicators after pulse t_5 are _____. [4 bits, with bit A on the left and bit D on the right]

24. _____

25. Refer to Fig. 9-3. The logic states appearing at the output indicators after pulse t_6 are _____. [4 bits, with bit A on the left and bit D on the right]

25. _____

26. Refer to Fig. 9-3. The logic states appearing at the output indicators after pulse t_7 are _____. [4 bits, with bit A on the left and bit D on the right]

26. _____

27. Refer to Fig. 9-4. The logic states appearing at the output indicators after pulse t_1 are _____. [8 bits, with bit Q_0 on the left and bit Q_7 on the right]

27. _____

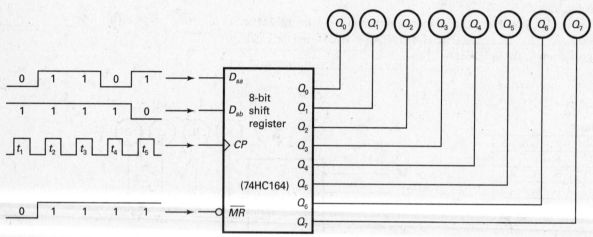

Fig. 9-4 Pulse-train problem.

28. Refer to Fig. 9-4. The logic states appearing at the output indicators after pulse t_2 are _____. [8 bits, with bit Q_0 on the left and bit Q_7 on the right]

28. _____

29. Refer to Fig. 9-4. The logic states appearing at the output indicators after pulse t_3 are _____. [8 bits, with bit Q_0 on the left and bit Q_7 on the right]

29. _____

30. Refer to Fig. 9-4. The logic states appearing at the output indicators after pulse t_4 are _____. [8 bits, with bit Q_0 on the left and bit Q_7 on the right]

30. _____

31. Refer to Fig. 9-4. The logic states appearing at the output indicators after pulse t_5 are _____. [8 bits, with bit Q_0 on the left and bit Q_7 on the right]

31. _____

32. Refer to Fig. 9-5. The 74HC164 shift register is wired as a(n) _____ in this circuit.

32. _____

33. Refer to Fig. 9-5. Resistor R_7 and capacitor C_4 are used during the power-up sequence first to _____ (clear, set) all outputs of the register and second to place a HIGH at the $\overline{MR}$ input of the 74HC164 IC.

33. _____

34. Refer to Fig. 9-5. The four NAND gates are wired as a(n) _____ (R-S, J-K) flip-flop and assist in loading only a single HIGH into the ring counter.

34. _____

35. Refer to Fig. 9-5. After the power is turned on and *before* the first clock pulse, the display reads _____. [8 bits, with bit 0 on the left and bit 7 on the right]

35. _____

36. Refer to Fig. 9-5. After pulse t_1, which LED is lit?

36. _____

37. Refer to Fig. 9-5. After pulse t_6, which LED is lit?

37. _____

38. Refer to Fig. 9-5. After pulse t_9, which LED is lit?

38. _____

260

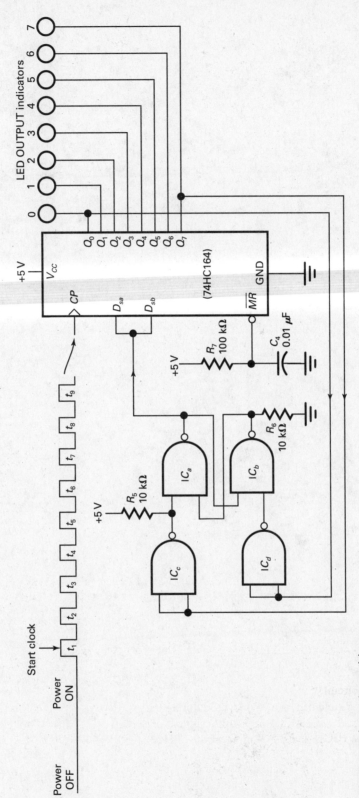

Fig. 9-5 Pulse-train problem.

9-1 LAB EXPERIMENT: SERIAL-LOAD SHIFT REGISTER

OBJECTIVES

1. To wire and test a 4-bit serial-load shift register using 7474 TTL ICs (D flip-flops).
2. *OPTIONAL:* Electronics Workbench or Multisim. To use electronic circuit simulation software to draw and operate an 8-bit shift register.

MATERIALS

Qty.

2 7474 dual D flip-flop ICs
1 clock (single pulses)
1 5-V dc regulated power supply

Qty.

2 logic switches
4 LED indicator-light assemblies
1 electronic circuit simulation software (optional)

SYSTEM DIAGRAM

The wiring diagram for the 4-bit serial-load shift register is drawn in Fig. 9-6. The circuit is wired using two 7474 TTL ICs. This register could be used as a serial-in parallel-out unit.

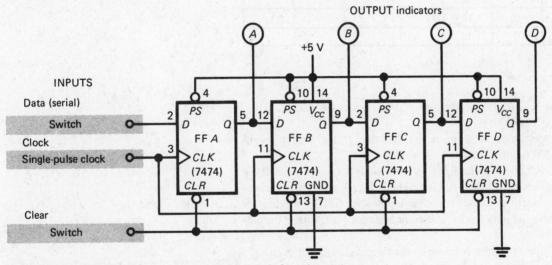

Fig. 9-6 Wiring diagram for a 4-bit serial-load shift-right register.

PROCEDURE

1. Insert two 7474 ICs into the mounting board.
2. Power OFF. Connect power (V_{CC} and GND) to the 7474 ICs.
3. Refer to Fig. 9-6 for the wiring diagram. Wire the 4-bit serial-load shift-right register. Use input switches for the data (serial load) and *CLR* (clear) inputs. Connect a single-pulse clock to the *CLK* (clock) input. Wire the 7474 ICs and the four LED output indicator-light assemblies.
4. Power ON. Clear data from the shift register by placing the *CLR* switch to 0 and then back to 1. This is shown on lines 1 and 2 of Table 9-1.
5. Operate the shift register according to Table 9-1. Observe and record the results in the Outputs columns. Read the table from left to right. Go through the sequence on the chart several times to make sure it is correct.

TABLE 9-1 Operation of a Serial-Load Shift Register

	INPUTS			OUTPUTS			
				LED Indicators			
Line	Clear	Data	Clock pulse number	A	B	C	D
1	0	1	0	0	0	0	0
2	1	1	0	0	0	0	0
3	1	1	1				
4	1	0	2				
5	1	0	3				
6	1	0	4				
7	1	0	5				
8	1	1	6				
9	1	1	7				
10	1	0	8				
11	1	0	9				
12	1	0	10				
13	1	0	11				
14	1	1	12				
15	1	1	13				
16	1	1	14				
17	1	1	15				
18	1	1	16				
19	1	0	17				
20	1	0	18				

6. Try clearing the register and loading several combinations, such as 0101, 0110, 0001, 1101, and 0011.
7. *OPTIONAL:* Electronics Workbench or Multisim. If assigned by your instructor, use EWB software to:
   a. Draw a logic diagram of an 8-bit serial-load shift register something like the one shown in Fig. 9-6. Use eight D flip-flops with asynchronous preset (deactivate these) and clear inputs. Use SPDT switches for the data, clock, and data inputs. Use eight logic probes (probes) for output indicators.
   b. Show your instructor your 8-bit serial-load parallel-out shift register.
8. Power OFF. Take down the circuit, and return all equipment to its proper place.

QUESTIONS

Complete questions 1 to 3.

1. Draw a logic symbol diagram for an 8-bit serial-load shift-right register. Use eight D flip-flops. Label the inputs *CLR,* data, and *CLK;* label the outputs *A, B, C, D, E, F, G,* and *H.*
2. What two characteristics of the D flip-flop do we take advantage of in constructing a shift register?

2. _____ _____

3. The circuit constructed in this experiment (Fig. 9-6) could be described as a _____ (parallel, serial)-in _____ (parallel, serial)-out register.

3. _____ _____

264

9-2 LAB EXPERIMENT: PARALLEL-LOAD SHIFT REGISTER

OBJECTIVES

1. To wire and test a 4-bit parallel-load recirculating shift-right register using 7476 TTL ICs (J-K flip-flops).
2. *OPTIONAL:* Electronics Workbench or Multisim. To use EWB software to draw and operate a 4-bit parallel-load recirculating shift register.

MATERIALS

Qty.		Qty.	
2	7476 dual J-K flip-flop ICs	5	logic switches
1	clock (single pulses)	4	LED indicator-light
1	5-V dc regulated power supply		assemblies

SYSTEM DIAGRAM

A wiring diagram for a parallel-load recirculating shift register is drawn in Fig. 9-7. This register is wired using two 7476 ICs. As shown in Fig. 9-7, this unit is a parallel-in parallel-out register. If serial data were taken from output D only, the unit could operate as parallel-in serial-out register.

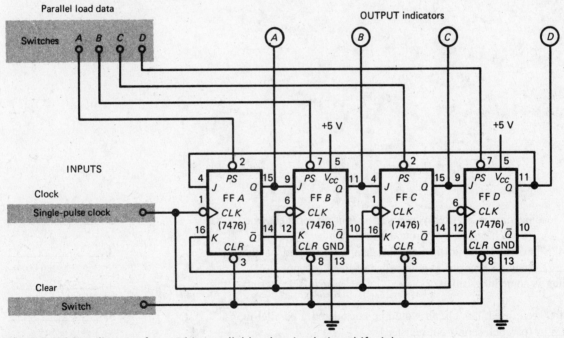

Fig. 9-7 Wiring diagram for a 4-bit parallel-load recirculating shift-right register.

PROCEDURE

1. Insert two 7476 ICs into the mounting board.
2. Power OFF. Connect power (V_{CC} and GND) to the 7476 ICs.
3. Refer to Fig. 9-7 for the wiring diagram. Wire the 4-bit parallel-load recirculating shift-right register. Use five input switches for the *CLR* (clear) and data (parallel-load) inputs. Connect the single-pulse clock to the *CLK* (clock) inputs. Wire the 7476 ICs and the four LED output indicator-light assemblies.

4. Power ON. Clear data from the shift register by placing the *CLR* switch to 0 and then back to 1. Load register with 1000. Repeat pulse clock input and observe the results.
5. Have your instructor approve your operating shift register. Be prepared to demonstrate and answer questions about the circuit.
6. *OPTIONAL:* Electronics Workbench or Multisim. If assigned by your instructor, use EWB software to:
   a. Draw a logic diagram of a 4-bit parallel-load recirculating shift register something like the one shown in Fig. 9-7. Use four J-K flip-flops (74LS112 ICs) with asynchronous preset and clear inputs. Use SPDT switches for the clear, clock, and four parallel data inputs. Use logic probes for the four output indicators.
   b. Show your instructor your 4-bit parallel-load parallel-out recirculating shift register.
7. Power OFF. Take down the circuit, and return all equipment to its proper place.

QUESTIONS

Complete questions 1 to 5.

1. Draw a logic symbol diagram for a 5-bit parallel load recirculating shift-right register. Use five J-K flip-flops. Label the inputs *CLR, CLK,* and parallel data (*A, B, C, D,* and *E*). Label the outputs *A, B, C, D,* and *E*.

2. The parallel loading on the register you wired in this experiment was _____ (asynchronous, synchronous) loading.

2. _____

3. Using the 7476 or 74LS112 J-K flip-flop, the data were shifted on the _____ (negative, positive)-going edge of the clock pulse.

3. _____

4. Parallel loading is something called _____ (broadside, serial) loading.

4. _____

5. If just output *D* were used, this circuit would be considered a parallel-in _____ (parallel, serial)-out register.

5. _____

9-3 LAB EXPERIMENT: THE UNIVERSAL SHIFT REGISTER

OBJECTIVES

1. To construct and test the 74194 TTL IC register wired as a serial-load shift-right register.
2. To construct and test the 74194 TTL IC register wired as a serial-load shift-left register.
3. To construct and test the 74194 TTL IC register wired as a parallel-load shift register.
4. To construct and test 74194 ICs wired together as an 8-bit shift register.

MATERIALS

Qty.   **Qty.**

2   74194 4-bit bidirectional   11   logic switches
    universal shift register ICs   8   LED indicator-light
1   clock (single pulses)   assemblies
1   5-V dc regulated power
    supply

SYSTEM DIAGRAMS

The wiring diagram for a 4-bit serial-load shift register is detailed in Fig. 9-8. The 74194 universal shift register IC is used in this circuit. Figure 9-9 shows how the 74194 IC would be wired as a serial-load *shift-left* register. Note the changes that must be made at the mode control inputs from Fig. 9-8 to 9-9. Also notice that serial data enter the serial (shift-right) input (pin 2) in Fig. 9-8, while the serial (shift-left) input (pin 7) is used in Fig. 9-9.

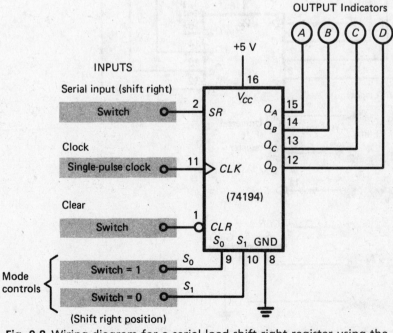

Fig. 9-8 Wiring diagram for a serial-load shift-right register using the 74194 TTL IC.

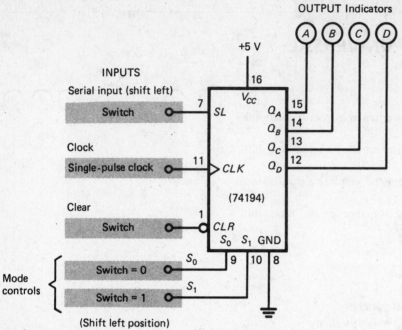

Fig. 9-9 Wiring diagram for a serial-load shift-left register using the 74194 TTL IC.

The circuit in Fig. 9-10 is an experimental circuit that uses all of the modes of operation of the 74194 universal shift register. Especially note the addition of the four parallel-load inputs.

The wiring diagram in Fig. 9-11 is for an 8-bit parallel-load recirculating shift-right register using two 74194 ICs.

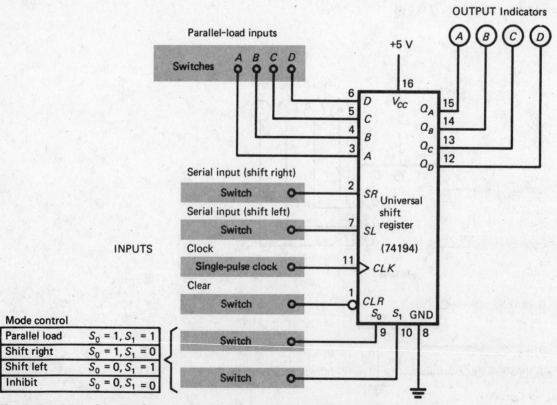

Fig. 9-10 Wiring diagram for a universal shift register circuit using the 74194 TTL IC.

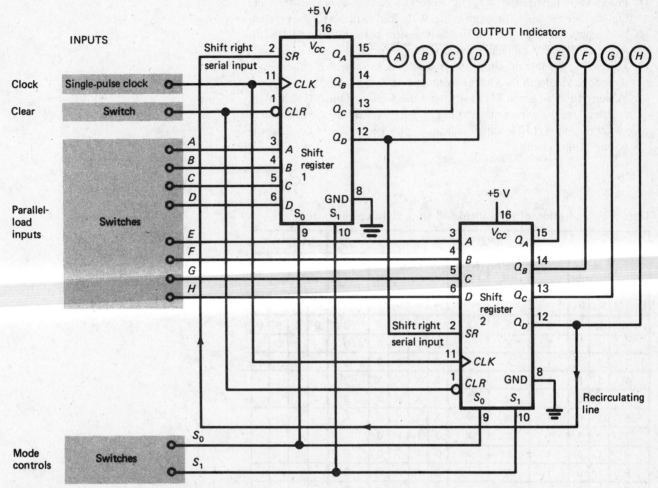

Fig. 9-11 Wiring diagram for an 8-bit parallel-load recirculating shift-right register using the 74194 TTL IC.

PROCEDURE

1. Insert the 74194 IC into the mounting board.
2. Power OFF. Connect power (V_{CC} and GND) to the 74194 IC.
3. Refer to Fig. 9-8 for the wiring diagram. Wire the 4-bit serial-load shift-right register. Use four input switches for the shift-right serial input, *CLR*, and mode control inputs (S_0, S_1). Connect a single-pulse clock to the *CLK* input. Wire the 74194 IC and the four LED output indicator-light assemblies.
4. Mode control to shift-right position ($S_0 = 1$, $S_1 = 0$).
5. Power ON. Clear data from the shift register by placing the *CLR* switch to 0 and then back to 1. Try loading 1000 into the register and pulse clock.
6. Demonstrate the operation of your serial-load shift-right register to your instructor.
7. Power OFF. Rewire the 74194 IC to form the serial-load shift-left register. The wiring diagram for this register is in Fig. 9-9. Change only *one wire*.
8. Mode control to the shift-left position ($S_0 = 0$, $S_1 = 1$).
9. Power ON. Clear data from the shift register by placing the *CLR* switch to 0 and then back to 1. Try loading 0001 and pulse clock.

10. Demonstrate the operation of the serial-load shift-left register to your instructor.
11. Power OFF. Rewire the 74194 IC to form a parallel-load shift-left/right register. See the wiring diagram in Fig. 9-10. Add switches for the parallel-load inputs (A, B, C, and D). Use switches for both shift-right and shift-left serial inputs. Complete the wiring of the circuit.
12. Power ON. Operate the shift register according to Table 9-2. Observe and record the results in the Outputs columns.
13. Power OFF. Using two 74192 ICs, wire the 8-bit recirculating shift-right register. Use the wiring diagram in Fig. 9-11. Notice the use of 11 input switches and 8 LED output indicator-light assemblies. Complete the wiring of this circuit.

TABLE 9-2 Operation of a Parallel-Load Shift-Right/Left Register

	INPUTS										OUTPUTS			
Line	Mode control		Clear (CLR)	Shift-left serial input	Shift-right serial input	Data Parallel inputs				Clock pulse number	4-bit shift register			
	S_0	S_1				A	B	C	D		A	B	C	D
1	X	X	0	X	X	X	X	X	X	0	0	0	0	0
2	1	1	1	X	X	0	1	0	0	0				
3	1	1	1	X	X	0	1	0	0	1				
4	1	0	1	X	0	X	X	X	X	2				
5	1	0	1	X	0	X	X	X	X	3				
6	1	0	1	X	0	X	X	X	X	4				
7	1	1	1	X	X	0	1	1	0					
8	1	0	1	X	X	0	1	1	0	5				
9	1	0	1	X	1	X	X	X	X	6				
10	1	0	1	X	1	X	X	X	X	7				
11	1	0	1	X	1	X	X	X	X	8				
12	1	0	1	X	1	X	X	X	X	9				
13	X	X	0	X	X	X	X	X	X					
14	1	1	1	X	X	1	0	1	0					
15	1	1	1	X	X	1	0	1	0	10				
16	1	1	1	X	0	1	0	1	0	11				
17	1	0	1	X	0	X	X	X	X	12				
18	1	0	1	X	0	X	X	X	X	13				
19	1	0	1	X	0	X	X	X	X	14				
20	1	0	1	X	0	X	X	X	X	15				
21	1	1	1	X	X	0	0	0	1	16				
22	0	1	1	0	X	X	X	X	X	17				
23	0	1	1	0	X	X	X	X	X	18				
24	0	1	1	0	X	X	X	X	X	19				
25	0	1	1	0	X	X	X	X	X	20				
26	1	1	1	X	X	0	1	1	0	21				
27	0	0	1	0	0	X	X	X	X	22				
28	0	0	1	0	0	X	X	X	X	23				
29	0	1	1	0	0	X	X	X	X	24				
30	1	0	1	0	0	X	X	X	X	25				

X - irrelevant (any input)

270

14. Power ON. Clear the register using the *CLR* input. Load 1000 0000 in the register. Shift to the right nine times. What is the result?

_____

15. Demonstrate the operation of the 8-bit parallel-load recirculating shift-right register to your instructor.

16. Power OFF. Take down the circuit, and return all equipment to its proper place.

QUESTIONS

Complete questions 1 to 10.

1. The *CLR* input on the 74194 IC is enabled by a logical _____ (0, 1).

1. _____

2. The data inputs (shift right, shift left, parallel load) on the 74194 IC are all _____ (asynchronous, synchronous) inputs.

2. _____

3. List the four modes of operation for the 74194 IC and the input conditions on S_0 and S_1 that produce these modes.

3. _____
_____
_____

4. Refer to Fig. 9-9. Explain how you would enter 0110 in a serial-load shift-right register (74194 IC).

4. _____
_____
_____

5. Refer to Fig. 9-10 and Table 9-2. Explain how your would enter a 0110 in a parallel-load shift register (74194 IC).

5. _____
_____
_____
_____

6. Explain how you would enter 0011 0110 in an 8-bit parallel-load shift register (see Fig. 9-11).

6. _____
_____
_____
_____

7. Refer to Table 9-2, line 1. Why do all inputs except the *Clear* input show an X (irrelevant)?

7. _____
_____
_____
_____

8. Refer to Table 9-2, lines 27 and 28. The mode control is in what position in these lines?

8. _____

9. Refer to Table 9-2. Why is there no shift from line 15 to 16 even if there is a clock pulse?

9. _____
_____

10. Refer to Table 9-2. The mode control is in what position in line 22?

10. _____

9-4 LAB EXPERIMENT: A CMOS 8-BIT SHIFT REGISTER

OBJECTIVES

1. To construct and test an 8-bit serial-in parallel-out shift register using the 74HC164 CMOS IC.
2. *OPTIONAL:* Electronics Workbench or Multisim. To use the electronic circuit simulation software to generate a logic diagram for an 8-bit shift register with a recirculating feature.

MATERIALS

Qty.   **Qty.**

1   74HC164 8-bit shift register CMOS IC
2   logic switches
1   clock (single pulses)

8   LED indicator-light assemblies
1   5-V dc regulated power supply
1   electronic circuit simulation software (optional)

SYSTEM DIAGRAM

A simple CMOS shift register test circuit is detailed in Fig. 9-12.

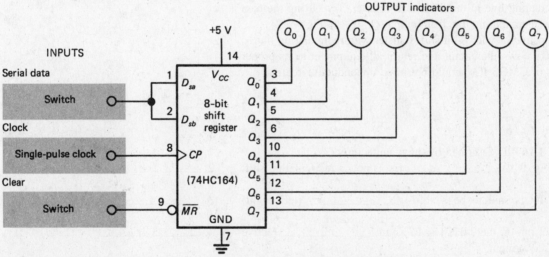

Fig. 9-12 Wiring diagram for an 8-bit shift register using the 74HC164 CMOS IC.

PROCEDURE

1. Insert the 74HC164 CMOS IC. Be careful about handling CMOS ICs.
2. Power OFF. Connect power (V_{CC} and GND) to the 74HC164 IC.
3. Wire the circuit in Fig. 9-12. Wire two logic switches and a single-pulse clock to the IC. Connect eight LED indicator-light assemblies to monitor the outputs.
4. Power ON. Operate the shift register according to Table 9-3. Observe and record the results in the Outputs columns. Go through the sequence on the chart several times to make sure it is correct.
5. *OPTIONAL:* Electronics Workbench or Multisim. If assigned by your instructor, use EWB software to:
   a. Draw a logic diagram of an 8-bit serial-load shift register something like the one shown in Fig. 9-12. Use the 74HC164 IC. Use SPDT

TABLE 9-3 Operation of the 74HC164 CMOS Shift Register

INPUTS			OUTPUTS							
Clear	Serial data	Clock pulse number	Q_0	Q_1	Q_2	Q_3	Q_4	Q_5	Q_6	Q_7
0	1	0	0	0	0	0	0	0	0	0
1	1	1								
1	1	2								
1	1	3								
1	0	4								
1	0	5								
1	0	6								
1	0	7								
1	1	8								
1	0	9								
1	0	10								
1	0	11								

switches for the clear, clock, and serial data input. Use a logic probe (probes) for the eight output indicators.

b. Operate the 8-bit serial-load shift register to make sure it is working properly.

c. Add a recirculating line to the 8-bit shift register. Try ORing the data input and the recirculating line from output Q_H (Q_7 in Fig. 9-12).

d. Show your instructor your 8-bit shift register with recirculating line.

6. Power OFF. Take down the circuit and return all equipment to its proper place. The pins of CMOS ICs should be stored in conductive foam.

QUESTIONS

Complete questions 1 to 6.

1. How is the 74HC164 IC described by the manufacturer?

1. _____

2. The circuit in Fig. 9-12 _____ (is, is not) a recirculating shift register.

2. _____

3. The 74HC164 IC is said to be a shift- _____ (left, right) register.

3. _____

4. The master reset pin on the 74HC164 IC is an _____ (active HIGH, active LOW) input.

4. _____

5. Data inputs D_{sa} and D_{sb} to the 74HC164 IC are ANDed together inside the chip. (T or F)

5. _____

6. The 74HC164 IC is manufactured using _____ (CMOS, TTL) technology.

6. _____

9-5 LAB EXPERIMENT: DIGITAL ROULETTE WHEEL CIRCUIT

OBJECTIVES

1. To construct and test an 8-bit electronic version of the roulette wheel.
2. To use an oscilloscope to observe the output of the voltage-controlled oscillator circuit.

MATERIALS

Qty.		Qty.	
1	555 timer IC	1	150-Ω, ¼-W resistor
1	74HC00 quad two-input NAND gate CMOS IC	2	10-kΩ, ¼-W resistors
1	74HC164 8-bit shift register CMOS IC	3	100-kΩ, ¼-W resistors
		2	0.01-μF capacitors
1	2N3904 NPN transistor	1	10-μF capacitor
1	keypad (N.O. push-button switch)	1	47-μF capacitor
		1	5-V dc regulated power supply
8	LED indicator-light assemblies	1	oscilloscope

SYSTEM DIAGRAMS

A simplified 8-bit digital roulette wheel circuit is detailed in Fig. 9-13. You will wire and test this circuit.

PROCEDURE

1. Power OFF. Construct the simplified 8-bit roulette wheel circuit in Fig. 9-13. Only a single N.O. switch on the keypad is used.
2. Power ON. Operate the 8-bit roulette wheel circuit.
3. If assigned by your instructor, use an oscilloscope to observe the waveform at the output of the VCO (voltage-controlled oscillator—pin 3 of 555 timer IC). Expect a frequency of about 100 to 500 Hz, which will decrease when you release the switch.
4. Demonstrate the operation of the 8-bit roulette wheel circuit you wired to your instructor. Be prepared to answer questions about the circuit.
5. Power OFF. Take down the circuit, and return all equipment to the proper place. The pins of CMOS ICs should be stored in conductive foam.

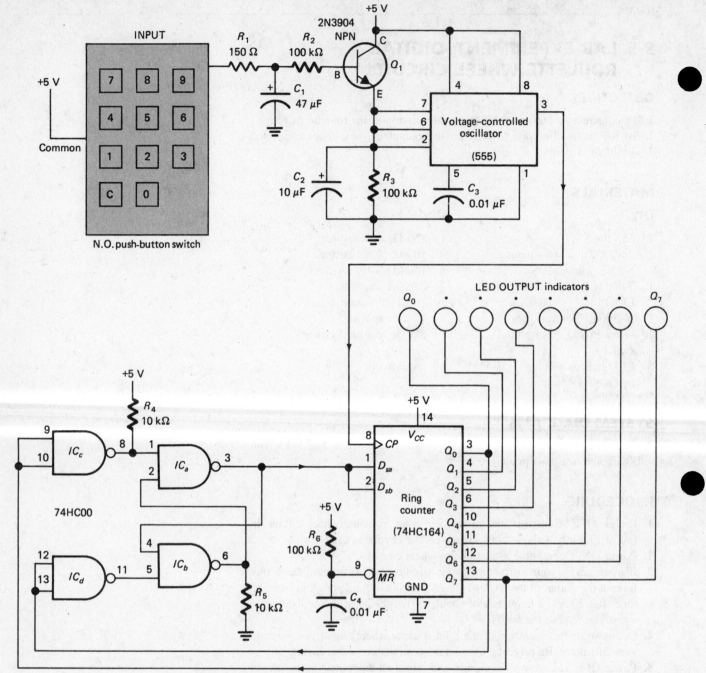

Fig. 9-13 Wiring diagram for a simplified electronic roulette wheel circuit using CMOS ICs.

CHAPTER 10

Arithmetic Circuits

TEST: ARITHMETIC CIRCUITS

Answer the questions in the spaces provided.

1. The most significant bit of the binary number 100000 is
   a. 0 **c.** Not determinable
   b. 1

   1. _____

2. What is the *sum* of the binary numbers 1111 and 0011?

   2. _____

3. What is the *difference* between the binary numbers 1110110 and 1010?

   3. _____

4. The binary number 1110110 equals _____ in decimal.

   4. _____

5. A half-adder circuit contains an AND gate and a(n) _____ gate.
   a. OR **c.** XNOR
   b. NOR **d.** XOR

   5. _____

6. A full-adder circuit contains two _____ circuits.
   a. Half-adder **c.** Subtractor
   b. NOR **d.** Summing

   6. _____

7. A full adder has _____ inputs and two outputs.
   a. One **c.** Three
   b. Two **d.** Four

   7. _____

8. To construct a 4-bit parallel adder you might use
   a. Two full adders and   **c.** Four summing circuits
        two XOR gates   **d.** Four half adders and
   b. Three full adders and   four OR gates
        one half adder

   8. _____

9. We need only add a single _____ gate to a half adder to make a half-subtractor circuit.
   a. AND **c.** NOR
   b. OR **d.** NOT (inverter)

   9. _____

10. Full-adder and -subtractor circuits are classified as _____ logic circuits.
    a. Combinational **c.** IIL
    b. ECL **d.** Sequential

    10. _____

11. Refer to Fig. 10-1. The sum output from the full-adder circuit for problem *a* will be _____. [1 bit]

    11. _____

12. Refer to Fig. 10-1. The C_o output from the full-adder circuit for problem *a* will be _____. [1 bit]

    12. _____

13. Refer to Fig. 10-1. The sum output from the full-adder circuit for problem *b* will be _____. [1 bit]

    13. _____

14. Refer to Fig. 10-1. The C_o output from the full-adder circuit for problem *b* will be _____. [1 bit]

    14. _____

15. Refer to Fig. 10-1. The sum output from the full-adder circuit for problem *c* will be _____. [1 bit]

    15. _____

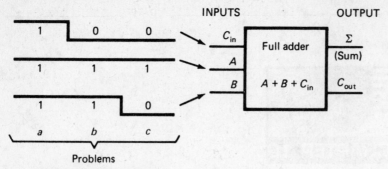

Fig. 10-1 Pulse-train problem.

16. Refer to Fig. 10-1. The C_o output from the full-adder circuit for problem c will be _____. [1 bit]

16. _____

17. Refer to Fig. 10-2. The sum at the output of the 4-bit parallel adder circuit at t_1 is _____.

17. _____

18. Refer to Fig. 10-2. The sum at the output of the 4-bit parallel adder circuit at t_2 is _____.

18. _____

19. Refer to Fig. 10-2. The sum at the output of the 4-bit parallel adder circuit at t_3 is _____.

19. _____

20. Refer to Fig. 10-2. The sum at the output of the 4-bit parallel adder circuit at t_4 is _____.

20. _____

21. Refer to Fig. 10-2. The sum at the output of the 4-bit parallel adder circuit at t_5 is _____.

21. _____

22. Refer to Fig. 10-2. The sum at the output of the 4-bit parallel adder circuit at t_6 is _____.

22. _____

23. Two 7483 TTL adder ICs can be _____ to form an 8-bit parallel binary adder.

23. _____

    a. Cascaded   **c.** Gated

    b. Demodulated   **d.** Multiplexed

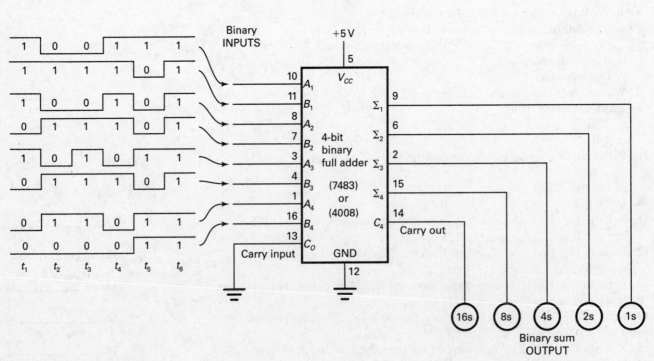

Fig. 10-2 Pulse-train problem.

278

24. What is the binary product of 1111 × 0101?

25. Binary multiplication can be performed using repeated addition or the _____ method.

26. When microprocessors process both positive and negative numbers, _____ representations are used.

27. The 4-bit 2s complement number 0101 represents a positive _____ in decimal.

28. The 4-bit 2s complement number 1111 represents a negative _____ in decimal.

29. In 2s complement representation, the MSB is called the _____ bit.

30. The decimal number −3 equals _____ in 2s complement. [4 bits]

31. The sum of the 2s complement numbers 1011 and 1110 equals _____ in 2s complement. [4 bits]

32. Subtract 2s complement 0011 from 2s complement 1011. (Give answer in 4-bit 2s complement.)

33. The 8-bit 2s complement number 11110000 equals _____ in signed decimal.

34. The 8-bit 2s complement number 00010000 equals _____ in signed decimal.

35. What is the 8-bit 2s complement sum of 00001111 (2s C) and 00110000 (2s C)?

36. What is the 8-bit 2s complement sum of 11111111 (2s C) and 00001000 (2s C)?

37. Refer to Fig. 10-3 for the results of logic probe checks on a faulty half-adder circuit. Which input(s) or output(s) appear to be in error in the half adder?

24. _____
25. _____
26. _____
27. _____
28. _____
29. _____
30. _____
31. _____
32. _____
33. _____
34. _____
35. _____
36. _____
37. _____

Measured Inputs		Measured Outputs	
B	A	Sum	C_{out}
L	L	H	L
L	H	H	L
H	L	H	L
H	H	H	H

H = HIGH as measured with a logic probe
L = LOW as measured with a logic probe

Fig. 10-3 Table of troubleshooting results from checks on a faulty half-adder circuit.

38. On the basis of the measured logic levels recorded for the faulty half adder in Fig. 10-3, the _____ gate within the circuit appears to be faulty.

    a. AND   **c.** OR

    b. NOT (inverter)   **d.** XOR

38. _____

10-1 LAB EXPERIMENT: HALF AND FULL ADDERS

OBJECTIVES

1. To draw, wire, and test a half-adder circuit.
2. To draw, wire, and test a full adder using XOR and NAND gates.
3. *OPTIONAL:* Electronics Workbench or Multisim. To use the electronic circuit simulation software to draw and test a 2-bit parallel adder circuit using one half adder and one full adder.

MATERIALS

Qty.

1	7400 two-input NAND gate IC
1	7486 two-input XOR gate IC
2	LED indicator-light assemblies

Qty.

1	7408 two-input AND gate IC
3	logic switches
1	5-V dc regulated power supply
1	electronic circuit simulation software (optional)

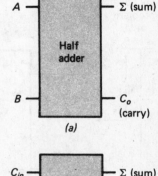

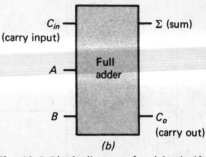

Fig. 10-4 Block diagram for *(a)* a half adder and *(b)* a full adder.

SYSTEM DIAGRAM

Figure 10-4 shows the block symbols for the half adder and full adder used in this experiment.

PROCEDURE

1. Draw a logic symbol diagram of the half adder illustrated in Fig. 10-4(*a*). Use XOR and AND gates.

TABLE 10-1 Truth Table for Half Adder

INPUTS		OUTPUTS	
B	**A**	**Σ**	**C_o**
0	0		
0	1		
1	0		
1	1		

2. Insert the 7408 and 7486 ICs into the mounting board.
3. Power OFF. Connect power (V_{CC} and GND) to both ICs.
4. Wire the half-adder circuit you drew in step 1. Use two input switches for inputs A and B. Use two LED indicator-light assemblies for the outputs.
5. Power ON. Operate the half adder. Observe and record the results in Table 10-1.
6. Power OFF. Leave the 7486 IC on the mounting board. Remove the 7408 IC.
7. Draw a logic symbol diagram of the full adder illustrated in Fig. 10-4(*b*). Use XOR and NAND gates only.

TABLE 10-2 Truth Table for Full Adder

INPUTS			OUTPUTS	
C_{in}	**B**	**A**	**Σ**	**C_o**
0	0	0		
0	0	1		
0	1	0		
0	1	1		
1	0	0		
1	0	1		
1	1	0		
1	1	1		

8. Power OFF. Insert and connect power (V_{CC} and GND) to the 7400 and 7486 ICs.
9. Wire the full adder you drew in step 7. Use three input switches for C_{in}, A, and B. Wire the 7486 and 7400 ICs. Connect the outputs to two LED indicator-light assemblies.
10. Power ON. Operate the full adder. Observe and record the results in Table 10-2.

281

11. *OPTIONAL:* Electronics Workbench or Multisim. If assigned, use EWB to:

    a. Draw a logic diagram for a 2-bit parallel adder circuit using one half adder and one full adder as diagrammed in Fig. 10-5. You may use individual gates or HA and FA blocks furnished by EWB.

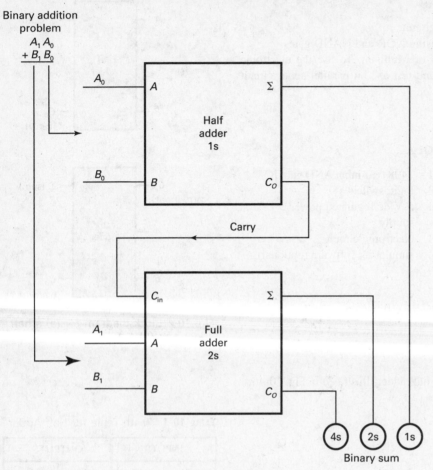

Fig. 10-5 Block diagram for a 2-bit adder circuit.

    b. Operate and test the 2-bit binary adder circuit.

    c. Show the instructor your adder circuit and test results. Be prepared to answer questions about your circuit.

12. Power OFF. Leave the full adder wired for the next experiment.

QUESTIONS

Complete questions 1 to 5.

1. Where can the half adder be used?

2. Where can the full adder be used?

3. Why is the C_o output needed on a half adder?

4. Why is the "extra" C_{in} input needed on a full adder?

5. Draw a logic symbol diagram of a full adder using AND, OR, and XOR gates.

1. _____

2. _____

3. _____

4. _____

282

10-2 LAB EXPERIMENT: 3-BIT PARALLEL ADDER

OBJECTIVE

To wire and operate a 3-bit parallel adder using AND, NAND, and XOR gates.

MATERIALS

Qty.   **Qty.**

2   7400 two-input NAND gate   1   7408 two-input AND gate IC
    ICs   6   logic switches
2   7486 two-input XOR gates ICs   1   5-V dc regulated power
4   LED indicator-light   supply
    assemblies

SYSTEM DIAGRAM

The wiring diagram for a 3-bit parallel adder using gates is drawn in Fig. 10-6. It will add two 3-bit binary numbers ($A_2A_1A_0 + B_2B_1B_0$) and show the sum on the display at the lower right. The top XOR and AND gates form a half adder for adding the two inputs from the 1s column. The middle five gates (two XORs and three NANDs) form a full adder, used to add the carry in (C_{in}) plus the two inputs from the 2s column. The bottom five gates also form a full adder, used to add the inputs from the 4s column plus the carry in (C_{in}) from the 2s full-adder circuit. The carry out (C_o) from the lower-right full adder is the overflow or carry into the 8s position of the sum.

PROCEDURE

1. Insert the two 7400, one 7408, and two 7486 ICs into the mounting board.
2. Power OFF. Connect power (V_{CC} and GND) to each of the five ICs.
3. Wire the 3-bit adder shown in Fig. 10-6. Use six switches for the input numbers $A_2A_1A_0$ and $B_2B_1B_0$. Wire the five ICs. Connect the outputs to the four LED indicator-light assemblies.
4. Power ON. Operate the 3-bit parallel adder. Try adding binary 111 to 111. The answer should be 1110 (decimal 14). Try five binary addition problems of your choice (not over 3 bits long). Record your *inputs* and *outputs*.
5. Show your instructor your 3-bit parallel adder circuit. Be prepared to demonstrate the circuit and answer questions about the circuit's operation.
6. Power OFF. Take down the circuit, and return all equipment to its proper place.

QUESTIONS

Complete questions 1 to 5.

1. Refer to Fig. 10-6. If inputs A_2 and B_2 and C_{in} of the bottom full adder are all 1, then indicators _____ and _____ [1s, 2s, 4s, 8s] are sure to be lit.

   1. _____   _____

2. Refer to Fig. 10-6. Indicator _____ [1s, 2s, 4s, 8s] could be considered an overflow or carry indicator.

   2. _____

3. Refer to Fig. 10-6. If A_0, B_0, A_1, and B_1 are all at 1, then the 1s indicator will read _____ (0, 1) and the 2s indicator _____ (0, 1).

   3. _____   _____

4. Refer to Fig. 10-6. What is the highest binary sum this adder will handle? This equals what in decimal?

   4. _____   _____

5. What do LSB and MSB mean in relation to a binary number?

   5. _____

   _____

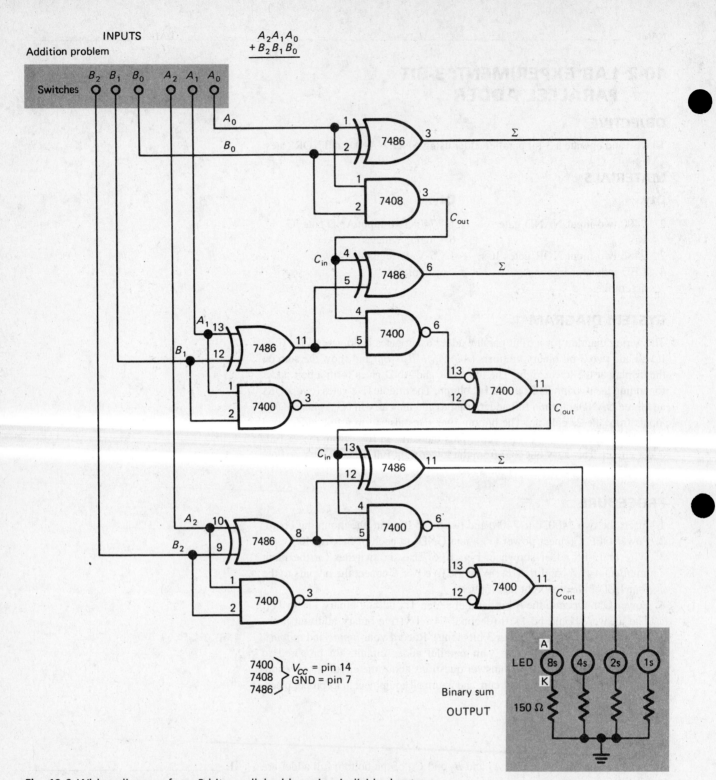

INPUTS

Addition problem

$A_2 A_1 A_0$
$+ B_2 B_1 B_0$

Switches B_2 B_1 B_0 A_2 A_1 A_0

A_0

B_0

7486

7408 C_{out}

C_{in} 7486

7486

7400

7400 C_{out}

A_1 7486

B_1

7400

C_{in} 7486

7486

7400

A_2

B_2

7400

7400 C_{out}

7400
7408 } V_{CC} = pin 14
7486   GND = pin 7

Binary sum

OUTPUT

A
LED   8s   4s   2s   1s
K
150 Ω

Fig. 10-6 Wiring diagram for a 3-bit parallel adder using individual gates.

10-3 LAB EXPERIMENT: USING THE 7483 TTL IC ADDER

OBJECTIVES

1. To wire and operate the 7483 TTL IC as a 4-bit parallel binary adder.
2. To convert the 7483 TTL IC adder into a 4-bit parallel binary subtractor using inverters. To wire and operate this 4-bit parallel subtractor.

MATERIALS

Qty.		Qty.	
1	7404 hex inverter IC	1	7483 4-bit binary adder IC
8	logic switches	5	LED indicator-light assemblies
1	5-V dc regulated power supply		

SYSTEM DIAGRAMS

A wiring diagram of a 4-bit parallel adder is shown in Fig. 10-7. The 7483 adder IC is used to perform the binary addition. The two numbers to be added

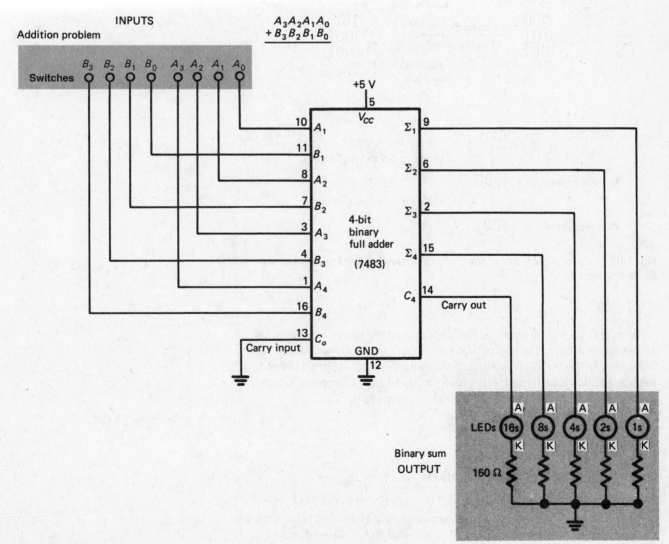

Fig. 10-7 Wiring diagram for a 4-bit parallel binary adder using the 7483 TTL IC.

$(A_3A_2A_1A_0$ and $B_3B_2B_1B_0)$ are applied to the input switches at the upper left. The sum will appear almost immediately on the LEDs at the lower right.

The 7483 4-bit adder IC has been rewired in Fig. 10-9 to perform *binary subtraction*. Notice that (1) four inverters have been added to the B inputs of the 4-bit adder and (2) a HIGH is placed on the Carry input (pin 13). The mathematical technique used to make an adder IC do binary subtraction is shown as an example in Fig. 10-8(a). On the left, Fig. 10-8(a) shows the decimal version of the subtraction problem $(12 - 7 = 5)$. At the center, Fig. 10-8(a) shows the binary version of the subtraction $(1100 - 0111 = 0101)$. At the right, Fig. 10-8(a) shows subtraction being performed using addition. Before performing the addition at the right in Fig. 10-8(a), the subtrahend must be converted from binary to its 2s complement form. Details of this conversion are shown in 10-8(b) where the binary number is first converted to its 1s complement form and then to its 2s complement form $(0111_2 \rightarrow 1000_{1sc} \rightarrow 1001_{2sc})$. The 2s complement form of binary 0111 is 1001. Back to the right in Fig. 10-8(a) shows addition with any overflow beyond 4 bits discarded. In this example, $1100 + 1001 = 1\ 0101$ with the MSB discarded yielding 0101.

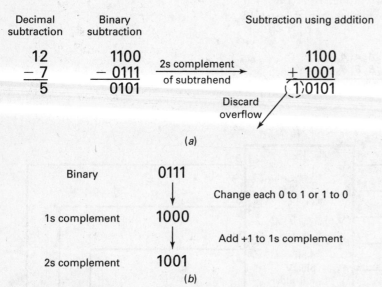

Fig. 10-8 (a) Subtraction using addition. (b) Changing binary number to its 2s complement.

In review, the four inverters in Fig. 10-9 convert the binary number to its 1s complement form. The logical 1 at carry input (pin 13 of the 7483 IC) is like adding $+1$ to the 1s complement yielding the 2s complement subtrahend. Regular addition is performed on the binary minuend and 2s complement subtrahend yielding the difference (the overflow is discarded).

PROCEDURE

Part I: 4-Bit Parallel Binary Adder

1. Power OFF. Insert and connect power (V_{CC} and GND) to the 7483 IC.
2. Refer to Fig. 10-7. Wire the 4-bit binary adder. Use eight switches for the binary numbers to be added ($A_3A_2A_1A_0$ and $B_3B_2B_1B_0$). Ground input C_o (carry input). Connect the five LED indicator-light assemblies to the five outputs.

286

3. Power ON. Operate the 4-bit binary adder. Try adding binary 1111 and 1111. The answer should be 11110 (decimal 15 + 15 = 30). Try five different binary addition problems of your choice (not over 4 bits long). Record your *inputs* and *outputs*.

4. Power OFF. Leave the adder wired for the next section.

Part II: 4-Bit Parallel Binary Subtractor

5. Insert and connect power (V_{CC} and GND) to the 7404 and 7483 ICs.

6. Wire the 4-bit parallel binary subtractor shown in Fig. 10-9. Use eight input switches. Wire the 7404 and 7483 ICs. Use four LED indicator-light assemblies for outputs.

7. Operate the 4-bit parallel binary subtractor. Try subtracting binary 0110 from 1111 ($A_3A_2A_1A_0 = 1111$, $B_3B_2B_1B_0 = 0110$). The answer on the LED indicators should be 1001 (decimal 15 − 6 = 9). Try five binary subtraction problems of your choice (not over 4 bits long, with the larger number being in the $A_3A_2A_1A_0$ position). Record your *inputs* and *outputs*.

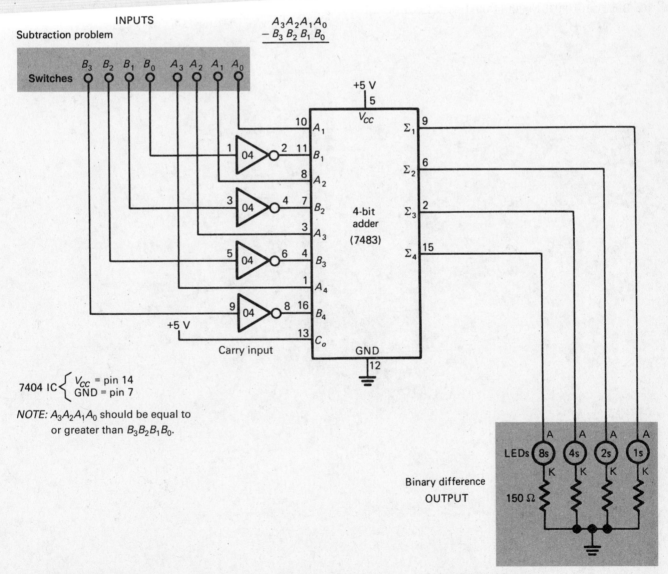

Fig. 10-9 Wiring diagram for a 4-bit parallel subtractor using inverters and the 7483 adder IC.

8. Show your instructor your 4-bit subtractor circuit. Be prepared to answer questions about the circuit's operation.
9. Power OFF. Take down the circuit, and return all equipment to its proper place.

QUESTIONS

Complete questions 1 to 10.

1. Describe *two* things you must do to a 4-bit adder to convert it to a binary subtractor.
2. What two inputs of the 7483 IC are the 1s digits?
3. Inside the 7483 IC we would find circuitry equal to _____ (one half adder and three full adders, four full adders).
4. Refer to Fig. 10-9. The 1s complement of the subtrahend is generated by what electronic components(s)?
5. Binary 1101 + 0111 = _____.
6. Binary 1001 + 1110 = _____.
7. Binary 1111 − 0110 = _____.
8. Binary 1011 − 0110 = _____.
9. Binary 1100 0011 + 0111 0111 = _____.
10. Binary 1010 0101 − 0011 0011 = _____.

1. _____
   _____
2. _____ _____
3. _____
4. _____
5. _____
6. _____
7. _____
8. _____
9. _____
10. _____

10-4 DESIGN PROBLEM: 2s COMPLEMENT ADDER/SUBTRACTOR

OBJECTIVE

To design, draw, construct, test, and demonstrate a 2s complement adder/subtractor system using only 7483 and 7486 ICs.

MATERIALS

Qty.

1 7483 4-bit adder IC
1 7486 quad two-input XOR gate IC
9 logic switches

Qty.

4 LED indicator-light assemblies
1 5-V dc regulated power supply

SYSTEM DIAGRAM

Figure 10-10 is a diagram of a 2s complement adder/subtractor system to be designed and constructed in this activity. The four full adders (FA) are housed in the 7483 4-bit adder IC. The 4 two-input XOR gates are housed in the 7486 IC.

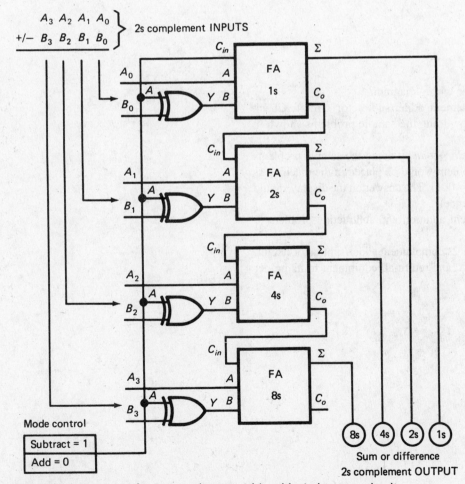

Fig. 10-10 Diagram of a 2s complement 4-bit adder/subtractor circuit used in the design problems.

PROCEDURE

1. Design the 2s complement adder/subtractor system shown in Fig. 10-10. Use a 7486 IC for the four XOR gates. Use a single 7483 adder IC for the four full adders (FAs).
2. Draw a wiring diagram of your adder/subtractor circuit. It is suggested that you use one 7486 IC, one 7483 IC, nine input switches, and four LED output indicator-light assemblies.

3. Have your instructor approve your wiring diagram.
4. Wire and operate the 2s complement adder/subtractor circuit. Check the 2s complement *adder mode* by doing the sample problems shown in Fig. 10-11.
5. Check the 2s complement *subtractor mode* by doing the sample problems shown in Fig. 10-12. The first problem would be placed in the machine as $A_3A_2A_1A_0 = 0111$ and $B_3B_2B_1B_0 = 0011$. The answer on the display should read 0100 (2s complement difference).
6. Demonstrate both a 2s complement addition and subtraction problem to your instructor.
7. Have your instructor approve your 2s complement adder/subtractor circuit.
8. Power OFF. Take down the circuit, and return all equipment to its proper place.

$$
\begin{array}{rr}
(+4) & 0100 \\
+\ (+3) & +\ 0011 \\
\hline
+7_{10} & 0111 \text{ (2s complement SUM)}
\end{array}
$$

(a)

$$
\begin{array}{rr}
(-1) & 1111 \\
+\ (-2) & +\ 1110 \\
\hline
-3_{10} & \textbf{1}\,1101 \text{ (2s complement SUM)}
\end{array}
$$

Discard

(b)

$$
\begin{array}{rr}
(+1) & 0001 \\
+\ (-3) & +\ 1101 \\
\hline
-2_{10} & 1110 \text{ (2s complement SUM)}
\end{array}
$$

(c)

$$
\begin{array}{rr}
(+5) & 0101 \\
+\ (-4) & +\ 1100 \\
\hline
+1_{10} & \textbf{1}\,0001 \text{ (2s complement SUM)}
\end{array}
$$

Discard

(d)

Fig. 10-11 Solved 4-bit 2s complement addition problems with signed numbers.

$$
\begin{array}{r}
(+7) \\
-\ (+3) \\
\hline
+4_{10}
\end{array} = 0011 \xrightarrow[\text{and ADD}]{\text{Form 2s complement}}
\begin{array}{r}
0111 \\
+\ 1101 \\
\hline
\textbf{1}\,0100 \text{ (2s complement DIFFERENCE)}
\end{array}
$$

Discard

(a)

$$
\begin{array}{r}
(-8) \\
-\ (-3) \\
\hline
-5_{10}
\end{array} = 1101 \xrightarrow[\text{and ADD}]{\text{Form 2s complement}}
\begin{array}{r}
1000 \\
+\ 0011 \\
\hline
1011 \text{ (2s complement DIFFERENCE)}
\end{array}
$$

(b)

$$
\begin{array}{r}
(+3) \\
-\ (-3) \\
\hline
+6_{10}
\end{array} = 1101 \xrightarrow[\text{and ADD}]{\text{Form 2s complement}}
\begin{array}{r}
0011 \\
+\ 0011 \\
\hline
0110 \text{ (2s complement DIFFERENCE)}
\end{array}
$$

(c)

$$
\begin{array}{r}
(-4) \\
-\ (+2) \\
\hline
-6_{10}
\end{array} = 0010 \xrightarrow[\text{and ADD}]{\text{Form 2s complement}}
\begin{array}{r}
1100 \\
+\ 1110 \\
\hline
\textbf{1}\,1010 \text{ (2s complement DIFFERENCE)}
\end{array}
$$

Discard

(d)

Fig. 10-12 Solved 4-bit 2s complement subtraction problems with signed numbers.

10-5 TROUBLESHOOTING PROBLEM: FULL-ADDER CIRCUIT

OBJECTIVES

1. To troubleshoot a full-adder circuit.
2. To determine which gate within an IC is faulty and whether it has a "stuck LOW," "stuck HIGH," or other problem.

MATERIALS

Qty.

1 *faulty* 7400 two-input NAND gate IC (get from instructor)
1 *faulty* 7486 two-input XOR gate IC (get from instructor)
3 logic switches

Qty.

2 LED indicator-light assemblies
1 5-V dc regulated power supply

SYSTEM DIAGRAM

You will troubleshoot the full-adder circuit shown in Fig. 10-13. Your instructor will furnish you with a faulty (or perhaps good) 7400 or 7486 IC. Your task will be to troubleshoot the circuit. You will determine which, if any, gate within the faulty IC is bad and the nature of the fault (output "stuck LOW," output "stuck HIGH," or floating input).

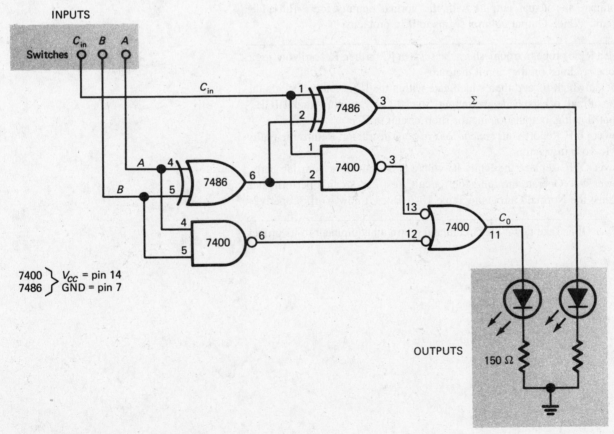

Fig. 10-13 Full-adder troubleshooting circuit.

PROCEDURE

1. Pick up a *faulty* 7400 or 7486 IC from your instructor. Do not mix the faulty unit with the good ICs.
2. Look up the truth table for the full-adder circuit. Fill out the Normal Outputs columns (Σ and C_o) on Table 10-3.

TABLE **10-3** Full-Adder Circuit

INPUTS			Normal outputs		Actual outputs	
C_{in}	B	A	Σ	C_o	Σ	C_o
0	0	0				
0	0	1				
0	1	0				
0	1	1				
1	0	0				
1	0	1				
1	1	0				
1	1	1				
Carry + B + A			Sum	Carry out	Sum	Carry out

3. Power OFF. Wire the full-adder circuit shown in Fig. 10-13.
4. Power ON. Operate the full-adder circuit, and record your observed results under the Actual Outputs columns in Table 10-3.
5. Compare the actual outputs with the normal outputs for a full-adder circuit. Which Output column seems to be a problem?

_____

6. Use a logic probe to troubleshoot the suspect IC further. Record your logic probe readings on the circuit diagram.
7. Decide which IC and then which gate within the IC is faulty. Decide the type of fault in each bad gate (output "stuck LOW," output "stuck HIGH," input floating, output floating, or short circuit in IC).
8. Power OFF. Report your *conclusions* to your instructor. Return the faulty IC to your instructor.
9. Power OFF. Replace the faulty IC with a good unit from your lab parts.
10. Power ON. Operate the full-adder circuit. Test the circuit's performance against the Normal Outputs in Table 10-3. Does it now work properly?

_____

11. Power OFF. Take down the circuit, and return all equipment to its proper place.

Memories

TEST: MEMORIES

Answer the questions in the spaces provided.

1. A personal computer contains a section called the _____ (three letters or words) that contains the arithmetic, logic, and control sections.

2. In a PC, two one-way buses that direct memory, storage and peripheral devices, and which does what and when, are the control bus and _____ bus.

3. Three types of semiconductor memory that are commonly found in a personal computer are NVRAM, RAM, and _____ (three letters).

4. Modern personal computers commonly feature bulk-storage devices such as a hard disk drive, a(n)_____ (optical CD/DVD, ZRAM) drive, and USB ports to support USB flash drives.

5. A hard disk drive (HDD) with 1 TB of storage capacity is considered to have a capacity of about _____ (1 million, 1 trillion) bytes of data.

6. One gigabyte of memory equals _____ bytes of storage capacity.

7. A semiconductor read/write memory is commonly referred to as a _____. [three letters]

8. To copy information into a storage location is called _____ into memory.

9. To copy information from a storage location is called _____ from memory.

10. A semiconductor unit referred to as a(n) _____ is a volatile memory. [three letters]

11. A semiconductor unit referred to as a(n) _____ is a permanent memory. [three letters]

12. A semiconductor dynamic RAM _____ (must, need not) be refreshed many times per second.

13. The computer term _____ is used to describe programs that are permanently held in ROM in a microcomputer system.
    a. Firmware
    b. Hardware

14. The _____ performs the task of programming a ROM.
    a. Distributor
    b. Manufacturer
    c. User

15. A 256- × 4-bit RAM would contain a total of _____ bits of memory.

1. _____
2. _____
3. _____
4. _____
5. _____
6. _____
7. _____
8. _____
9. _____
10. _____
11. _____
12. _____
13. _____
14. _____
15. _____

16. An EEPROM is a field-programmable ROM that
    a. Can be erased by turning off the power
    b. Can be erased by using ultraviolet light
    c. Can be erased electrically
    d. Cannot be erased

16. _____

17. A 1K × 8 EPROM would contain 1024 words, each _____ bits wide.

17. _____

18. External computer bulk storage devices are classified as _____ in nature.
    a. Mechanical or chemical
    b. Mechanical or solar
    c. Mechanical, magnetic, optical, or semiconductor

18. _____

19. Magnetic disks have an advantage over magnetic tapes in that they are _____-access bulk storage devices.

19. _____

20. An 8-bit group of data is called a _____.

20. _____

21. The _____ storage device has the fastest access time.
    a. SRAM
    b. Floppy disk
    c. Magnetic tape

21. _____

22. The _____ (floppy, hard) disk has the greater storage capacity.

22. _____

23. The _____ (floppy, hard) disk drive has the fastest access time.

23. _____

24. An EPROM with the window on the top can be erased _____ (electrically, by using ultraviolet light).

24. _____

25. A flash EEPROM is erased _____ (electrically, by ultraviolet light).

25. _____

26. A RAM with 1024 memory locations (such as the 2114 IC) would need _____ address pins. [number]

26. _____

27. Refer to Fig. 11-1. The output of the diode ROM during pulse *a* will be _____. [4 bits].

27. _____

28. Refer to Fig. 11-1. The output of the diode ROM during pulse *b* will be _____. [4 bits]

28. _____

29. Refer to Fig. 11-1. The output of the diode ROM during pulse *c* will be _____. [4 bits]

29. _____

30. Refer to Fig. 11-1. The output of the diode ROM during pulse *d* will be _____. [4 bits]

30. _____

31. Refer to Fig. 11-1. The output of the diode ROM during pulse *e* will be _____. [4 bits]

31. _____

32. Flash EEPROM, battery backup SRAM, and FeRAM are all examples of devices that may be used to implement nonvolatile read/write memory. (T or F)

32. _____

33. Ferroelectric RAM is a _____ (nonvolatile, volatile) RAM whose memory cells are based on a ferroelectric capacitor and MOS transistor.

33. _____

34. Magnetoresistive RAM (MRAM) is a newer nonvolatile RAM with excellent access speed and high density. (T or F)

34. _____

35. FeRAM semiconductor memory is nonvolatile and can be updated electrically. (T or F)

35. _____

36. SDRAM semiconductor memory is commonly packaged in _____ (DIMM, FIP) cards which slide into sockets in modern desktop PCs.

36. _____

37. In computer memory jargon, RIMM stands for a radio implanted memory module. (T or F)

37. _____

38. A modern laptop computer might typically use a _____ (30-pin SIMM, 200-pin SO-DIMM) module to implement SDRAM.

38. _____

39. Like magnetic tapes, floppy disks are recorded in a spiral track and are sequential-access devices. (T or F)

39. _____

40. PCMCIA describes the physical and electrical characteristics or standards for a _____ (2 words).

40. _____

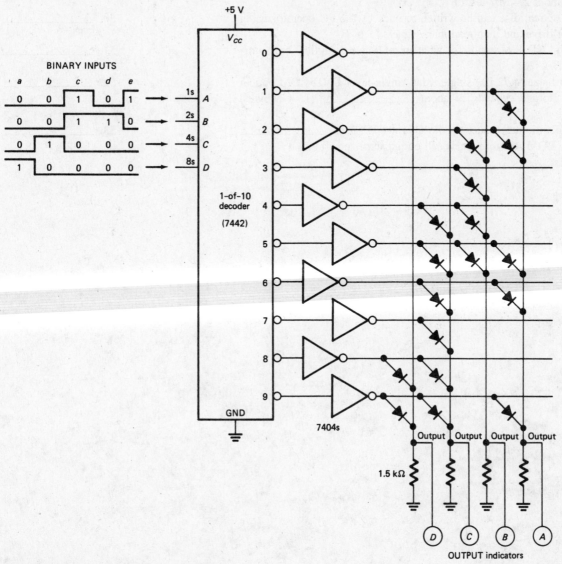

Fig. 11-1 Diode ROM pulse-train problem.

41. The newer "solid-state disk" in portable computers might use _____ (flash EEPROMs, ROMs) as the memory cell.

41. _____

42. If a bulk storage device is a called a Winchester drive, it is a _____ (hard, solid-state) disk drive.

42. _____

43. The hard disk drive in a modern computer, may have a spindle speed of _____ (7200 rpm, 600 rpm) and a storage capacity of 500 GB.

43. _____

44. When you purchase a compact computer, the laptop may feature a _____ (solid-state disk drive, magnetic tape drive) instead of a traditional HDD.

44. _____

45. A modern PC might implement the dynamic RAM within the computer using a _____ (DIMM or RIMM, SIP) memory module.

45. _____

46. The acronym *WORM* stands for what in reference to bulk storage devices?

46. _____

47. Modern personal computers do not use optical discs because they are very fragile, extremely expensive, and have low storage capacities. (T or F)

47. _____

48. The typical optical CD or DVD is 1.2 mm thick and has a diameter of _____ (20, 120) mm.

48. _____

49. If a CD-ROM drive with a 1X designation has a maximum data transfer rate of 150 Kbytes per second, then a 16X drive would have a data transfer rate of about 2.4 Mbytes. (T or F)

50. The CD-R optical disc can be written to once by the PC operator using a CD-writer drive and then read many times. (T or F)

51. The acronym CD-RW stands for what in reference to optical bulk storage devices?

52. The standard size (4.72 in.) single-sided single-layer DVD-ROM optical disc has a storage capacity of about _____ (1.44 Mbytes, 4.7 Gbytes).

53. A digital potentiometer, such as the DS1804 IC, uses_____ (EEPROM, ROM) to store and recall output wiper positions.

49. _____

50. _____

51. _____

52. _____

53. _____

298

11-1 LAB EXPERIMENT: RANDOM-ACCESS MEMORY

OBJECTIVES

1. To wire and read the memory contents (without programming) of a 7489 TTL read/write random-access memory (RAM).
2. To program the 7489 RAM by writing in the Gray code.
3. To operate the 7489 RAM as a binary-to-Gray code converter after the memory is programmed.
4. *OPTIONAL:* Reprogram the RAM to perform as a Gray-to-binary code converter.

MATERIALS

Qty.		Qty.	
1	7404 hex inverter IC	4	LED indicator-light assemblies
9	logic switches		
1	5-V dc regulated power supply	4	1-kΩ resistors
		1	single-pulse clock (negative pulse)
1	logic probe		
1	7489 read/write RAM IC		

SYSTEM DIAGRAM

The wiring diagram for the 16- $\times$ 4-bit RAM is shown in Fig. 11-2. The address switches select which 4-bit memory word will be retrieved from memory (read operation) or at which location new data will be stored

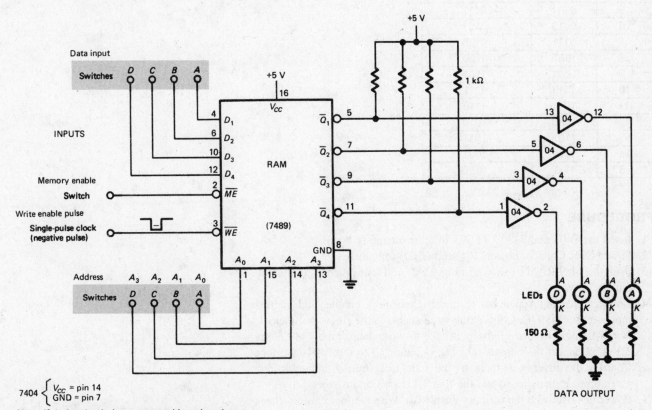

7404 $\begin{cases} V_{CC} = \text{pin 14} \\ \text{GND} = \text{pin 7} \end{cases}$

Note: If single-pulse clock puts out a positive pulse only, put an extra 7404 inverter between single-pulse positive clock and pin 3 ($\overline{WE}$) of 7489 IC.

Fig. 11-2 Wiring diagram for the 7489 TTL SRAM IC.

299

(write operation). The memory enable switch must be LOW for either the read or write operations. A *negative pulse* at the write-enable input $\overline{WE}$ transfers data from the data inputs to the location addressed by the address inputs. Inverted data appear at the complementary outputs $\overline{Q}_{1-4}$ of the 7489 IC during the read operation.

The 7489 TTL IC has *open collector* outputs. For proper operation in this circuit (Fig. 11-2), external pull-up resistors are required to hold the output HIGH until they are activated. Most TTL ICs use *totem-pole* outputs. TTL ICs with totem-pole outputs provide an internal transistor connecting from the output to +5 V of the power supply. This internal transistor acts like the external pull-up resistors shown in Fig. 11-2.

In Fig. 11-2, the small bubbles at the outputs of the 7489 IC symbol suggest the RAM features active LOW outputs. To display the true data held in the RAM, four 7404 inverters complement the output data from the IC for a true output shown on the LEDs.

During the lab, you will program the 7489 RAM IC with the 4-bit Gray code. The Gray code (also called the binary-reflected Gray code) is commonly associated with the rotary encoder used to determine the angular position of a shaft or disk. It is also used in some modern digital communications systems.

A listing of the 4-bit Gray code is shown in Table 11-1 along with decimal and binary equivalents.

TABLE 11-1 Gray Code

Decimal	Binary	4-bit Gray code
0	0000	0000
1	0001	0001
2	0010	0011
3	0011	0010
4	0100	0110
5	0101	0111
6	0110	0101
7	0111	0100
8	1000	1100
9	1001	1101
10	1010	1111
11	1011	1110
12	1100	1010
13	1101	1011
14	1110	1001
15	1111	1000

PROCEDURE

1. Insert the 7404 and 7489 TTL ICs into the mounting board.
2. Power OFF. Connect power (V_{CC} and GND) to both ICs.
3. Wire the 7489 RAM as shown in Fig. 11-2. Add four inverters to the outputs as shown.
4. Connect nine input switches to the data, memory enable, and address inputs of the 7489 IC. Connect the write enable to the single-pulse clock. *Note:* A single *negative pulse* is needed to write data in the 7489 memory IC. Connect the outputs ($\overline{Q}_1$, $\overline{Q}_2$, $\overline{Q}_3$, and $\overline{Q}_4$) to the 7404 inverters. Connect the inverter outputs to the four LED output indicator-light assemblies. Remember to add the four 1-kΩ pull-up resistors.
5. Power ON. Activate the memory enable (0). With the logic probe, check the write-enable input to make sure it is HIGH. This is the *read* position.
6. *Read* the random contents of the 7489 memory IC using the LED output as you address each of the 16 word locations in the storage unit.

300

7. Observe the outputs of the 7489 memory with no program in it. Record them. Turn the power off and on again. Try reading again. Are the memory contents the same?

8. Write in or program the 7489 IC memory with the 4-bit Gray code. *Do not turn off the power at any time* or you will lose the Gray code program from the memory.

9. Power ON. Set the memory enable switch to 0. For each address, set the address and data inputs and press the write-enable pulse push button (single negative clock pulse). The Gray code contents should appear almost immediately on the output indicators.

10. Memory enable to 0. Read the Gray code for each address, and check against Table 11-1 for accuracy.

11. Show your instructor your programmed SRAM IC. Be prepared to answer question about the circuit.

12. *OPTIONAL:* Your instructor may ask you to reprogram the 7489 RAM chip to perform as a Gray-to-binary decoder.

13. Power OFF. Take down the circuit, and return all equipment to its proper place.

QUESTIONS

Complete questions 1 to 10.

1. The device you ended up with in step 10 of the procedure was a _____ (counter, decoder).

    1. _____

2. What happens to the stored program in the 7489 RAM if the power is turned off?

    2. _____

3. The 7489 IC was a _____ (nonvolatile, volatile) memory.

    3. _____

4. The 7489 IC was organized as a _____- × _____- bit memory.

    4. _____ _____

5. The 7489 memory contained _____ word locations. Each word contained _____ bits.

    5. _____ _____

6. The memory enable was activated by a logical _____ on the 7489 IC.

    6. _____

7. Which things listed below are *true* about the 7489 RAM IC?

    7. _____

a. 64-bit memory	i. PROM
b. 256-bit memory	j. Inverted outputs
c. 16 × 4 memory	k. Volatile memory
d. 32 × 2 memory	l. Read-only memory
e. Scratchpad memory	m. Read/write memory
f. Permanent memory	n. Open collector outputs
g. RAM	o. Dynamic RAM
h. ROM	p. Static RAM

8. To *write* a 0111 in word location 3, what are the logic levels needed at the following inputs?

    8. *Place answers below.*

  a. Address (D, C, B, A)

    a. _____

  b. Data (D_4, D_3, D_2, D_1)

    b. _____

  c. Memory enable

    c. _____

  d. Write enable

    d. _____

9. To *read* the contents of word location 13 in the 7489 memory, what are the necessary logic levels at the following inputs?

    9. *Place answers below.*

  a. Address (D, C, B, A)

    a. _____

  b. Data (D_4, D_3, D_2, D_1)

    b. _____

  c. Memory enable

    c. _____

  d. Write enable

    d. _____

10. The 7489 RAM IC is a _____ (CMOS, TTL) device.

    10. _____

11-2 LAB EXPERIMENT: READ-ONLY MEMORY

OBJECTIVE

To wire and operate a random-counter circuit using a ROM to store the counting sequence.

MATERIALS

Qty.

- 4   7447 BCD-to-seven-segment decoder ICs
- 1   logic switch
- 4   LED indicator-light assemblies
- 21   150-Ω, ¼-W resistors
- 1   7493 binary counter IC

Qty.

- 1   clock (single pulses)
- 3   seven-segment LED displays or commercial display board (DB-1000 by Dynalogic Concepts)
- 1   5-V dc regulated power supply

SYSTEM DIAGRAM

You will construct a random counter that will generate the decimal count 1, 117, 22, 6, 114, 44, 140, 17, 0, 14, 162, 146, 134, 64, 160, 177, and then back to 1. A wiring diagram of the random-counter circuit is drawn in Fig. 11-3. You will also use Table 11-2.

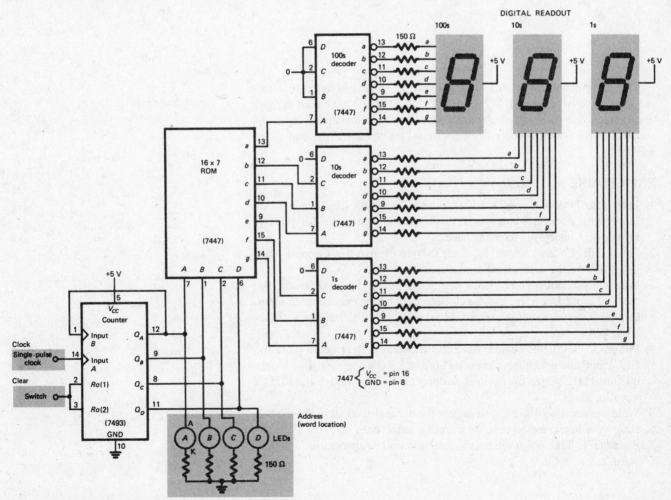

Fig. 11-3 Wiring diagram for a "random-counter" circuit.

TABLE 11-2   Random Counter

| Word location in ROM | | | | | Counting sequence | | |
| Decimal | Binary | | | | Decimal | | |
	D	C	B	A	100s	10s	1s
0							
1							
2							
3							
4							
5							
6							
7							
8							
9							
10							
11							
12							
13							
14							
15							

Custom-programmed ROMs are expensive in small quantities. For this experiment we have used a 7447 decoder TTL IC that you have used in other experiments. The 7447 decoder is not a ROM but a combinational logic circuit. This off-the-shelf decoder IC demonstrates the characteristics and use of a ROM.

PROCEDURE

1. Insert four 7447 ICs, one 7493 IC, and three 7-segment LED displays into the mounting board. A commercial display board such as the DB-1000 by Dynalogic Concepts is recommended.
2. Power OFF. Connect power (V_{CC} and GND) to the five ICs. Connect +5 V (V_{CC}) only to the 3 seven-segment displays.
3. Wire the random-counter circuit in Fig. 11-3. Use an input switch for the clear input. Use a single-pulse clock (positive pulse) for the clock input. Complete the wiring of the circuit. Make sure all 150-Ω limiting resistors are in place.
4. Power ON. Clear the counter (to 1 and back to 0). Repeatedly pulse the *CLK* input, and notice the address inputs to the ROM increase from binary 0000 to 1111. Notice the decimal readouts as you go through the ROM's word locations.
5. Write the address or word location and the decimal readout in Table 11-2.
6. Have your instructor approve the counting sequence.
7. Power OFF. Take down the circuit, and return all equipment to its proper place.

QUESTIONS

Complete questions 1 to 9.

1. What do the following abbreviations stand for?
   a. RAM
   b. ROM
   c. PROM
   d. EPROM

1. *Place answers below.*
   a. _____
   b. _____
   c. _____
   d. _____

2. Explain the difference between the four types of semiconductor memories in question 1.

2. _____
_____
_____
_____
_____

3. What is a disadvantage of a ROM?

3. _____

4. The ROM is an example of _____ (nonvolatile, volatile) memory.

4. _____

5. What is the purpose of the 7493 counter in Fig. 11-3?

5. _____
_____

6. A logical _____ is needed to activate the clear input on the 7493 IC. A logical _____ is therefore needed to permit the counter to operate.

6. _____ _____

7. The 7447 IC was used as a _____-bit ROM in this experiment. The 7447 ROM is organized with _____ words. Each word contains _____ bits.

7. _____ _____
_____

8. What do the following abbreviations stand for?
   a. SRAM
   b. DRAM
   c. EEPROM
   d. NVSRAM

8. *Place answers below.*
   a. _____
   b. _____
   c. _____
   d. _____

9. Refer to Fig. 11-3. One 7447 IC is used like a 16 × 7 ROM in this circuit even though internally the 7447 IC is a _____ (combinational, sequential) logic circuit.

11-3 LAB EXPERIMENT: DIGITAL POTENTIOMETER—USING EEPROM

OBJECTIVES

1. To wire and test the DS1804 digital potentiometer.
2. To wire and test the DS1804 IC as a voltage divider.

MATERIALS

Qty.   **Qty.**

1   1804-100 digital   1   5-V dc regulated power
    potentiometer IC   supply
2   logic switches   1   DMM
1   single-pulse clock
    (negative pulse)

SYSTEM DIAGRAMS

The first circuit used to test the operation of the *DS1804 NV trimmer poten-tiometer* is shown in Fig. 11-4. Two logic switches and a single-pulse clock (negative pulse) form the inputs to the digital potentiometer. An ohmmeter will be used to measure the resistance between the wiper (W) and bottom end of resistance (L). The chip used will be the DS1804-100. The output resistance is divided, having 100 separate taps. Therefore, if the total resistance is 100 kΩ, then the output resistance can vary from 0 to 100 kΩ in discrete steps of 1 kΩ, depending on the position of the wiper.

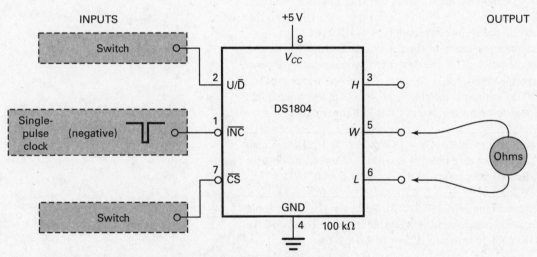

Fig. 11-4 Testing the DS1804-100 digital potentiometer.

The chip select ($\overline{CS}$) input to the DS1804 IC is like a master switch al-lowing inputs from the U/$\overline{D}$ and $\overline{INC}$ pins when it is activated (LOW). If the $\overline{CS}$ input is HIGH, the inputs are deactivated; however, the IC still operates as a one-time programmable (OTP) trimmer potentiometer. Its memory holds one resistance value only.

The second circuit will test the DS1804-100 digital potentiometer as a voltage divider, which is drawn in Fig. 11-5. The measured output voltage should increase or decrease in small discrete amounts of about 0.05 V per step (5 V divided by 99 steps equals about 0.05 V/step). An accurate DMM would be useful for measuring output voltage in this circuit.

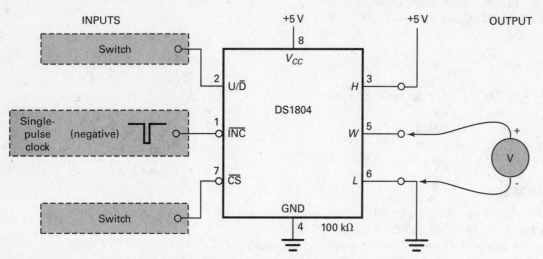

Fig. 11-5 Testing the DS1804-100 digital potentiometer wired as a voltage divider.

PROCEDURE

1. Power OFF. Wire the circuit shown in Fig. 11-4 using two logic switches, a single-pulse clock (negative pulses), and a DMM to measure electrical resistance.

2. Power OFF. Deactivate the chip select input ($\overline{CS}$ = HIGH).

3. Power ON. Observe the resistance of the potentiometer (outputs W to L resistance). Repeat the power OFF and power ON sequence *several times slowly*, and observe the resistance each time. The stored wiper position is read from EEPROM each time the chip is powered up, and you should read the same resistance each time. Record this initial resistance.

-----

4. Power ON. Activate the chip select ($\overline{CS}$ = LOW). Set the Up/Down input to (U/$\overline{D}$ = HIGH). Inject negative pulse(s) into the $\overline{INC}$ input, and observe the output results. Record your results at the output ($\overline{CS}$ = 0, U/$\overline{D}$ = 1).

-----

5. Power ON. Activate the chip select ($\overline{CS}$ = 0). Reset the Up/Down input to down (U/$\overline{D}$ = 0). Inject negative pulse(s) into the $\overline{INC}$ input, and observe the output results. Record your results ($\overline{CS}$ = 0, U/$\overline{D}$ = 0).

-----

6. Power ON. Inputs: $\overline{CS}$ = 0, U/$\overline{D}$ = 0 or 1 as needed. Inject single pulses $\overline{INC}$ input until the resistance reading is close to 50% of total (50 kΩ for a 100-kΩ pot).

7. Deactivate chip select input ($\overline{CS}$ = 1). Observe output resistance. Turn power OFF and then ON again *slowly*. Observe the output resistance. Record your results. (*Hint:* This is a newly programmed wiper position.)

-----

8. Power OFF. Rewire the DS1804-100 IC as shown in Fig. 11-5 using an accurate dc voltmeter (DMM) to measure the output voltage.

9. Power ON. Test the DS1804 digital potentiometer wired as a voltage divider. Observe and record your results.

_____

_____

_____

10. Show your instructor your circuit (Fig. 11-5), and prepare to answer questions about testing the DS1804-100 digital potentiometer IC wired as a voltage divider.

11. Power OFF. Take down the circuit, and return all equipment to its proper place.

QUESTIONS

Complete questions 1 to 6.

1. Refer to Fig. 11-4. The small bubbles on the logic diagram at pins 1 and 7 mean these are _____ (active HIGH, active LOW) inputs.

    1. _____

2. Draw a simple block diagram of the DS1804-100 digital potentiometer. Include inputs labeled $\overline{CS}$, $U/\overline{D}$, and $\overline{INC}$. Include outputs labeled H, L, and W. Include internal blocks labeled control logic, EEPROM, multiplexer, and series resistors. Refer to the Dallas Semiconductor or Maxim Integrated Products data sheet or your textbook.

3. The purpose of the internal EEPROM section of the DS1804 IC is to serve as nonvolatile storage of the wiper position during power down and recalling that position to the control logic during power up. (T or F)

    3. _____

4. Refer to Fig 11-4. With the chip select ($\overline{CS}$) input _____ (HIGH, LOW), all other inputs are disabled.

    4. _____

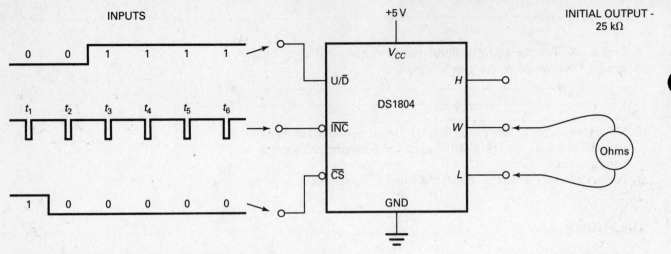

Fig. 11-6 DS1804 digital potentiometer pulse-train problem.

5. Refer to Fig. 11-6. Give the expected resistance reading immediately following these input pulses:

$t_1 =$ _____

$t_2 =$ _____

$t_3 =$ _____

$t_4 =$ _____

$t_5 =$ _____

$t_6 =$ _____

6. Refer to Fig. 11-5. Calculate the smallest voltage change when the wiper moves one position in the digital potentiometer. (*Hint:* 100 steps, 5 V)

6. _____

Digital Systems

TEST: DIGITAL SYSTEMS

Answer the questions in the spaces provided.

1. Encoders and decoders would both be classified as
   - **a.** Analog subsystems
   - **b.** Analog systems
   - **c.** Digital subsystems
   - **d.** Digital systems

   1. _____

2. A _____ would be classified as a large-scale integration.
   - **a.** Decoder IC
   - **b.** Flip-flop IC
   - **c.** Logic-gate IC
   - **d.** Calculator IC

   2. _____

3. A microcontroller is a system on a chip and might be described by the manufacturer as a(n)
   - **a.** Discrete component
   - **b.** MSI
   - **c.** SSI
   - **d.** VLSI

   3. _____

4. The 4000 series ICs were operated on a _____ (5-V, 12-V) power supply in the digital dice game circuit.

   4. _____

5. The 4016 IC used in the digital dice game circuit contains _____ which can conduct either ac or dc.
   - **a.** Bilateral switches
   - **b.** NAND gates
   - **c.** Multiplexers
   - **d.** Relays

   5. _____

6. A(n) _____ (known, unknown) frequency is the main input into a digital clock.

   6. _____

7. A digital clock makes extensive use of _____ subsystems.
   - **a.** Counter
   - **b.** Encoder
   - **c.** ROM
   - **d.** Shift register

   7. _____

8. Frequency division in a digital clock is accomplished by
   - **a.** Counters
   - **b.** Demodulators
   - **c.** Multiplexers
   - **d.** Multipliers

   8. _____

9. Most LSI digital clock chips are manufactured using low-power _____ (bipolar, MOS) technology.

   9. _____

10. The National Semiconductor MM5314 clock chip can produce a _____ display.
    - **a.** Direct drive hours and minutes
    - **b.** Multiplexed hours, minutes, and seconds

    10. _____

11. Multiplexed multidigit displays are
    - **a.** All left on continuously
    - **b.** All turned on and off together to save power
    - **c.** All turned on most of the time and turned off when they overheat
    - **d.** Turned on and off, one at a time, in rapid succession

    11. _____

12. The experimental frequency counter in Chap. 12 used _____ to convert the input signals to a square wave.
   a. AC/DC converters   **c.** Level adjusters
   b. Decoders   **d.** Schmitt trigger inverters

13. Frequency range, input sensitivity, input impedance, input protection, accuracy, gate intervals, and display time are important specifications of commercial
   a. Digital clocks   **c.** Frequency counters
   b. Digital multimeters   **d.** Oscilloscopes

14. A frequency counter counts the number of pulses from an unknown frequency source in
   a. 1 h   **c.** A short given amount of time
   b. 100 s   **d.** An unknown amount of time

15. The experimental frequency counter went through an entire _____ sequence each second.
   a. Input-countdown-output **b.** Reset-count-display

16. Whereas the experimental frequency counter used a 60-Hz known frequency, commercial frequency counters use a much
   a. Higher known frequency, produced by an accurate oscillator
   b. Lower known frequency, produced by a transformer

17. Liquid-crystal displays are commonly used with _____ ICs in low-power digital systems.
   a. CMOS   **c.** SSI
   b. IEEE   **d.** TTL

18. The accuracy of the _____ is critical in determining the accuracy of digital timer circuits.
   a. Comparator   **b.** Time-base clock

19. Two 74HC85 ICs were cascaded in the LCD timer circuit to form an 8-bit
   a. Latch   **c.** Magnitude comparator
   b. Counter   **d.** Data bus

20. The 74HC4543 was the IC used as a _____ in the LCD timer circuit.
   a. Clock   **b.** Decoder/driver

21. Two technologies used in simple low-cost distance sensors includes ultrasonic sound and _____ (blue-ray-, infrared-) light sensors.

22. The Pololu 1134 distance sensor module studied in this chapter was an infrared-light sensor that sensed objects in a range from about 1 to 4 inches and featured a(n) _____ (analog, digital) output.

23. The ultrasonic distance sensor depends on an attached microcontroller to precisely measure the time it takes in _____ (seconds, microseconds) for an ultrasonic sound burst to travel to and echo off the target. The microcontroller then calculates the distance to the target.

24. In common usage, boundary scan is also referred to as _____ (IEEE, JTAG) in the field of digital electronics.

25. A testing architecture which includes automated testing and test access points for miniaturized complex PC boards is covered by IEEE _____ (Standard 1149.1, Standard 2001) and is commonly referred to as JTAG.

12. _____
13. _____
14. _____
15. _____
16. _____
17. _____
18. _____
19. _____
20. _____
21. _____
22. _____
23. _____
24. _____
25. _____

312

12-1 LAB EXPERIMENT: DIGITAL DICE

OBJECTIVES

1. To wire and test an experimental digital dice game using TTL ICs.
2. *OPTIONAL:* To design a double digital dice game to simulate the toss of a pair of dice.
3. *OPTIONAL:* To wire and test your double digital dice game circuit.

MATERIALS

Qty.   **Qty.**

1	555 timer IC	1	seven-segment LED display
1	7408 two-input AND gate IC	7	150-Ω, ¼-W resistors
1	7410 three-input NAND gate IC	1	10-kΩ, ¼-W resistor
1	7447 BCD-to-seven-segment decoder/driver IC	1	27-kΩ, ¼-W resistor
1	74192 decade counter IC	1	0.033-μF capacitor
1	logic switch	1	5-V dc regulated power supply

SYSTEM DIAGRAM

The wiring diagram in Fig. 12-1 details a simple digital dice game. When the input switch is HIGH, the 7408 control gate permits the 600-Hz square-wave signal from the clock assembly to pass through to the counter. When the input control switch is moved to LOW, the clock signal is blocked and the counter stops. A random number from 1 through 6 appears on the seven-segment LED display.

PROCEDURE

1. Insert the five ICs and one 7-segment LED display into the mounting board.
2. Power OFF. Connect power (V_{CC} and GND) to the five ICs. Connect +5 V (V_{CC}) only to the anode of the seven-segment LED display.
3. Wire the experimental digital dice game shown in Fig. 12-1. Use an input switch for the roll/stop control.
4. Power ON. Move the roll/stop switch to a logical 1. The 74192 IC should be counting upward at about 600 digits per second. The output will appear to be an 8. Move the roll/stop switch to a logical 0. The output display should contain a random number from 1 through 6.
5. Show your instructor your digital dice circuit. Be prepared to answer questions about the circuit's operation.
6. *OPTIONAL:* Add a second clock/counter/decoder/driver and display to simulate the roll of a pair of dice.
7. On a separate sheet, design a circuit to be added to the one in Fig. 12-1. Use a three-input AND gate and a single roll/stop switch to control both circuits.
8. Power OFF. Wire your additional circuit.
9. Power ON. Test your expanded digital dice game.
10. Show your instructor your expanded digital dice game.
11. Power OFF. Take down the circuit, and return all equipment to its proper place.

QUESTIONS

Complete questions 1 to 8.

1. Refer to Fig. 12-1. Which IC is wired as a free-running MV?   1. _____
2. Refer to Fig. 12-1. Which ICs are wired as a mod-6 up counter?   2. _____
3. Refer to Fig. 12-1. Which IC is considered the input control gate?   3. _____

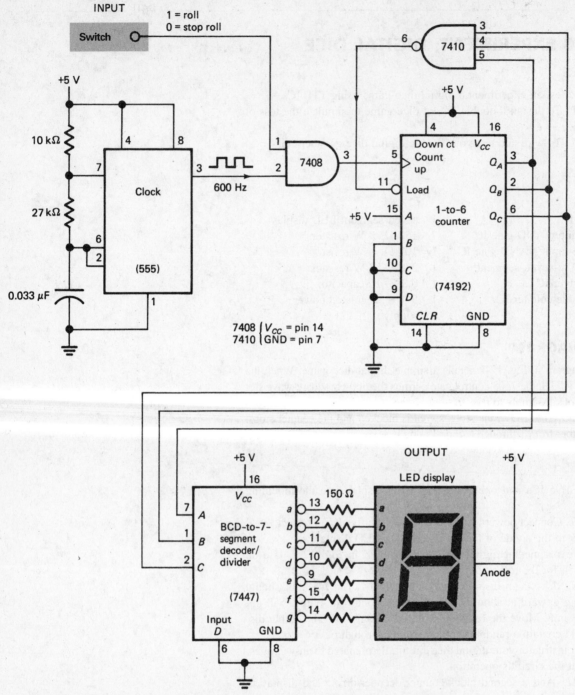

Fig. 12-1 Wiring diagram for a digital dice game using TTL ICs and an LED display.

4. Refer to Fig. 12-1. Which IC is considered the BCD-to-seven-segment decoder?

4. _____

5. Refer to Fig. 12-1. The 7447 IC has active _____ (HIGH, LOW) outputs.

5. _____

6. Refer to Fig. 12-1. What is the purpose of the 7410 IC in this circuit?

6. _____
_____

7. On a separate sheet, draw a block diagram of the digital dice game in Fig. 12-1.

8. On a separate sheet, draw a block diagram of a *double* digital dice game which simulates the toss of a *pair* of dice.

12-2 LAB EXPERIMENT: DIGITAL STOPWATCH SYSTEM

OBJECTIVE

To wire and test a simple digital stopwatch system using various TTL sub-systems.

MATERIALS

Qty.

1 7400 two-input NAND gate IC
1 7408 two-input AND gate IC
1 7414 Schmitt trigger inverter IC
2 7447 BCD-to-seven-segment decoder ICs
2 7476 J-K flip-flop ICs
1 7493 binary counter IC
1 74192 decade counter IC

Qty.

2 seven-segment LED displays (commercial display board is recommended)
1 logic switch
14 150-Ω, ¼-W resistors
1 5-V dc regulated power supply
1 60-Hz square-wave source such as a function generator or clock on trainer

SYSTEM DIAGRAM

Counters serve as the heart of any digital clock system. Figure 12-2 is a wiring diagram for a simple digital timepiece. The digital stopwatch you will build in this experiment works just like a regular digital clock, but it is much less complex. The elapsed time counted on the display in Fig. 12-2 will be in tenths of a second. The maximum time that can be measured by this simple system will be 9.9 s. More counters, decoders, and displays would be needed to display minutes and hours as on a regular digital clock.

You should be aware that the purpose for including this simple circuit is to study the fundamental operation of digital clock operation and to demonstrate how SSI and MSI ICs may be used to form digital systems. Real-world digital clocks and timepieces are based on highly complex MOS LSI chips.

PROCEDURE

1. Insert the nine ICs and two 7-segment LED displays into the mounting board.
2. Power OFF. Connect power (V_{CC} and GND) to the nine ICs. Connect +5 V (V_{CC}) only to the two common-anode seven-segment LED displays.
3. Wire the digital stopwatch system shown in Fig. 12-2. Use a logic switch for the start/stop control input. Use a 60-Hz source for the known frequency input to the 7414 Schmitt trigger inverter. Your instructor will provide information on which 60-Hz square-wave source to use.
4. Power ON. Move the start/stop control switch to a logical 1 or the *start* (*count*) position. The digital displays should count in tenths of a second. Check the approximate accuracy with a regular clock or wristwatch.
5. Have your instructor approve your completed circuit. Be able to explain the function or job of each IC in the circuit.
6. Use your digital stopwatch for checking your dexterity against your classmates'. One test you can try is laying a small bolt and nut on the table next to your circuit. With the count at 0.0 s do the following:
   a. Start the counter (start/stop control to logical 1).
   b. Thread the nut on the bolt to the end of the threads.

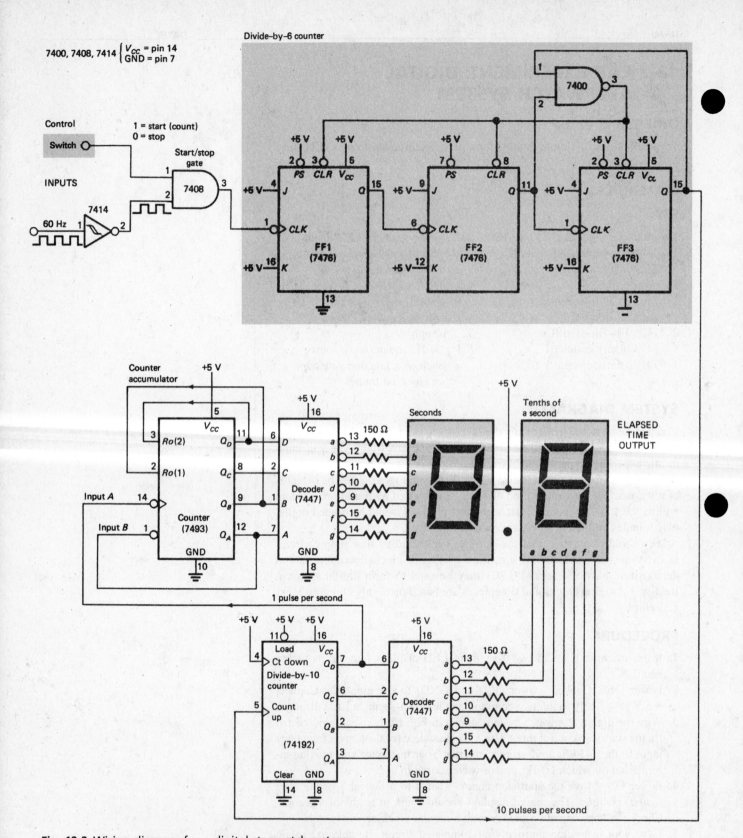

Fig. 12-2 Wiring diagram for a digital stopwatch system.

c. Place the bolt and nut in a small box on the other side of the circuit.

d. Stop the counter (start/stop control switch to logical 0).

e. Check your time.

7. If time permits, try devising your own test of dexterity or speed.

8. Power OFF. Take down the circuit, and return all equipment to its proper place.

QUESTIONS

Complete questions 1 to 12.

1. Draw a block diagram of the digital stopwatch constructed in this experiment. Include count accumulator, decoder, digital display, frequency divider, start/stop control, 60-Hz input, and waveshaping.

2. What IC(s) perform the following jobs in the digital stopwatch you constructed?

    a. Count accumulator (seconds)

    b. Count accumulator (tenths of a second)

    c. Decoder (seconds)

    d. Decoder (tenths of a second)

    e. Divide-by-6 counter

    f. Divide-by-10 counter

    g. Waveshaping circuit

    h. Start/stop control gate

3. Refer to Fig. 12-2. The count-up input to the 74192 IC sees _____ pulse(s) per second. [a number] The output of the 74192 IC at Q_D puts out _____ pulse(s) per second. [a number]

4. Refer to Fig. 12-2. A logical _____ (0, 1) at the start/stop control input causes the AND gate to allow the 60-Hz square-wave signal to pass through it to the divide-by-6 counter.

5. Refer to Fig. 12-2. The highest binary number the divide-by-6 counter gets to is _____. [a number]

6. The 7493 IC is wired as a mod-_____ counter in the stopwatch circuit.

7. What are the two jobs of the 74192 decade counter in the digital stopwatch circuit?

8. Which two ICs in the digital stopwatch you constructed act as storage units (or accumulators)?

9. Refer to Fig. 12-2. A jumper wire from the output of the Schmitt trigger inverter to the count-up input of the 74192 counter (pin 5) might serve what purpose (as it does on many digital clocks)?

2. *Place answers below.*

    a. _____

    b. _____

    c. _____

    d. _____

    e. _____

    f. _____

    g. _____

    h. _____

3. _____ _____

4. _____

5. _____

6. _____

7. _____

    _____

8. _____

9. _____

10. Refer to Fig. l2-3. This is a _____ (divide-by-6, divide-by-10) counter whose output is _____ Hz when the input control switch is HIGH.

11. The counter circuit drawn in Fig. 12-3 can replace the divide-by-6 counter used in the digital stopwatch circuit (Fig. 12-2). (T or F)

12. Refer to Fig. 12-3. What is the sequence of binary counts from this 74l92 IC starting at binary 001?

10. _____
_____

11. _____

12. _____
_____

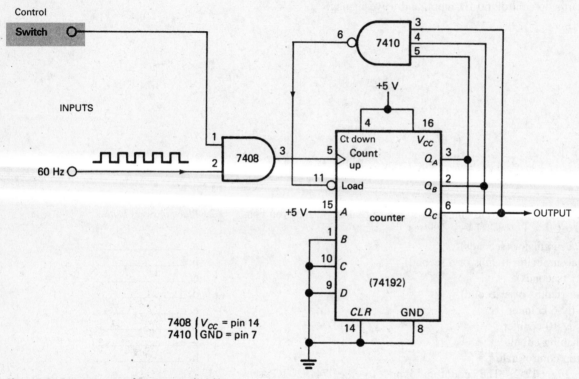

Fig. 12-3 Counter as a frequency divider.

12-3 LAB EXPERIMENT: MULTIPLEXED DISPLAYS

OBJECTIVES

1. To wire and test a system using multiplexed seven-segment LED displays.
2. To demonstrate the multiplexing action by slowing down the higher-frequency multiplex clock.

MATERIALS

Qty.		Qty.	
1	555 timer IC	2	seven-segment LED displays (commercial display board is recommended)
1	7404 inverter IC		
1	7447 BCD-to-seven-segment decoder/driver IC		
1	74157 quad 2-line to 1-line multiplexer IC	1	47-Ω, ¼-W resistor
		7	150-Ω, ¼-W resistors
2	74192 counter ICs	1	15-kΩ, ¼-W resistor
1	clock (free-running)	1	100-kΩ, ¼-W resistor
		1	0.1-μF capacitor
		1	10-μF electrolytic capacitor
		1	5-V dc regulated power supply

SYSTEM DIAGRAM

The circuit in Fig. 12-4 is a 0-to-99 counter. The 1s count is generated by the top 74192 decade counter and the 10s count by the bottom 74192 counter. The 74157 multiplexer *alternately selects* the count from the 1s and 10s counters and passes it on to the decoder/driver. The 7447 decoder drives the segments of both displays alternately. The digit drive pulses are timed to turn on either the 1s or 10s decimal display. The 1s 74192 counter is driven by a low-frequency clock (about 1 Hz). The multiplex clock is set at a higher frequency (about 50 to 75 Hz).

The output of the 555 timer IC is routed to both the *select input* of the 74157 multiplexer and the displays. If a LOW (0) is present at the output of the 555, the 74157 multiplexer passes the BCD from the 1s counter to the decoder/driver. When the digit drive line is LOW (0), the 10s display is disabled, while the 1s seven-segment decimal LED display is activated by the 7404 inverter. When the digit drive line is LOW (0), the 1s decimal count is lit. When the output of the 555 timer goes HIGH (1), the opposite happens; the 74157 multiplexer passes the BCD from the 10s counter on to the decoder/driver. The HIGH (1) on the digit driver line activates the 10s display. The 1s display is disabled by the LOW generated by the 7404 inverter. The 10s decimal count display is lit when the digit drive line is HIGH (1).

PROCEDURE

1. Insert the six ICs and two seven-segment LED displays into the mounting boards. Arrange the displays with the 10s display on the left and 1s display on the right for easy reading. A commercial display board is recommended, such as the DB1000A by Dynalogic Concepts.
2. Power OFF. Connect power (V_{CC} and GND) to the six ICs.
3. Wire the multiplexed-display counter system shown in Fig. 12-4. Use the free-running clock to generate the slow pulses into the 1s counter.

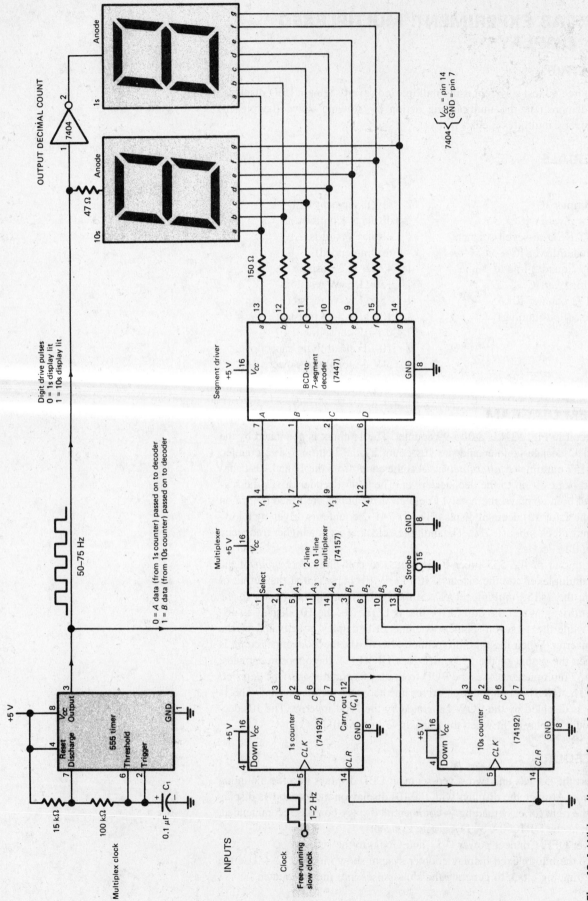

Fig. 12-4 Wiring diagram for a 0-to-99 counter circuit using LED display multiplexing.

320

4. Power ON. The counter should be counting upward within the range of 0 to 99. Can you observe any flickering of the output decimal displays?

_____

5. Power OFF. Remove the 0.1-μF capacitor C_1, and replace it with a 10-μF electrolytic capacitor.

6. Power ON. Observe the slow multiplexing action on the seven-segment displays.

7. Have your instructor approve your operating counter system with "slow" multiplexed displays. Be able to explain the function or job of each IC in the circuit.

8. Power OFF. Leave the counter circuit (Fig. 12-4) wired for use in the next lab activity.

QUESTIONS

Complete questions 1 to 8.

1. Refer to Fig. 12-4. The multiplex clock uses the 555 _____ IC to generate a _____ (higher, lower) frequency than the counter clock.

2. Refer to Fig. 12-4. If the frequency of the multiplex clock is too low, the displays tend to _____ (flicker, heat up).

3. Refer to Fig. 12-4. When the output of the multiplex clock is at logical 0, BCD data from the _____ (1s, 10s) counter is passed to the decoder and the _____ (1s, 10s) decimal display is lit.

4. Refer to Fig. 12-4. When the output of the multiplex clock is at logical 1, BCD data from the _____ (1s, 10s) counter is passed to the decoder and the _____ (1s, 10s) decimal display is lit.

5. Refer to Fig. 12-4. When the digit drive pulse is at a logical 1, the 1s seven-segment display is _____ (lit, not lit).

6. Refer to Fig. 12-4. The _____ [number] IC is used as the segment driver in this circuit, while the _____ [number] IC serves as the digit driver.

7. Refer to Fig. 12-4. The 74157 multiplexer IC operates something like a(n) _____ (SPST, 4PDT) mechanical switch.

8. Refer to Fig. 12-4. The two 74192 ICs are said to be _____ (cascaded, decoupled) to form a 0-to-99 counter.

1. _____ _____

2. _____

3. _____ _____

4. _____ _____

5. _____

6. _____ _____

7. _____

8. _____

12-4 TROUBLESHOOTING PROBLEM: COUNTER WITH MULTIPLEXED DISPLAYS

OBJECTIVES

1. To wire and operate a normal 0-to-99 counter with multiplexed LED displays.
2. To introduce faults in counter circuit and troubleshoot the circuit.
3. To report your observations and troubleshooting procedure to your instructor.

MATERIALS

Same parts used in Lab Experiment 12-3. See Fig. 12-5.
Test instruments such as logic probe, DMM, and oscilloscope.

SYSTEM DIAGRAM

You will operate the 0-to-99 counter circuit from the Lab Experiment 12-3 and observe its normal operation. The wiring diagram for this counter circuit has been reproduced in this lab as Fig. 12-5.

In troubleshooting knowing the *normal operation* of the circuit and your *powers of observation* is extremely important. After a fault is introduced into the counter circuit, you begin the troubleshooting procedure. This might include:

1. Use your senses. *Listen* to descriptions of problems from others. *Look* for problems. *Feel* the top of ICs for excessive heating. *Smell* for odd odors.
2. Check documentation/manuals and with co-workers for troubleshooting ideas.
3. Check for power. Check the main power supply using a DMM and/or battery checker. Check the power to individual ICs with a logic probe or DMM.
4. Check for signals. Check signal sources. This includes the 555 timer clock and slow clock inputs in the counter circuit drawn in Fig. 12-5. A logic probe or oscilloscope may be used for this step.
5. Replace suspect plug-in parts (such as boards or ICs if in sockets).
6. Verify proper operation after correcting a fault in the circuit.

In real-world troubleshooting you might work in one of two different types of jobs. In engineering/development, in education, or in research, you might troubleshoot down to the IC and component level. However, in a repair job dealing with consumer, business, military, or industrial equipment, you may only replace circuit boards, plug-in cards, or modules.

Many modern electronic devices are programmable. This means you also need good computer skills.

PROCEDURE

1. Power OFF. Use the 0-to-99 counter circuit with multiplexed LED displays from Lab Experiment 12-3.
2. Power ON. Operate the counter circuit (wiring diagram reproduced in Fig. 12-5). Observe its normal operation.
3. Power OFF. **Introduce fault 1** into the counter circuit. Ground output pin 14 of the 7447 decoder IC. See Fig. 12-5.
4. Power ON. Carefully observe the seven-segment LED outputs of the counter. Notice the effect of the "stuck LOW" output of the 7447 decoder IC. You will report to your instructor your observations and how you might track down this problem using your knowledge of circuit operation and test

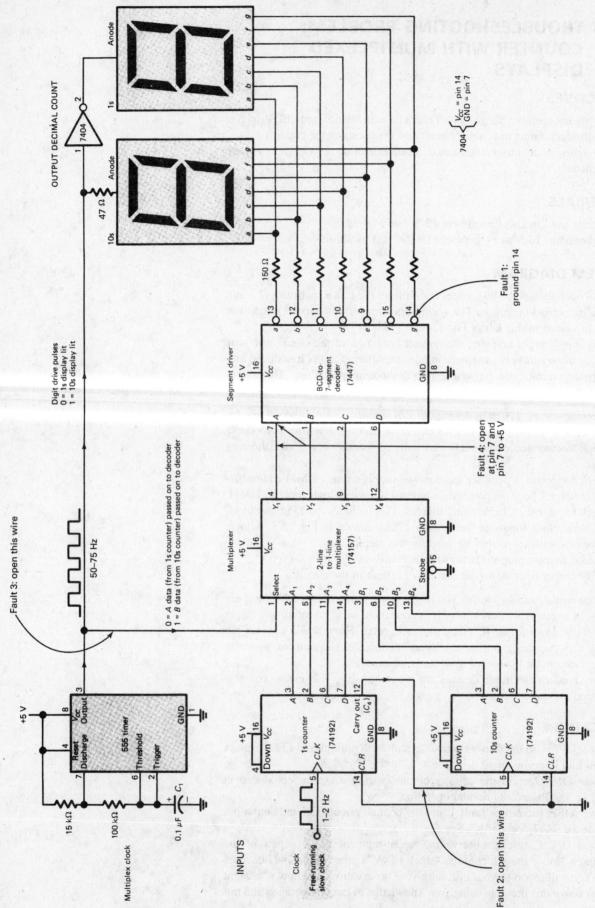

Fig. 12-5 Wiring diagram for a 0-to-99 counter circuit. Faults are shown for trouble shooting activity.

instruments, such as the logic probe or DMM. What action might you take to repair a problem like this in a real circuit (clear shorts, replace the 7447 IC)?

5. Power OFF. *Report you results* to your instructor, and be prepared to answer questions about your observations and actions.

6. Power OFF. Clear fault 1.

7. Power ON. Operate the counter circuit. Observe its normal operation.

8. Power OFF. **Introduce fault 2** into the counter circuit. Open the wire between output pin 12 of the ls counter and CLK input pin 5 of the 10s counter. See Fig. 12-5.

9. Power ON. Carefully observe the seven-segment LED outputs of the counter. Notice the effect of the "open" circuit. You will report to your instructor your observations and how you might track down this problem using your knowledge of circuit operation and test instruments, such as the logic probe or DMM. What action might you take to repair a problem like this in a real circuit (repair circuit board, replace a 74l92 IC)?

10. Power OFF. *Report you results* to your instructor, and be prepared to answer questions about your observations and actions.

11. Power OFF. Clear Fault 2.

12. Power ON. Operate the counter circuit. Observe its normal operation.

13. Power OFF. **Introduce fault 3** into the counter circuit. Open the wire leading to pin 1 of the 74157 IC from the clock. Do not open wire from the clock to the seven-segment LEDs and 7404 IC. See Fig. 12-5.

14. Power ON. Carefully observe the seven-segment LED outputs of the counter. Notice the effect of the "open" circuit. You will report to your instructor your observations and how you might track down this problem using your knowledge of circuit operation and test instruments, such as the logic probe, DMM, or oscilloscope. What action might you take to repair a problem like this in a real circuit (repair circuit board, replace ICs such as 74157 or 555)?

15. Power OFF. *Report you results* to your instructor, and be prepared to answer questions about your observations and actions.

16. Power OFF. Clear Fault 3.

17. Power ON. Operate the counter circuit. Observe its normal operation.

18. Power OFF. **Introduce fault 4** into the counter circuit. Open the wire leading to pin 7 of the 7447 decoder IC and tie input pin 7 of 7447 IC to +5 V. This will simulate data input *A* of the 7447 decoder being "stuck HIGH." See Fig. 12-5.

19. Power ON. Carefully observe the seven-segment LED outputs of the counter. Notice the effect of the "stuck HIGH" input to the 7447 IC on circuit operation. You will report to your instructor your observations and how you might track down this problem using your knowledge of circuit operation and test instruments, such as the logic probe and DMM. What action might you take to repair a problem like this in a real circuit (repair the circuit board, replace ICs such as 7447)?

20. Power OFF. *Report you results* to your instructor, and be prepared to answer questions about your observations and actions.

21. Power OFF. Clear fault 4.

22. Power OFF. Take down the circuit, and return all equipment to its proper place.

12-5 LAB EXPERIMENT: LCD TIMER WITH ALARM

OBJECTIVES

1. To wire and test a CMOS timer circuit with an LCD display and alarm.
2. To calibrate the time-base circuit in the digital timer circuit.

MATERIALS

Qty.		Qty.	
2	555 timer ICs	1	1-kΩ, ¼-W resistor
2	74HC85 4-bit magnitude comparator CMOS ICs	1	2.2-kΩ, ¼-W resistor
2	74HC192 decade up/down counter CMOS ICs	1	74HC393 dual 4-bit binary counter CMOS IC
1	2N3904 general-purpose NPN transistor	2	74HC4543 BCD-to-seven-segment latch/decoder/driver CMOS ICs
1	1N4001 silicon diode	3	10-kΩ, ¼-W resistors
9	logic switches	1	100-kΩ, ¼-W resistor
1	two-digit seven-segment LCD (commercial display board is recommended)	1	0.033-μF capacitor
		1	1-μF electrolytic capacitor
		1	5-V dc regulated power supply
1	piezo (or electronic) buzzer	1	frequency counter (*optional*)

SYSTEM DIAGRAMS

A block diagram of the digital timer with alarm is sketched in Fig. 12-6. To operate the timer, first set the *load/start control* to 0 (load mode). Second, set the two groups of four switches each to the BCD equivalent of the decimal numbers you want to appear on the LCD display. Third, move the load/start control to a logical 1 (start mode). The timer will start counting downward by seconds. The LCD display shows the number of seconds (99 down to 00) that remain before the alarm sounds. When the timer's count on the LCD display decreases to 00, an alarm will sound. Finally, temporarily turning off the 5-V dc power supply will stop the alarm from sounding.

The time-base clock shown in Fig. 12-6 should generate a 256-Hz square-wave signal which is fed into a divide-by-256 counter. This block reduces the frequency to 1 Hz. The 1-Hz signal enters the *count down* clock input of the 1s down counter. Each decrement of the 1s down counter is latched and decoded and displayed as the right digit on the LCD. The *borrow line* between the 1s and the 10s counters is activated when the 10s down counter must be decremented. An example would be when the timer counts from 40 to 39. Each decrement of the 10s down counter is latched, decoded, and displayed as the left digit on the LCD.

The *load/start* input switch is connected to the asynchronous parallel-load inputs of the 74HC192 down counter ICs. When the load input is activated by a LOW, parallel BCD data are immediately loaded into both the 1s and 10s counters.

The alarm shown in Fig. 12-6 will be activated when the BCD count from the 1s and 10s counters equals 0000 0000$_{BCD}$. An 8-bit magnitude comparator checks the BCD outputs of both counters. When the count reaches 0000 0000, the comparator does two things. First, it activates a buzzer which

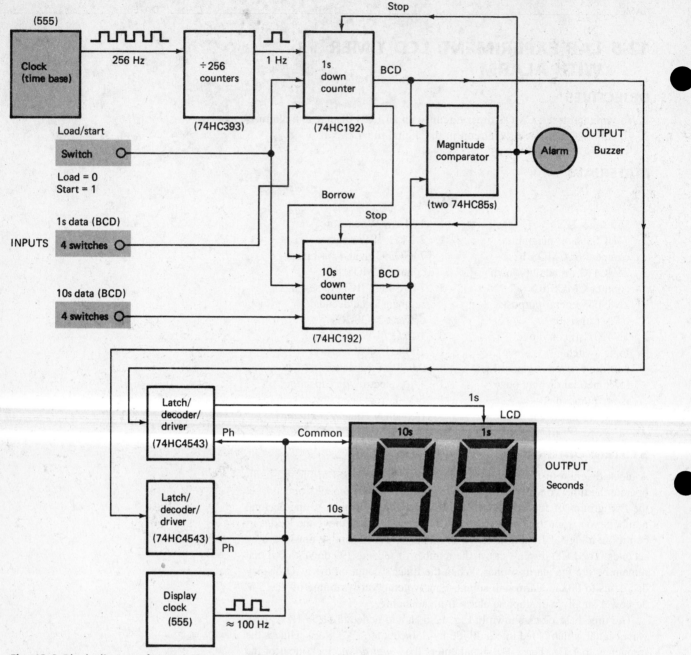

Fig. 12-6 Block diagram for an LCD timer with alarm.

serves as the timer alarm in this circuit. Second, the comparator sends a HIGH *stop signal* back to the clear inputs of both counters so they will stop their down counting at 0000. The liquid-crystal displays will read 00 when the alarm sounds.

The clock shown at the lower left in Fig. 12-6 generates a square-wave signal of about 100 Hz. The 100-Hz signal is used for driving the liquid-crystal displays. It is fed to the common (backplane) of the LCD and the *Ph* inputs of the two driver chips (74HC4543).

A detailed wiring diagram of the LCD timer circuit is drawn in Fig. 12-7. The ICs are positioned as in the block diagram. This is a *difficult wiring problem* and should be *reserved for the more advanced students.*

PROCEDURE

1. Insert the nine ICs in the mounting boards. A commercial display board is recommended for the liquid-crystal display (LCD).

 **CAUTION** CMOS ICs can be damaged by static electricity.

2. Power OFF. Connect power (V_{CC} and GND) to the nine ICs. It is suggested that you color-code the wires (red = V_{CC}, black = GND).
3. Power OFF. Wire the LCD timer circuit detailed in Fig. 12-7. Use nine logic switches for the BCD data and load/start control at the left. On complex circuits, color coding of wires is highly recommended.
4. Power OFF. Resistor R_1 should have an initial value of about 20 kΩ (two 10-kΩ resistors in series). This can be changed later when calibrating the *time-base* clock for maximum accuracy.
5. Power ON. Move the load/start switch to 0 (load mode). Manipulate the load 1s counter and load 10s counter data switches for some number between 99 and 01.
6. Power ON. Move the load/start switch to 1 (count down mode). The LCD should count downward in seconds until it reaches 00. The piezo buzzer should then sound.
7. Power OFF. This turns off the alarm (buzzer).
8. Demonstrate the operation of your LCD timer to your instructor. Be prepared to answer questions about the function of each IC in the LCD timer circuit.
9. *OPTIONAL:* Calibrate the time-base clock for greatest accuracy.
10. Power ON. Set the LCD for 99 seconds and start the timer, checking its accuracy against a clock or wristwatch.
11. Power OFF. Experiment with changing the values of resistor R_1 connected to the 555 timer IC to get the greatest accuracy.
12. Power ON. Check the frequency of the newly calibrated time-base clock using a frequency counter. When the calibration is accurate, test your new LCD timer design.
13. Show your instructor your calibrated LCD timer.
14. Power OFF. Take down the circuit and return all equipment to its proper place. The pins of CMOS ICs should be stored in conductive foam.

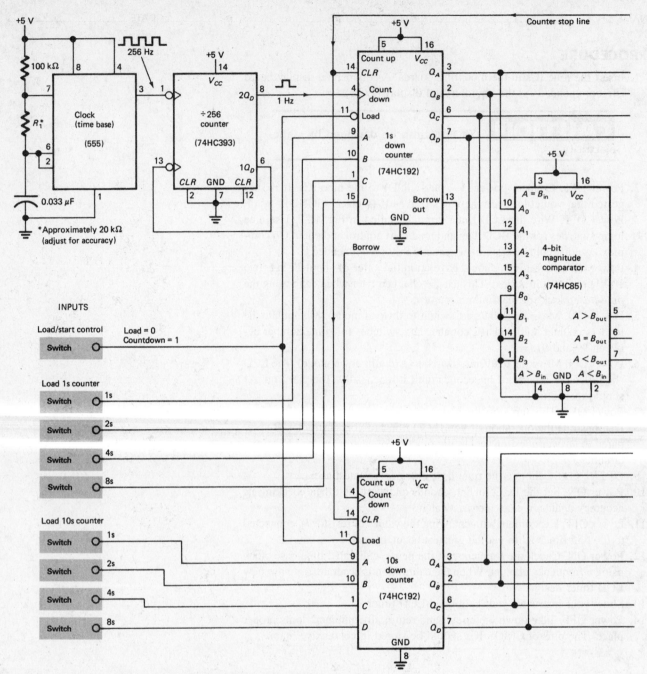

Fig. 12-7 Wiring diagram for the LCD timer with alarm system.

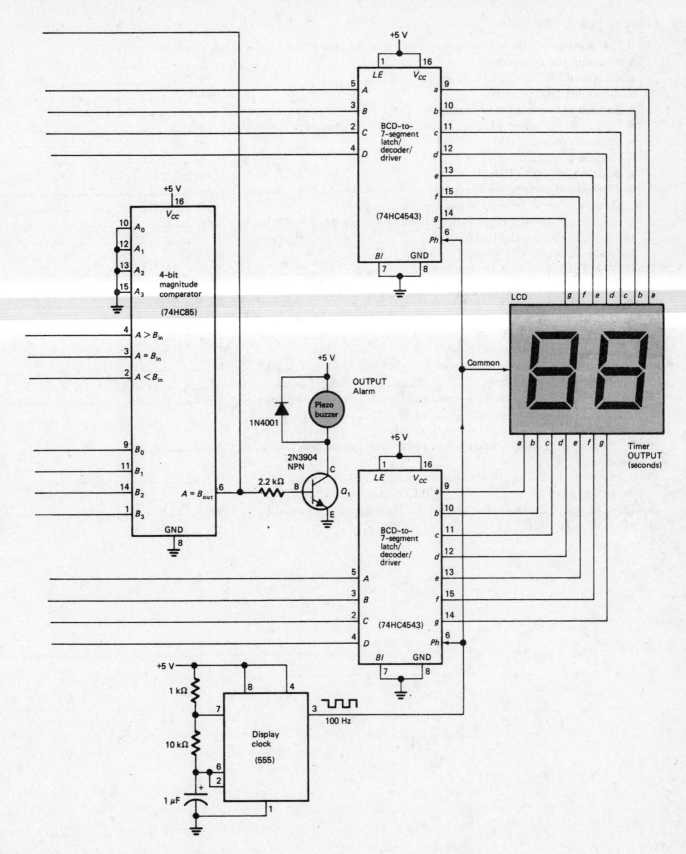

QUESTIONS

Complete questions 1 to 10.

1. Which IC(s) or transistor(s) performed the following jobs in the LCD timer circuit you constructed?
   a. Generated 256-Hz square-wave signal
   b. Generated 100-Hz signal
   c. Latched and decoded the 1s count
   d. Turned on the buzzer when the output of the comparator went HIGH
   e. Counted downward (1s)
   f. Counted downward (10s)
   g. Reduced 256 Hz to 1 Hz
   h. Latched and decoded the 10s count
   i. Checked to see if the timer output was down to 00
   j. Sent the counter stop signal back to 1s and 10s counters

2. The 74HC393 IC actually contains _____ units wired as two divide-by-16 counters. [a number]

3. The 74HC192 IC's *load* input is activated with a _____ (HIGH, LOW).

4. The two 74HC192 ICs are wired as mod-_____ [a number] _____ (down, up) counters.

5. The two 74HC85 ICs are wired as a(n) _____-bit _____ comparator.

6. Refer to Fig. 12-7. When the output of the right 74HC85 ($A = B_{out}$) goes HIGH, the LCD output reads _____ in decimal.

7. Refer to Fig. 12-7. When the output of the right 74HC85 comparator ($A = B_{out}$) goes HIGH, transistor Q_1 is turned _____ (off, on) and the alarm _____ (is off, sounds).

8. Refer to Fig. 12-7. After sounding, how is the alarm buzzer turned off in this circuit?

9. Refer to Fig. 12-7. The 100-Hz display clock provides the square wave required to drive the _____ (LCD, transistor).

10. The accuracy of the LCD timer depends on the frequency generated by which IC in Fig. 12-7?

---

1. *Place answers below.*
   a. _____
   b. _____
   c. _____
   d. _____
   e. _____
   f. _____
   g. _____
   h. _____
   i. _____
   j. _____

2. _____

3. _____

4. _____

5. _____

6. _____

7. _____ _____

8. _____

9. _____

10. _____

CHAPTER 13

Computer Systems

TEST: COMPUTER SYSTEMS

Answer the questions in the spaces provided.

1. The organization of subsystems in a computer would most commonly be called the _____ of the computer.
   a. Architecture   c. Dimensions
   b. Blueprint   d. Plan

    1. _____

2. Devices, such as tape and disk drives, located outside the computer's CPU are often called _____ devices.
   a. Neighboring   c. Peripheral
   b. Outside   d. Surrounding

    2. _____

3. An IC called a(n) _____ forms the CPU of a microcomputer.
   a. ALU   c. Full adder
   b. Digital clock   d. Microprocessor

    3. _____

4. The parts of a microcomputer system are connected by control lines, an address bus, and a two-way _____ bus.
   a. Accumulator   c. Information
   b. Data   d. Program

    4. _____

5. Program instructions that tell the MPU what to do can reside in either RAM or _____ in a microcomputer. [three letters]

    5. _____

6. Typically microcomputer instructions are divided into two parts: the action part is called the _____ and the second part is called the *operand.*
   a. Directive   c. Function
   b. Effector   d. Operation

    6. _____

7. The microcomputer's MPU follows a(n) _____ sequence when running a program.
   a. Encode-decode-execute   c. Find-do-transfer
   b. Fetch-decode-execute   d. Find-transfer-do

    7. _____

8. A typical small microcomputer system would be composed of a power supply, MPU, some ROM and RAM, I/O ports, address bus, control lines, data bus, and
   a. Address decoder   c. Control latch
   b. Bus gater   d. Data encoder

    8. _____

9. All devices connected to a microcomputer's data bus must have _____ between them and the shared bus.
   a. High-impedance RTL   c. Three-state buffers
   b. Low-impedance latches   d. Three-state gates

    9. _____

10. A microcomputer's address decoder helps select one of several
    a. Instructions to be executed next
    b. Paths for data to follow
    c. Cloud devices to be placed online
    d. RAMs, ROMs, or I/O devices to connect to the data bus

11. A digital signal processor (DSP) is a specialized microprocessor IC designed for high-speed data manipulation. (T or F)

12. The main use for DSP devices is to replace existing audio amplifiers in low-cost consumer products. (T or F)

13. In a digital signal processor (or most microprocessors), starting a task before the previous one is finished is referred to as _____.
    a. Aliasing   c. Product-of-sums referencing
    b. Pipelining

14. A common type of calculation performed at various speeds by a digital signal processor is known as _____ (repeated differential subtraction, sum of products).

15. The transfer of eight or more binary digits at one time over a data link is called _____ data transmission.
    a. Parallel   c. Register
    b. Parity   d. Serial

16. Peripheral interface adapters (such as the Motorola 6820 PIA or the Intel 8255 PPI) are used for interfacing a microcomputer's _____ with a peripheral device such as a printer.
    a. CPU   c. Keyboard
    b. Graphics tablet   d. Mouse

17. The speed at which serial data is transmitted over a data link is called the _____ rate.
    a. Address   c. Cycle
    b. Baud   d. Pulse

18. Parity bit generators and detectors make extensive use of _____ logic gates.
    a. AND   c. OR
    b. NAND   d. XOR

19. Errors in transmission can be detected by sending an extra _____ bit with the data over the transmission lines.
    a. Correcting   c. Parity
    b. Error   d. Redundant

20. In transmission devices such as modems, cyclic redundancy checks (CRC) create a _____ on data that allows them to detect almost all errors in data transmission.
    a. Gray code
    b. Smart code
    c. Checksum
    d. Parity bit

21. A piece of equipment that can transmit and receive data at the same time is referred to as a _____ (full-, half-) duplex device.

22. A _____ (flag, ping) is a single bit in a register used to signal a status or error condition.

23. A _____ (BEL, PLC) is a specialized computer used to replace relays in industrial process control systems.

24. A _____ (personal computer system, programmable logic controller) is a heavy-duty computer system used for machine control in factories, warehouses, and chemical plants.

25. A Boolean expression that describes the relay schematic in Fig. 13-1 is _____.

26. A _____ (microcontroller, PLC) may be described as a "computer on a chip" because it contains a CPU, RAM, read-only memory, a clock, and I/O pins within a single IC.

10. _____
11. _____
12. _____
13. _____
14. _____
15. _____
16. _____
17. _____
18. _____
19. _____
20. _____
21. _____
22. _____
23. _____
24. _____
25. _____
26. _____

334

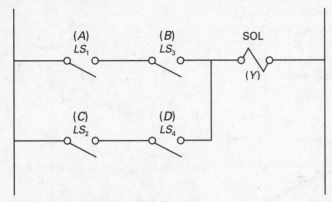

Fig. 13-1 Relay schematic diagram.

27. The microcontroller (computer on a chip) is small in size but very expensive. (T or F)

27. _____

28. Microcontrollers can address _____ (very large, very small) amounts of RAM as compared to microprocessors.

28. _____

29. Microcontrollers are commonly used as the CPU section of a personal computer system. (T or F)

29. _____

30. Microcontrollers are manufacturered in huge quantities at very low cost. (T or F)

30. _____

31. A BASIC Stamp 2 module contains a _____ (microcontroller, microprocessor) along with other associated components mounted on a small PC board using a 24-pin DIP package.

31. _____

32. The BS2-IC uses the _____ (PIC16C57, TI7404) microcontroller, its EEPROM will hold about 500 instructions, and it has _____ (8, 16) general-purpose I/O ports.
    a. PIC16C57, 8   c. TI7404, 8
    b. PIC16C57, 16   d. TI7407, 16

32. _____

33. The V_{dd} pin of a BASIC Stamp module is connected to _____ (+5 V, ground), while a pin labeled P_3 will be a(n) _____ (I/O port, reset input).
    a. +5 V, I/O port   c. ground, I/O port
    b. +5 V, reset input   d. ground, reset input

33. _____

34. Once programmed, microcontrollers, like the BASIC Stamp 2 modules, are _____ (single-purpose, general-purpose) devices that are commonly embedded in products.

34. _____

35. The PBASIC program statement **output 7** might be explained with the remark statement _____.
    a. 'Title of PBASIC program
    b. 'Configure I/O port P7 as an output
    c. 'Configure I/O port P0 as an input

35. _____

36. The PBASIC program statement **pause 1000** causes the microcontroller from a BASIC Stamp 2 IC to do nothing for about 1000 _____ (milliseconds, seconds).

36. _____

37. Refer to Fig. 13-2. The light-sensitive image sensor in a digital camera may be either a CMOS image sensor or a _____ (CCD, cascaded amplifier) type device.

37. _____

38. Refer to Fig. 13-2. The special colored filter located between the camera lens and the image sensor is commonly referred to as a _____ (Bayer, Kirchhoff) filter.

38. _____

39. Refer to Fig. 13-2. The job of the block labeled *A* in the digital camera is _____ (digital signal processing, multiplexing).

40. Refer to Fig. 13-2. The removable media block used for storing photographs produced by the digital camera would probably use a _____ (ROM, flash EEPROM) semiconductor memory device.

41. In a digital camera, colored images are stored as digital values for the three primary colors, which are _____ (orange, green, and violet; red, green, and blue).

39. _____

40. _____

41. _____

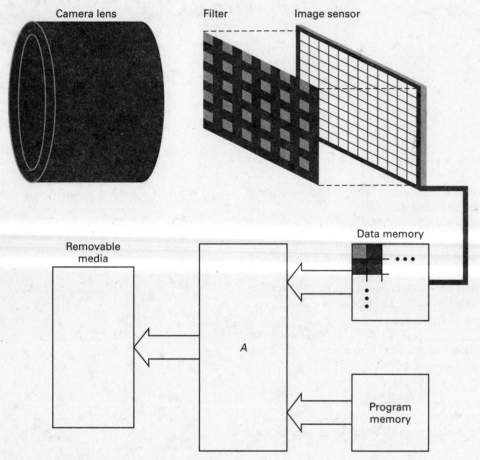

Fig. 13-2 Block diagram of a digital camera.

42. Hobby-type servo motors are commonly rotated using a _____ (10 to 30-Hz audio, pulse-width modulated) signal.

43. The CdS photocell is a sensor that changes resistance as more or less light strikes its surface. (T or F)

44. The PBASIC statement **ckphotocell:** with the colon (:) after the title is called a *label* that usually is the first statement in a main routine (or subroutine) and is a target point for other PBASIC statements (such as GOTO). (T or F)

45. The PBASIC statement **PULSOUT 12,500** causes the BS2 microcontroller to generate a pulse of 1 ms at _____ (input port 500, output port 12).

42. _____

43. _____

44. _____

45. _____

13-1 LAB EXPERIMENT: MICROCOMPUTER MEMORY ADDRESS DECODING

OBJECTIVES

1. To wire and test a memory address decoding circuit similar to those used in microcomputer systems.
2. To use three-state buffers to send information over a common data bus.

MATERIALS

Qty.   **Qty.**

1   7404 inverter IC   1   clock (single pulses)
2   7432 two-input OR gate ICs   4   LED indicator-light
1   7475 4-bit latch IC   assemblies
2   7489 RAM ICs   1   5-V dc regulated power
2   74125 three-state buffer   supply
    ICs   1   logic probe
10 logic switches

SYSTEM DIAGRAM

The circuit in Fig. 13-3 on the next page uses two 16- $\times$ 4-bit RAMs. The outputs of these RAMs are connected in parallel to the same 4-bit data bus as they might be in a microprocessor-based system. Three-state buffers will isolate the outputs of one of the RAMs from the data bus when the other is sending data. Like a very small microprocessor-based system, the circuit uses an 8-bit address bus. Address decoding is accomplished by the two groups of OR gates and inverter at the left in Fig. 13-3. The address decoding is handled as it would be in a microprocessor-based system. *Unlike* a microprocessor-based system, the loading of data in the RAMs will be done directly from logic switches. The data bus *will not* be a two-way bus, as it would be in a microprocessor-based system. The 4-bit latch at the right will be used to latch data off the data bus. The latch also inverts data back to their original form, loaded and stored in RAM. The LEDs at the lower right show the data stored in the RAM location pointed to by the 8-bit address.

PROCEDURE

1. Insert the eight ICs into the mounting board.
2. Power OFF. Connect power (V_{CC} and GND) to the eight ICs.
3. Wire the memory address decoding circuit shown in Fig. 13-3. Use five input switches for the *E, D, C, B,* and *A* address inputs. GND address inputs are *H, G,* and *F.* Use five input switches for the Load Data-RAM 0 inputs. *Temporarily* leave the Load Data–RAM 1 inputs unconnected. Use a single-pulse clock (positive pulse) for the enable input to the 7475 latch IC. Use four LED indicator-light assemblies.
4. Power ON. Move all address switches (*E* through *A*) to 0. With a logic probe, check the logic level at the memory enable pins $\overline{ME}$ on both RAMs. Pin 2 on RAM 0 should be LOW, while pin 2 on RAM 1 should be HIGH.
5. Load in RAM 0. *Note: Do not turn off the power until the experiment is complete, or you will lose the contents programmed into the RAMs.* The top half of Table 13-1 shows the data to be loaded into RAM 0.

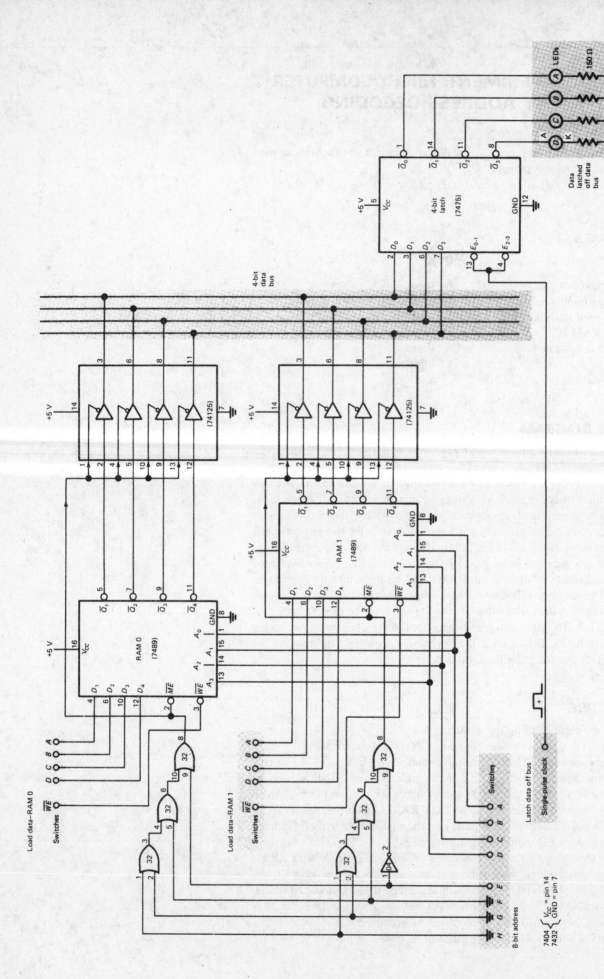

Fig. 13-3 Wiring diagram for a circuit with two 16- × 4-bit RAMs sharing a common data bus. Address decoding and three-state buffers are both used in this circuit.

6. Leave power ON. To program or load each location in the RAMs, follow this procedure.
   a. Write enable switch $\overline{WE}$ to 1 (read mode).
   b. Set five address switches to memory address.
   c. Set four data input switches to data listed under contents of RAM in Table 13-1.
   d. Move the write-enable switch $\overline{WE}$ to 0 (write mode) and then back to 1 (read mode).
   e. Verify the loaded data by pressing the clock push button. The correct data at this memory address should be latched off the data bus and should appear on the LEDs.

TABLE 13-1 Readouts From RAM Circuits

Memory address								Contents RAM 0				Output indicators			
H	G	F	E	D	C	B	A	D	C	B	A	D	C	B	A
0	0	0	0	0	0	0	0	0	0	0	0				
0	0	0	0	0	0	0	1	0	0	0	1				
0	0	0	0	0	0	1	0	0	0	1	0				
0	0	0	0	0	0	1	1	0	0	1	1				
0	0	0	0	0	1	0	0	0	1	0	0				
0	0	0	0	0	1	0	1	0	1	0	1				
0	0	0	0	0	1	1	0	0	1	1	0				
0	0	0	0	0	1	1	1	0	1	1	1				
0	0	0	0	1	0	0	0	1	0	0	0				
0	0	0	0	1	0	0	1	1	0	0	1				
0	0	0	0	1	0	1	0	1	0	1	0				
0	0	0	0	1	0	1	1	1	0	1	1				
0	0	0	0	1	1	0	0	1	1	0	0				
0	0	0	0	1	1	0	1	1	1	0	1				
0	0	0	0	1	1	1	0	1	1	1	0				
0	0	0	0	1	1	1	1	1	1	1	1				
								Contents RAM 1							
0	0	0	1	0	0	0	0	0	0	0	1				
0	0	0	1	0	0	0	1	0	0	1	0				
0	0	0	1	0	0	1	0	0	0	1	1				
0	0	0	1	0	0	1	1	0	1	0	0				
0	0	0	1	0	1	0	0	0	1	0	1				
0	0	0	1	0	1	0	1	0	1	1	0				
0	0	0	1	0	1	1	0	0	1	1	1				
0	0	0	1	0	1	1	1	1	0	0	0				
0	0	0	1	1	0	0	0	1	0	0	1				
0	0	0	1	1	0	0	1	1	0	1	0				
0	0	0	1	1	0	1	0	1	0	1	1				
0	0	0	1	1	0	1	1	1	1	0	0				
0	0	0	1	1	1	0	0	1	1	0	1				
0	0	0	1	1	1	0	1	1	1	1	0				
0	0	0	1	1	1	1	0	1	1	1	1				
0	0	0	1	1	1	1	1	0	0	0	0				

7. Leave power ON. Move the wires from the load data switches ($\overline{WE}$, *D*, *C*, *B*, and *A*) to the inputs of RAM 1.

8. Leave power ON. Program (load) RAM 1 with the data shown in Table 13-1. Use the procedure in step 6.

9. Leave power ON. Check the contents of memory locations 0000 0000 through 0001 1111, and record them in the right column, Table 13-1. Remember to press the clock push button at each step to latch the most up-to-date data off the data bus.

10. Leave power ON. Set the address switches at memory address 0001 0000. With a logic probe, check pin 2 $\overline{ME}$ of both RAMS. Which RAM is enabled?

_____

11. Leave power ON. Have your instructor check your memory address decoding circuit.

12. Power OFF.

13. Power ON. Check the contents of several memory locations in both RAMs. Are the contents the same as in Table 13-1?

_____

14. Power OFF. Take down the circuit, and return all equipment to its proper place.

QUESTIONS

Complete questions 1 to 10.

1. The 7489 RAMs are _____ (nonvolatile, volatile) memory devices.

2. The 7432 and 7404 ICs are used as _____ _____ in the circuit in Fig. 13-3.

3. The 74125 buffer ICs have _____-_____ outputs.

4. Refer to Fig. 13-3. With the address 0000 0000, RAM _____ is enabled as is the _____ (bottom, top) buffer.

5. Refer to Fig. 13-3. Inverted data leave the RAM outputs; this is complemented to its correct form by the _____ IC.

6. Refer to Fig. 13-3. With an address of 0000 0000, pin 2 of RAM _____ goes LOW and enables that memory. Pin 2 of RAM _____ goes HIGH and disables that memory and its output buffer.

7. Refer to Fig. 13-3. With an address of 0001 1111, RAM _____ is enabled. The outputs of the _____ (bottom, top) 74125 buffer are in their high-impedance state.

8. Refer to Fig. 13-3. Memory could be expanded by adding another RAM, a buffer, and _____ _____ circuitry.

9. The data bus in Fig. 13-3 is a(n) _____-way unit, while on a real microcomputer it would be a(n) _____-way path for data.

10. The eight address lines in Fig. 13-3 would be connected to the _____ bus of a real microprocessor-based system.

1. _____

2. _____ _____

3. _____ _____

4. _____ _____

5. _____

6. _____ _____

7. _____ _____

8. _____ _____

9. _____ _____

10. _____

13-2 LAB EXPERIMENT: DATA TRANSMISSION

OBJECTIVE

To wire and test a serial data transmission system using a 74150 multiplexer (MUX) IC and a 74154 demultiplexer (DEMUX) IC.

MATERIALS

Qty.   **Qty.**

3 7404 hex inverter ICs   1 7447 BCD-to-seven-
1 7493 binary counter IC   segment decoder IC
1 74154 demultiplexer/   1 74150 multiplexer/data
   distributor IC   selector IC
1 clock (single pulses)   1 logic switch
1 seven-segment LED display   16 LED indicator-light assemblies
1 5-V dc regulated power supply   7 150-Ω, ¼-W resistors

SYSTEM DIAGRAM

Figure 13-4 is the wiring diagram of the simplified data transmission system you will study in this activity.

PROCEDURE

1. Insert the seven ICs and the single seven-segment LED display into the mounting board. A commercial display board is recommended.
2. Power OFF. Connect power (V_{CC} and GND) to the seven ICs. Connect +5 V (V_{CC}) only to the seven-segment LED display.
3. Wire the data transmission system shown in Fig. 13-4. Use an input switch for the reset input. Let the parallel inputs to the 74150 IC float HIGH. Connect a single-pulse clock to the *CLK* input of the 7493 counter. Finish the wiring of the entire circuit.
4. Power ON. *Clear* the 7493 counter by enabling the reset switch (1) and then disabling (0). The decimal readout should now read 0.
5. Repeatedly pulse the clock input and observe the results. Remember you are applying logical 1s to all the inputs of the 74150 IC by leaving them float.
6. Have your instructor approve your completed circuit.
7. Power OFF. Change the inputs of the 74150 IC so that inputs 1, 3, 5, 7, 9, 11, 13, and 15 are at a logical 0 (tie to ground).
8. Power ON. Clear the counter. Operate the data transmission system by repeatedly pulsing the clock input.
9. Have your instructor approve your completed circuit. Be able to explain what is happening in the circuit.
10. Power OFF. Take down the circuit and return all equipment to its proper place.

QUESTIONS

Complete questions 1 to 8.

1. A multiplexer is also called a(n) _____.   1. _____
2. A demultiplexer is also called a(n) _____.   2. _____
3. Refer to Fig. 13-4. If the data select inputs of the 74150 IC were at 0110, what input will be connected to output *W?*   3. _____
4. For the 74150 IC to operate, the enable (strobe) input must be at a logical _____.   4. _____

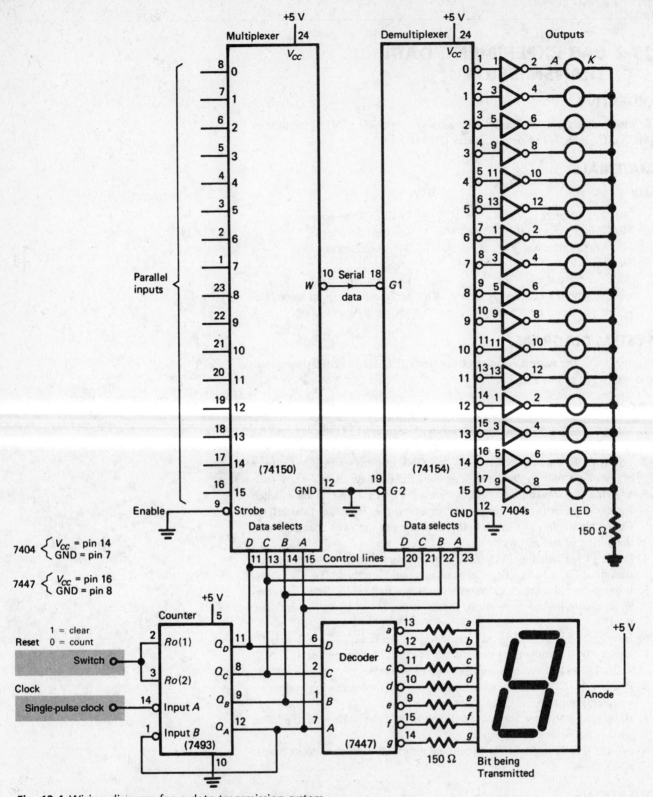

Fig. 13-4 Wiring diagram for a data transmission system.

5. Refer to Fig. 13-4. Inputs G_1 and G_2 of the 74154 DEMUX must be at a logical _____ to activate the IC.

5. _____

6. The inputs of the 74150 IC appear to float at a logical _____ when not connected.

6. _____

7. The outputs of the 74154 IC appear to float at a logical _____.

7. _____

8. Refer to Fig. 13-4. What is the meaning of the small bubbles on the 74150 output and the 74154 inputs and outputs?

8. _____
_____

342

13-3 BASIC STAMP EXPERIMENT: SIMPLE INPUT AND OUTPUT

OBJECTIVES

1. To use a microcontroller (BASIC Stamp 2 module) as a control device.
2. To use a BASIC Stamp 2–based development board (like Board of Education by Parallax) and personal computer to program in PBASIC, download the program, and wire and test several simple circuits.

MATERIALS

Qty.		**Qty.**	
1	BASIC Stamp 2 development board (such as Board of Education or HomeWork board, both available from Parallax)	1	Parallax manual *What Is a Microcontroller?*
		2	light-emitting diodes
1	BASIC Stamp 2 module (this IC may come mounted on development board)	1	pushbutton switch, N.O. contacts
		1	piezo buzzer
1	PC with MS Windows	2	150-Ω resistors
1	PBASIC editor from Parallax	1	10-kΩ resistor
1	serial cable (for downloading to BASIC Stamp 2 module) (see instructor for using a USB port)		

Note: BASIC Stamp 2 module, development board, PBASIC software, and both experiment and reference manuals are available from Parallax, Inc., 599 Menlo Drive, Rocklin, CA 95765. Most of the manuals and software can be downloaded free from Parallax's websites. General website: *www.parallaxinc.com*. Telephone: 916-624-8333.

SYSTEM DIAGRAMS

The first circuit to be wired to the BASIC Stamp 2 module (BS2 IC) is shown in Fig. 13-5. Two LEDs with limiting resistors connected between both port 1 (P_1) and port 0 (P_0) I/O pins of the BASIC Stamp 2 module and V_{dd} (+5 V) of the power supply. The BS2 IC will use ports 0 and 1 as outputs to light the LEDs. Notice that a logical 0 or LOW at either P_0 or P_1 is needed to light that respective LED. PBASIC (Parallax BASIC) programming is accomplished on a MS Windows–based personal computer using the PBASIC editor STAMPW.EXE. The program is downloaded using a serial cable between the PC's serial port and the BASIC Stamp 2 development board. The BS2 module stores the program in EEPROM, and it is interpreted and executed by the microcontroller. A USB port can also be used with instructor approval.

The following PBASIC program will cause the BS2 module to flash the LEDs alternately. Statements starting with an apostrophe ('), such as '**Two LEDs flashing,** are titles or remarks that make the program more understandable for humans. They are not executed by the microcontroller. Statements that end with a colon (:), such as **flash:**, are called *labels* and are reference points in the program used by statements such as **goto** to locate the beginning of the flash routine. The early statements (**output 0, output 1, out1 = 1**) after the title configure and initialize ports. The alternate flashing routine (**flash:**) turns first LED 0 on and off, then secondly turns LED 1 on and off, and then repeats. The **flash:** routine will continue until the power to the BS2 IC is turned off. When power to the BS2 IC is turned off and then back on, the program will start from the beginning. The remark statements (they begin with apostrophes) explain the detailed operation of the program statements.

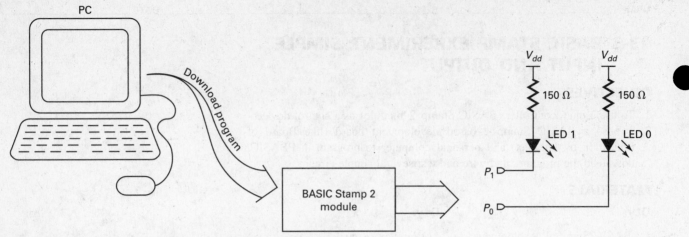

Fig. 13-5 BASIC Stamp 2 module programmed to light LEDs.

```
'Two LEDs flashing        'Title of PBASIC program (see Fig. 13-5)

output 0                  'Configure I/O port P0 as an output
output 1                  'Configure I/O port P1 as an output
out1=1                    'Initialize--Port 1 goes HIGH, LED 1 is off

flash:                    'Label for alternate flashing routine
  out0=0                  'Port 0 goes LOW lighting LED 0
  pause 1000              'Pause for 1000 ms (1 second)
  out0=1                  'Port 0 goes HIGH, LED 0 turned off
  pause 500               'Pause for 500 ms
  out1=0                  'Port 1 goes LOW lighting LED 1
  pause 1000              'Pause for 1000 ms
  out1=1                  'Port 1 goes HIGH, LED 1 turned off
  pause 500               'Pause for 500 ms
goto flash
```

The second circuit to be wired to the BASIC Stamp 2 module is drawn in Fig. 13-6. The N.O. push-button switch will serve as an input, while the output device is a piezo buzzer. The advantage of the BASIC Stamp (or any microcontroller) is that the 16 I/O pins can be configured as either inputs or outputs during programming. In this circuit, port 12 (P_{12}) of the BASIC Stamp is configured as an input, while port 2 (P_2) will be used as the output. The input at P_{12} is wired as an active LOW switch with a 10-kΩ pull-up resistor. The output at P_2 is wired so that a LOW will activate and sound the low-current piezo buzzer.

The following PBASIC program will cause the BS2 module to sound the piezo buzzer when the input switch is pressed. Early in the program, the I/O ports to be used are configured as either inputs or outputs (**input 12, output 2**). The **out2=1** statement initializes port 2 to HIGH, which ensures the buzzer will be off. The **n var nib** line of code declares that the program will use a variable called *n*. The **nib** (stands for nibble) part of the declaration line means the variable can be any number from 0 to 15 in decimal (0000–1111 in binary).

```
'Input switch/output buzzer   'Title of PBASIC program (see Fig. 13-6)

input 12                  'Configure I/O port P12 as input
output 2                  'Configure I/O port P2 as output
out2=1                    'Initialize P2 HIGH (buzzer off)
n    var    nib           'Declare variable n as nibble (0-15)
```

344

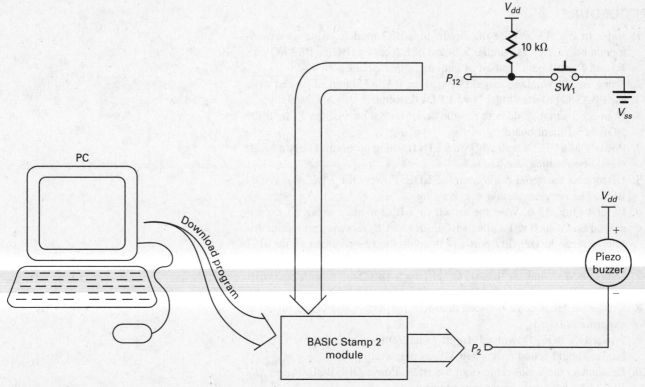

Fig. 13-6 BASIC Stamp 2 module programmed to accept switch input and piezo buzzer output.

```
checkswitch:              'Label for switch checking routine
  if in12=0 then buzzer   'IF SW1 closed THEN go to buzzer routine
goto checkswitch          'Send back to check switch again

buzzer:                   'Label for buzzer sounding routine
  for n=1 to 5            'Begin FOR-NEXT loop, 5 loops
    out2=0                'Port 2 goes LOW activating buzzer
    pause 75              'Pause 75 milliseconds
    out2=1                'Port 2 goes HIGH turning off buzzer
    pause 75              'Pause 75 ms
  next                    'End or FOR-NEXT loop, 5 loops only
goto checkswitch          'Send back to checkswitch routine
```

The **checkswitch:** line of PBASIC code is a label for the routine for switch checking. The **if in12=0** part of the IF-THEN statement is the condition that must be true before the microcontroller will proceed to the **then buzzer** section of line of code. If the condition is false (in this example if input 12 is HIGH), then the microcontroller will cause the next line of code to be executed which is **goto checkswitch.**

The **buzzer:** routine in the PBASIC program below will cause the buzzer to sound and then return to the **checkswitch:** routine. The **buzzer:** routine contains a FOR-NEXT loop, which is used as a counting loop. The microcontroller will cycle through the loop five times based on the instruction **for n = 1 to 5.** The **for n = 1 to 5** makes the variable $n = 1$ the first time through the loop and then increments the variable n by 1 each time though the cycle. The **next** statement checks the value stored in variable n. If the value of n is from 1 to 5, the **next** statement causes a jump back to the for **n = 1 to 5** line of code. However, if the value of n is greater than 5, then the **next** statement sends the program to the

345

goto checkswitch line of code. In summary, the **buzzer:** routine will cause the buzzer to beep five times and return to the **checkswitch:** routine.

PROCEDURE

1. Refer to Fig. 13-5. Wire the circuit to a BS2 module using an experimental board such as Parallax's Board of Education (BOE). BS2 I/O pins P_0 and P_1 will both be used as outputs in this circuit.
2. Using an MS Windows–based PC, start the BASIC Stamp editor and write the PBASIC program titled **'Two LEDs flashing.**
3. Connect a serial cable between the serial output port of the PC and the BOE experiment board.
4. Power ON BOE. Download **'Two LEDs flashing** program. The two LEDs should be flashing alternately.
5. Disconnect the serial cable from the BOE. Power OFF BOE. Power ON BOE. The program should start running.
6. Refer to Fig. 13-6. Wire the circuit to a BS2 module using an experimental board such as Parallax's BOE. I/O port P_2 serves as an output driving a piezo buzzer. I/O port P_{12} is configured as an input to the BS2 module and is connected to an active LOW push-button switch.
7. Using an MS Windows–based PC, start the BASIC Stamp editor and write the PBASIC program titled **'Input switch/output buzzer.**
8. Connect a serial cable between the serial output of the PC and the BOE experiment board.
9. Power ON BOE. Download **'Input switch/output buzzer** program. The buzzer should sound only when SW_1 is pressed.
10. Disconnect the serial cable from the BOE. Power OFF BOE. Power ON BOE. The program should start running.
11. Have your instructor check the operating switch input/buzzer output circuit, and be prepared to answer questions about both the circuit and program.
12. Power OFF. Take down the circuit and return all equipment to its proper place.

QUESTIONS

Answer questions 1 to 9.

1. At the heart of the BASIC Stamp 2 module is a programmable device called a _____ (microcontroller, multiplexer).

2. The high-level language used to program a BS2 IC is _____ (LISP, PBASIC).

3. The BASIC Stamp 2 module has _____ (4, 16) I/O ports that can be configured as either inputs or outputs by the programmer.

4. Refer to Fig. 13-5. A logical 0 or LOW at input P_1 of the light-emitting diode circuit will cause the LED 1 to _____ (light, not light).

5. Refer to the program titled **'Two LEDs flashing.** Lines 2 and 3 configure I/O ports of the BASIC Stamp 2 as _____ (inputs, outputs).

6. Refer to the program titled **'Two LEDs flashing.** The program statement **out1=0** causes the BS2 microcontroller to drive _____ (I/O port 1 LOW, I/O port 0 HIGH) lighting _____ (LED 0, LED 1).

7. Refer to Fig. 13-6. Pressing push-button switch SW_1 will input a _____ (HIGH, LOW) to I/O port 12 of the BASIC Stamp 2 module.

8. Refer to the program titled **'Input switch/output buzzer.** The program line _____ (for n=1 to 5, n var nib) declares the variable named *n* and allows it to range in value from 0 to 15 (0000 to 1111 in binary).

9. Refer to the program titled **'Input switch/output buzzer.** The program line **if in12=0 then buzzer** will cause the program to jump to the **buzzer:** subroutine only with input switch SW_1 is _____ (closed, open) generating a LOW at I/O _____ (port 2, port 0).

1. _____

2. _____

3. _____

4. _____

5. _____

6. _____

7. _____
   _____

8. _____

9. _____
   _____

346

13-4 BASIC STAMP EXPERIMENT: PHOTO INPUT AND SERVO MOTOR OUTPUT

OBJECTIVES

1. To use a microcontroller (BASIC Stamp module) as a control device.
2. To use a BASIC Stamp 2–based development board (like Board of Education or HomeWork board by Parallax) and personal computer to program in PBASIC, download the program, and wire and test several simple circuits.
3. To test and use a CdS photocell as an input to a BS2 IC.
4. To test and use both LEDs and a hobby servo motor as outputs from a BS2 IC.

MATERIALS

Qty.		Qty.	
1	BASIC Stamp 2 development board (such as Board of Education or HomeWork boards both available from Parallax)	1	Parallax manual *What Is a Microcontroller?*
1	BASIC Stamp 2 module (this IC may come mounted on development board)	2	LEDs (one red, one green)
		1	photoresistive cell, CdS
		1	Hobby servo motor
		2	150-Ω resistors
1	PC with MS Windows	1	2.2-kΩ resistor
1	PBASIC editor from Parallax	1	10-kΩ resistor
1	serial cable (for downloading to BASIC Stamp 2 module) (see instructor for using a USB port)	1	10-kΩ potentiometer
		1	logic probe or DMM

Note: BASIC Stamp module, development board, PBASIC software, and both experiment and reference manuals are available from Parallax, Inc., 599 Menlo Drive, Rocklin, CA 95765. Most of the manuals and software can be downloaded free from Parallax's websites. General website: *www.parallaxinc.com*. Telephone: 916-624-8333.

SYSTEM DIAGRAMS

The first circuit to be wired to the BASIC Stamp 2 module is shown in Fig. 13-7. The input device is a cadmium sulfide (CdS) photoresistive cell. The CdS photocell reacts to light by changing resistance. When bright light strikes the photocell, its resistance decreases. When no light strikes the photocell, its resistance increases and is high. The photocell is wired in series with a 10-kΩ resistor forming a voltage divider. When bright light strikes the photocell, its resistance is low and the voltage input to I/O P_7 is low (near ground voltage), which is interpreted by the microcontroller as a LOW logic level. When no light hits the photocell, its resistance increases and voltage input to I/O P_7 is high (nearing +5 V), which is interpreted by the BS2 IC as a HIGH logic level.

Two LEDs are used in the circuit shown in Fig. 13-7. The green LED (LED 3) glows when light strikes the CdS photocell and indicates a LOW input at P_7. The red LED (LED 5) glows when no light (less light) strikes the CdS photocell and indicates a HIGH input at P_7.

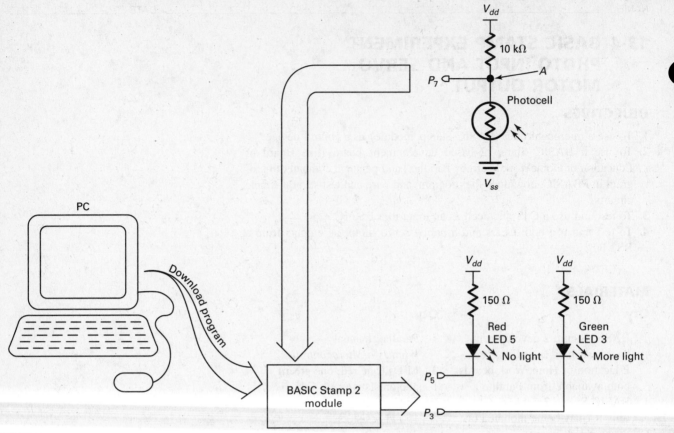

Fig. 13-7 CdS photocell input to BS2 IC with logic level output indicators (LEDs).

The following PBASIC program will cause the BS2 module to evaluate the amount of light striking the CdS photocell and respond by activating either a green LED (more light) or a red LED (less light). Notice that I/O port 7 (P_7) is configured as input (**input 7**). The **ckphotocell:** routine repeatedly checks the logic level at input P_7 and directs the program to jump to either the **green:** or **red:** subroutines. If the P_7 input is LOW, then the **in7=0 then green** statement directs the green LED to light. A glowing green LED means a high light level striking the input sensor (CdS photocell). However, if the P_7 input is HIGH then the **in7=1 then red** statement directs the red LED to light. A glowing red LED means a low light level striking the CdS photocell.

```
'Reading photocell input    'Title of PBASIC program (see Fig. 13-7)

output 3
output 5
input 7

Ckphotocell:               'Label for check photocell routine
  if in7=0 then green      'If input P7 is LOW then go to green subroutine
  if in7=1 then red        'If input P7 is HIGH then go to red subroutine
goto ckphotocell           'Go back and check photo cell again

Green:                     'Label for green subroutine (means more light)
  out3=0                   'Output P3 goes LOW, green LED glows
```

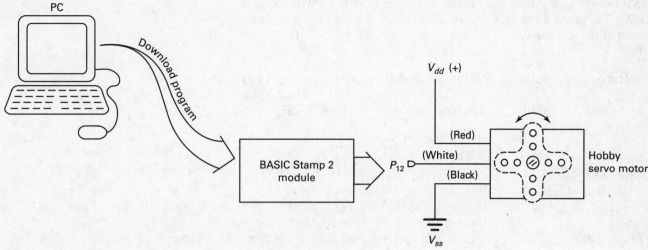

Fig. 13-8 BASIC Stamp 2 module driving a hobby servo motor using pulse-width modulation (PWM).

```
pause 250           'Pause 250 ms
out3=1              'Output P3 goes HIGH, green LED is off
pause 150           'Pause 150 ms
goto ckphotocell    'Go back to check photocell routine

Red:                'Label for red subroutine (means less light)
  out5=0
  pause 250
  out5=1
  pause 150
goto ckphotocell
```

The second circuit to be wired to the BASIC Stamp 2 module is sketched in Fig. 13-8. A hobby-type servo motor is to be repeatedly rotated clockwise (CW) and then counterclockwise (CCW) using pulse-width modulation (PWM). Recall that the hobby servo motor you may have already used will *rotate CW when a series of 1 millisecond pulses* enters its control wire. The servo will *rotate CCW when a series of 2 millisecond pulses* enters the control of the motor. Some servos rotate the opposite direction depending on the internal wiring of the motor.

The following program will cause the hobby servo motor to be rotated fully CW and then CCW. This sequence will repeat until power to the BS2 module is turned off. Port 12 (P_{12}) of the BASIC Stamp 2 module is configured as the output by the statement **output 12.** Variable x is declared by the statement **x   var   byte.** Variable x can be any number in a range from 0–255 (00000000–11111111 in binary). Each of the FOR-NEXT statements **(for x = 1 to 50)** defines that the loop will be executed 50 times. The **pulsout 12,500** statement allows the microcontroller to output a positive pulse to I/O port 12 that is 1000 microseconds (500 × 2 μs = 1000 μs) or 1 millisecond in duration. Immediately following is the **pause 10** statement pausing the 10 milliseconds before the next positive pulse. The second **pulsout 12, 1000** statement generates another positive pulse to I/O port 12 that is

349

2000 microseconds (1000 × 2 μs = 2000 μs) or 2 milliseconds long. Each **pulsout** statement is repeated 50 times to make sure the servo has time to move mechanically from one extreme to another (CW to CCW or CCW to CW). The **pause 1000** lines of code allow about 1s before the next movement of the servo motor.

```
'Rotate servo motor      'Title of PBASIC program (see Fig. 13-8)

x    var    byte          'Declare variable x (from 0-255)
output 12

servo:                    'Label for rotate servo CW
  for x=1 to 50           'Start loop, continue 50 times
    pulsout 12,500        'PWM of 1 millisecond to servo
    pause 10              'Pause 10 ms before next pulse
  next                    'Repeat loop till x > 50 then continue on
pause 1000                'Pause 1 s before changing direction of rotation
  for x=1 to 50           'Start loop, continue 50 times
    pulsout 12,1000       'PWM of 2 milliseconds to servo
    pause 10              'Pause 10 ms before next pulse
  next                    'Repeat loop till x > 50 then continue on
pause 1000                'Pause 1 s before changing direction of rotation
goto servo                'Go back to start servo routine again
```

The third circuit combines the former two using a photo input to control the direction of rotation of a servo motor. This BASIC Stamp system might be the basis for some device like an automatic shutter or door that responds to changes in light, causing opening and closing.

A system you will use is sketched in Fig. 13-9. The CdS photocell input (port P_7) is shown at the upper right. Notice that a potentiometer has been included to adjust the sensitivity of the light sensor (photocell). Two LED outputs (ports P_3 and P_5) are drawn to the right of the BS2 module and will verify to the operator when less light or more light is striking the photocell. The hobby servo motor is connected to output port 12 of the BS2 module. With more light striking the photocell, it will rotate fully CW. However, with less light hitting the light sensor, the servo will rotate fully in the opposite direction. Some servos may turn in the opposite direction due to the permanent internal wiring of the motor.

The following program will cause the servo motor to be rotated fully CW with more light striking the photocell. With less light striking the photocell, the servo will rotate fully CCW.

The **ckphotocell:** main routine checks the condition of input 7 generated by the photocell and series resistance. If the P_7 input is LOW, then the program jumps to the **CW:** subroutine, which rotates the servo fully CW using PWM. The **pulsout 12,500** command generates a 1 ms (500 × 2 μs = 1000 μs = 1 ms) positive pulse at port P_{12}. The FOR-NEXT loop generates 30 of these pulses. Thirty pulses are generated to make sure the servo rotor rotates all the way CW. This routine also blinks the green LED on and then off (**low 3** and **high 3**). The **low 3** is another command used to reset output 3 LOW. After the routine is finished, the program jumps back to the **ckphotocell:** routine.

If the P_7 input is HIGH, then the program jumps to the **CCW:** subroutine, which rotates the servo fully CCW using PWM. The **pulsout 12,1000** command generates a 2 ms (1000 × 2 μs = 2000 μs = 2 ms) positive pulse at port 12. The FOR-NEXT loop generates 30 pulses. This routine also blinks the red LED on and then off (**low 5** and **high 5**). The **high 5** is another command used to set output 5 HIGH. After the routine is finished, the program jumps back to the **ckphotocell:** routine.

350

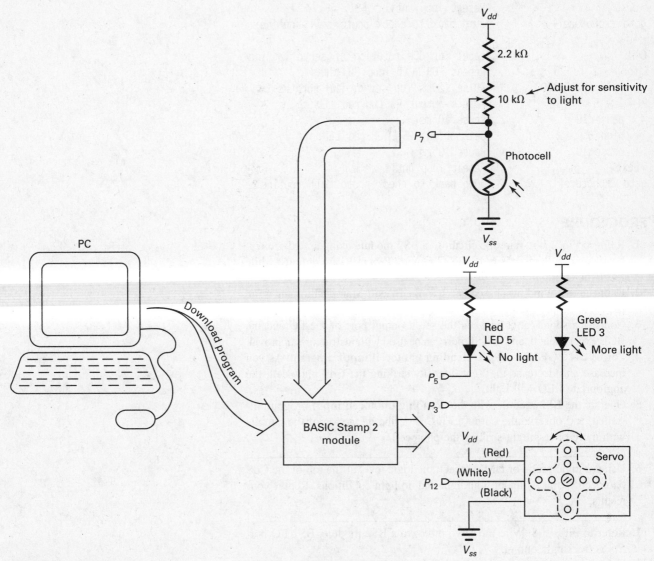

Fig. 13-9 Photocell reads light input causing servo motor to rotate either fully CW or CCW.

```
'Read light-rotate servo   'Title of PBASIC program (see Fig. 13-9)

x     var     byte         'Declare variable x; range from 0 thru 255
input 7                    'Configure I/O port P7 as an input (photocell)
output 12                  'Configure I/O port P12 as an output (servo motor)

ckphotocell:               'Label for checking photocell routine
 if in7=0 then CW          'If P7 input is LOW then jump to CW routine
 if in7=1 then CCW         'If P7 input is HIGH then jump to CCW routine
goto ckphotocell           'Jump back to check photo cell routine

CW:                        'Label for CW rotation of servo routine
 for x=1 to 30             'Repeat FOR-NEXT loop 30 times
  pulsout 12,500           'Pulse (1 ms) to servo, PWM rotates CW
  low 3                    'Resets output P3 LOW-green LED on
  pause 10                 'Pause 10 ms
```

```
  high 3              'Sets output P3 HIGH-green LED off
  pause 10            'Pause 10 ms
next                  'Repeat loop until x=31
goto ckphotocell      'Jump back to check photo cell routine


CCW:                  'Label for CCW rotation of servo routine
  for x=1 to 30       'Repeat FOR-NEXT loop 30 times
    pulsout 12,1000   'Pulse (2 ms) to servo, PWM rotates CCW
    low 5             'Resets output P5 LOW-red LED on
    pause 10          'Pause 10 ms
    high 5            'Sets output P5 HIGH-red LED off
    pause 10          'Pause 10 ms
  next                'Repeat loop until x=31
goto ckphotocell      'Jump back to check photo cell routine
```

PROCEDURE

1. Refer to Fig. 13-7. Wire the circuit to a BS2 module using an experiment board such as Parallax's BOE. BS2 I/O ports P_3 and P_5 are outputs, while I/O P_7 is an input.
2. Using an MS Windows–based PC, start the BASIC Stamp editor and write the PBASIC program titled **'Reading photocell input.**
3. Connect a serial cable between the serial output port of the PC and the BOE experiment board. A USB port can be used with instructor's approval.
4. Power ON BOE. Download **'Reading photocell input** program. As you increase and decrease the light intensity striking the CdS photocell, the appropriate LED will light.
5. Operate the BS2 module with the **'Reading photocell input** program installed, and observe the output LEDs. Describe the action of the outputs with more or less light striking the photocell.

_____

6. Using a logic probe or DMM, check the logic level at the top of the CdS photocell (point A) under high light and no light conditions. Report your results.

_____

7. Refer to Fig. 13-8. Wire the servo motor to a BS2 module. BS2 I/O port P_{12} is the single output.
8. Using an MS Windows–based PC, start the BASIC Stamp editor, and write the PBASIC program titled **'Rotate servo motor.**
9. Connect a serial cable between the serial output port of the PC and the BOE experiment board. A USB port can be used with instructor's approval.
10. Power ON BOE. Download **'Rotate servo motor** program. The servo motor should rotate first fully CW and then fully CCW. Describe your results.

_____

11. Refer to Fig. 13-9. Wire the circuit to a BS2 module using an experiment board such as Parallax's BOE. Three BS2 I/O ports serve as outputs (P_3, P_5, and P_{12}). One BS2 I/O port is a light sensor input (P_7).
12. Using an MS Windows–based PC, start the BASIC Stamp editor, and write the PBASIC program titled **'Read light-rotate servo.**
13. Connect a serial cable between the serial output port of the PC and the BOE experiment board. A USB port can be used with instructor's approval.
14. Power ON BOE. Download **'Read light-rotate servo** program. As you increase and decrease the light intensity striking the CdS photocell, the appropriate LED will light and the servo will rotate fully CW and then CCW. You may have to adjust the 10-kΩ potentiometer for the proper sensitivity to light. Describe your results.

_____

_____

15. Show instructor your operating photocell/servo motor circuit from Fig. 13-9, and be prepared to answer questions about both the program and circuit.

16. Power OFF. Take down the circuit, and return all equipment to its proper place.

QUESTIONS

Complete questions 1 to 14.

1. Refer to Fig. 13-7. When a bright light strikes the CdS photocell, its resistance is _____ (high, low) and input P_7 to the BS2 IC is at a _____ (HIGH, LOW) logic level.

2. Refer to Fig. 13-7 and PBASIC program **'Reading photocell input.** When no light strikes the photocell, its resistance is high and input P_7 to the BS2 IC is at a _____ (HIGH, LOW) logic level, which causes the BASIC Stamp module to light the _____ (green, red) LED.

3. Refer to Fig. 13-7. Under bright light conditions, I/O port P_3 is _____ (HIGH, LOW) and P_5 goes _____ (HIGH, LOW) causing the green LED to glow.

4. Refer to the PBASIC program titled **'Reading photocell input.** The two lines of code that configure the BASIC Stamp module I/O ports P_3 and P_5 as outputs are _____ and _____.

5. Refer to the PBASIC program titled **'Reading photocell input.** The purpose of the PBASIC statement **input 7** is to _____ (configure P_7 as an input, check input P_7 as to the exact analog voltage present).

6. Refer to the PBASIC program titled **'Reading photocell input.** The purpose of the PBASIC statement **out5=0** is to _____ (check I/O port P_5 as to the exact analog voltage present, cause output P_5 to go LOW).

7. Refer to the PBASIC program titled **'Reading photocell input.** The **pause 250** PBASIC statement causes the microcontroller to pause or do nothing for _____ (250 milliseconds, 250 seconds).

8. Refer to Fig. 13-8. The servo is caused to rotate either CW or CCW by the width of pulses entering the motor, which is called pulse-width modulation. (T or F)

9. Refer to the PBASIC program titled **'Rotate servo motor.** The **pulsout 12,500** PBASIC statement sends a positive pulse _____ (1 ms, 500 ms) in duration to output _____ (P_1 and P_2, P_{12})

10. Refer to the PBASIC program titled **'Rotate servo motor.** The PBASIC statement **for x=1 to 50** starts a FOR-NEXT loop that is executed 50 times before dropping out of the bottom of the loop. (T or F)

11. Refer to Fig. 13-9. The CdS photocell's sensitivity to light is adjusted by _____ (varying the resistance of the potentiometer, modifying the photocell).

12. Refer to the PBASIC program titled **'Read light-rotate servo.** The BS2 has I/O port _____ (P_7, P_{12}) configured as an input from the photocell and port _____ (P_7, P_{12}) set up as an output to drive the servo motor.

13. Refer to the PBASIC program titled **'Read light-rotate servo.** The PBASIC statement **low 3** is a shorthand method of setting I/O port _____ (P_3 HIGH, P_3 LOW).

14. Refer to the PBASIC program titled **'Read light-rotate servo.** Which PBASIC statement notifies the microcontroller that the program will use a variable with the name of *x* and that this variable will range from 0 through 255?

1. _____
   _____

2. _____
   _____

3. _____
   _____

4. _____
   _____

5. _____

6. _____

7. _____
   _____

8. _____

9. _____
   _____

10. _____

11. _____

12. _____
    _____

13. _____

14. _____

Connecting with Analog Devices

TEST: CONNECTING WITH ANALOG DEVICES

Answer the questions in the spaces provided.

1. An interface device that converts analog to digital information is called a(n) _____.

2. An interface device that converts digital to analog information is called a(n) _____.

3. A D/A converter consists of a(n)
   a. Encoder and decoder
   b. Op amp and two resistors
   c. Resistor network and summing amplifier
   d. Scaling amplifier and decoder

4. The name op amp stands for _____.

5. Op amps are characterized by high input impedance, low output impedance, and a variable voltage gain that is set _____.
   a. At the manufacturer
   b. With external resistors
   c. With internal resistors
   d. With external capacitors

6. Calculate the voltage gain for an op amp with a feedback resistor of 10 kΩ and an input resistor of 5 kΩ.

7. Calculate the output voltage from the op amp in question 6 if the input voltage is 0.5 V.

8. Refer to Fig. 14-1. The _____ (inverting, noninverting) input to the op amp is being used in this circuit.

9. Refer to Fig. 14-1. Calculate the combined resistance of parallel resistors R_1 and R_4 if both switches A and D are at logical 1.

10. Refer to Fig. 14-1. Calculate the voltage gain of the op amp when both switches A and D are at a logical 1. Use the resistance figure from question 9.

11. Refer to Fig. 14-1. Using the voltage gain from question 10, calculate the output voltage of the D/A converter when both switches A and D are at logical 1.

12. The _____ ladder-type D/A converter overcomes some of the disadvantages of the more basic unit shown in Fig. 14-1.

13. Refer to Fig. 14-2 for a block diagram of a counter-ramp-type A/D converter. The block labeled E would be a
   a. Counter   c. Voltage comparator
   b. D/A converter   d. Waveshaping circuit

1. _____

2. _____

3. _____

4. _____

5. _____

6. _____

7. _____

8. _____

9. _____

10. _____

11. _____

12. _____

13. _____

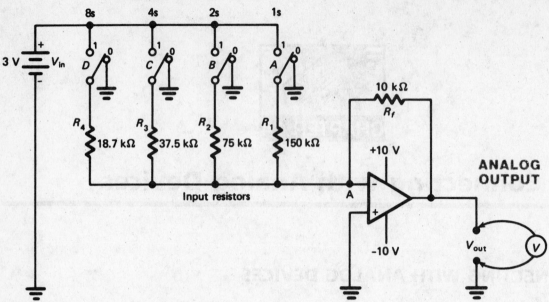

Fig. 14-1 A basic D/A converter circuit.

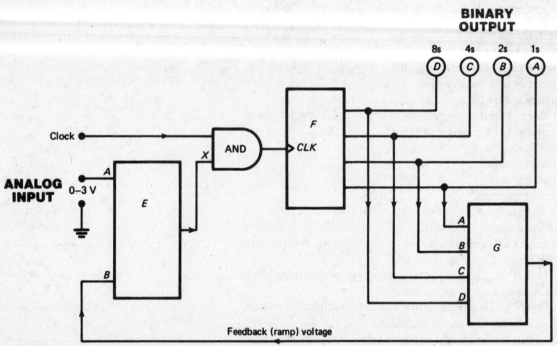

Fig. 14-2 A counter-ramp-type A/D converter circuit.

14. Refer to Fig. 14-2 for a block diagram of a counter-ramp-type A/D converter. The block labeled *F* would be a
    a. Counter
    b. D/A converter
    c. Decoder
    d. Shift register

14. _____

15. Refer to Fig. 14-2 for a block diagram of a counter-ramp-type A/D converter. The block labeled *G* would be a
    a. Counter
    b. D/A converter
    c. Decoder
    d. Voltage comparator

15. _____

16. Fig. 14-3 shows the schematic diagram for a _____ circuit using an op amp.
  a. Digital comparator
  b. Scaling amplifier
  c. Summing amplifier
  d. Voltage comparator

16. _____

17. Refer to Fig. 14-3. With 1.5 V at input *A* and 2.0 V at input *B*, the output will indicate a logical _____ (0, 1).

17. _____

18. Refer to Fig. 14-3. If input *B* drops to 1.0 V while input *A* remains at 1.5 V, the output will indicate a logical _____ (0, 1).

18. _____

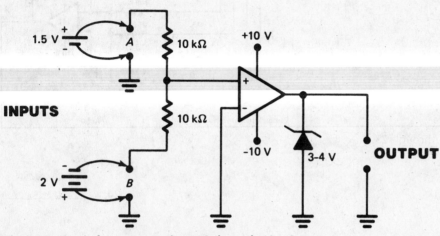

Fig. 14-3 Circuit for test questions 16 through 18.

19. The wiring diagram in Fig. 14-4 is for an experimental _____ circuit.

19. _____

20. The 741 op amp at the left in Fig. 14-4 is part of the _____ circuit.
  a. Ramp
  b. Successive-approximation
  c. Voltage comparator
  d. None of the above

20. _____

21. The 7493 IC in Fig. 14-4 is wired as a(n)
  a. A/D converter
  b. Counter
  c. Reset circuit
  d. Successive-approximation circuit

21. _____

22. The 7447 IC in Fig. 14-4 performs as a(n)
  a. A/D converter
  b. Counter
  c. D/A converter
  d. Decoder

22. _____

23. The 741 op amp at the right in Fig. 14-4 is part of the _____ circuit.
  a. A/D converter
  b. Counter
  c. D/A converter
  d. Voltage-sensing

23. _____

357

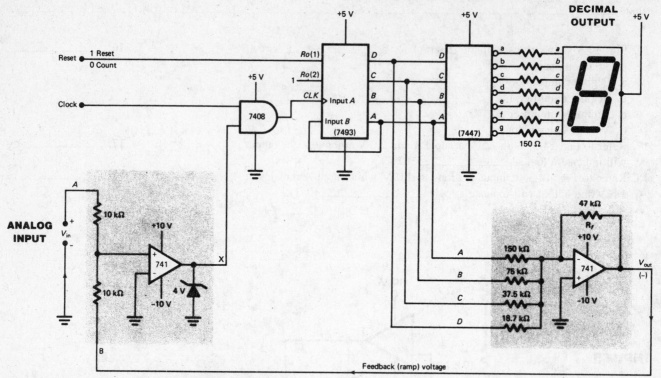

Fig. 14-4 Circuit for test questions 19 through 23.

24. A digital voltmeter is basically a(n)
    a. A/D converter
    b. Counter circuit
    c. D/A converter
    d. Voltage comparator

24. _____

25. The _____-type A/D converter is the fastest type unit.
    a. Counter-ramp
    b. Ramp
    c. Successive-approximation
    d. Varley loop

25. _____

26. The ADC0804 A/D converter IC has a resolution of _____ (4, 8, 16) bits.

26. _____

27. A 12-bit A/D converter has greater resolution than a(n) _____ (8, 16)-bit chip.

27. _____

28. The ADC0804 A/D converter IC has a typical conversion time of about _____ (2 ns, 100 μs).

28. _____

29. The ADC0804 IC is a _____ (meter, microprocessor)-type A/D converter.

29. _____

30. The conversion time for a microprocessor-type A/D converter is _____ (longer, shorter) than for a meter-type A/D converter IC.

30. _____

31. The light-dependent resistor (R_3) in Fig. 14-5 is probably a CdS photocell. (T or F).

31. _____

32. Refer to Fig. 14-5. Decreasing the light intensity striking R_3 will _____ (decrease, increase) the resistance of the photocell.

32. _____

33. Refer to Fig. 14-5. Increasing the light intensity striking the photocell _____ (decreases, increases) the decimal reading on the LED output display.

33. _____

34. Refer to Fig. 14-5. The A/D converter performs about _____ conversions per second.

34. _____

358

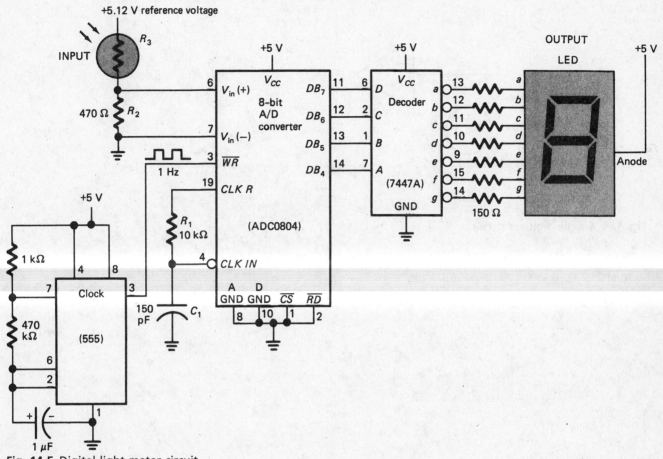

Fig. 14-5 Digital light-meter circuit.

35. Refer to Fig. 14-5. The analog input voltage between $V_{in}(+)$ and $V_{in}(-)$ _____ (decreases, increases) as the light intensity striking the photocell decreases.

36. Refer to Fig. 14-5. Only the four _____ (LSBs, MSBs) of the binary output of the ADC0804 A/D converter IC are being decoded to drive the seven-segment LED display.

37. Refer to Fig. 14-6. This component functions as the *temperature transducer* in this circuit.
    a. Potentiometer
    b. Schmitt trigger inverter
    c. Thermistor

38. Refer to Fig. 14-6. This component functions as a simple *digitizer* (A/D converter) in this circuit.
    a. Potentiometer
    b. Schmitt trigger inverter
    c. Thermistor

39. Refer to Fig. 14-6. If the temperature of the thermistor is greatly decreased, its resistance _____ (decreases, increases) and the input voltage to the inverter _____ (decreases, increases). The output of the Schmitt trigger inverter snaps to a _____ (HIGH, LOW) logic level.
    a. increases, increases, HIGH
    b. increases, decreases, HIGH
    c. decreases, increases, LOW
    d. decreases, decreases, LOW

35. _____

36. _____

37. _____

38. _____

39. _____

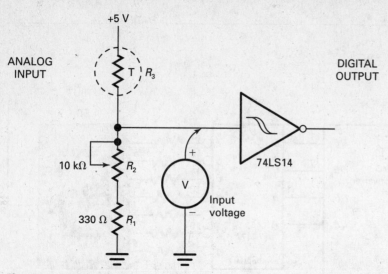

Fig. 14-6 Simple digitizer circuit.

14-1 LAB EXPERIMENT: VOLTAGE GAIN OF AN OPERATIONAL AMPLIFIER

OBJECTIVES

1. To construct and set the gain of an operational amplifier (op amp).
2. To use the op amp as a dc amplifier.
3. *OPTIONAL*: Use circuit simulation software (such as Multisim) to construct and test an op-amp circuit.

MATERIALS

Qty.   **Qty.**

1   741 op-amp (μA741) IC   1   dual 9-V, 12-V, or 15-V
3   10-kΩ, ¼-W resistors   dc power supply (or two
1   variable 0- to 10-V dc power   separate supplies)
    supply   1   voltmeter (such as DMM)
1   1-kΩ, ¼-W resistor   1   circuit simulation pro-
                              gram (*optional*)

SYSTEM DIAGRAM

Figure 14-7 is the circuit you will use to set the gain of the operational amplifier.

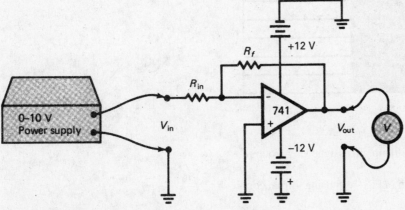

Fig. 14-7 Measuring dc gain in an op-amp circuit.

PROCEDURE

1. Insert the 741 IC into the mounting board.
2. Power OFF. Connect the +12- and −12-V dc power supplies to the op amp. Use a dual supply or two separate supplies. Dynalogic DT-1000 Trainer provides +12- and −12-V dc power.
3. Wire the op-amp circuit in Fig. 14-7. Use two 10-kΩ resistors for R_{in} and R_f.
4. Power ON. Apply the voltages shown in the left column of Table 14-1. Measure the output voltages (V_{out}) and record in Table 14-1. Record the polarity (+ or −) of the voltages.
5. Power OFF. Replace resistor R_{in} with a 1-kΩ resistor.
6. Power ON. Apply the voltages shown in the left column of Table 14-1. Measure the output voltages V_{out}, and record them in the center column of Table 14-1. Record the polarity of the voltages.
7. Power OFF. Replace resistor R_{in} with a 20-kΩ resistor (two 10 kΩs in series).
8. Power ON. Apply the voltages shown in the left column of Table 14-1. Measure the output voltages V_{out}, and record them in the right column. Record the polarity of the voltages.

361

TABLE 14-1 Op-Amp Voltage Gain

INPUT	OUTPUT		
V_{in} (volts)	R_f = 10k, R_{in} = 10k V_{out} (volts)	R_f = 10k, R_{in} = 1k V_{out} (volts)	R_f = 10k, R_{in} = 20k V_{out} (volts)
−0.4			
−0.8			
−1.2			
−1.6			
−2.0			
−2.4			
−2.8			
−3.2			
−3.6			
−4.0			
+0.4			
+0.8			
+1.2			
+1.6			
+2.0			
+2.4			
+2.8			
+3.2			
+3.6			
+4.0			

9. *Optional:* Electronics Workbench or Multisim. If assigned by your instructor, use circuit simulation software to:

   a. Construct the op-amp test circuit shown in Fig. 14-8 using virtual components when possible.

   b. Operate the circuit-changing input voltages (such as 100 mV, 200 mV, 500 mV, 1 V, 2 V).

   c. Observe the output voltages and *voltage gains* (see *Av* formulas in Fig. 14-8) for the op-amp test circuit.

   d. Change R_{in} and R_f values. Operate the circuit-changing input voltages. Observe the voltage gain of the op amp.

   e. Show your instructor your simulated circuit, and be prepared to answer questions about the op-amp test circuit (Fig. 14-8).

10. Power OFF. Leave the circuit set up for the next experiment.

QUESTIONS

Complete questions 1 to 9.

1. What is the formula for voltage gain A_v for this op-amp circuit (using R_f and R_{in})?

   1. _____

2. Calculate the voltage gain A_v of the op-amp circuit (use the formula from question 1) with:

   a. R_f = 10 kΩ, R_{in} = 10 kΩ

   b. R_f = 10 kΩ, R_{in} = 1 kΩ

   c. R_f = 10 kΩ, R_{in} = 20 kΩ

   2. *Place answers below.*

   a. _____

   b. _____

   c. _____

3. Write the formula for voltage gain A_v for the op-amp circuit (using V_{in} and V_{out}).

4. Calculate the voltage gain A_v for the op-amp circuit (use the formula from question 3) with (see data in Table 14-1):
  a. $R_f = 10 \text{ k}\Omega$, $R_{in} = 10 \text{ k}\Omega$
  b. $R_f = 10 \text{ k}\Omega$, $R_{in} = 1 \text{ k}\Omega$
  c. $R_f = 10 \text{ k}\Omega$, $R_{in} = 20 \text{ k}\Omega$

5. Refer to Fig. 14-7. If the input voltage V_{in} in this circuit is negative, the output voltage V_{out} is _____ ($-$, $+$).

6. Refer to Fig. 14-7. What is the purpose of resistors R_{in} and R_f in this op-amp circuit?

7. Resistor R_{in} is called the _____ resistor, and R_f is called the _____ resistor.

8. Refer to Fig. 14-9. What is the voltage gain of this op-amp circuit?

9. Refer to Fig. 14-9. If the input voltage were changed to 500 mV, what would be the output voltage?

3. _____

4. *Place answers below.*

a _____

b. _____

c. _____

5. _____

6. _____

7. _____ _____

8. _____

9. _____

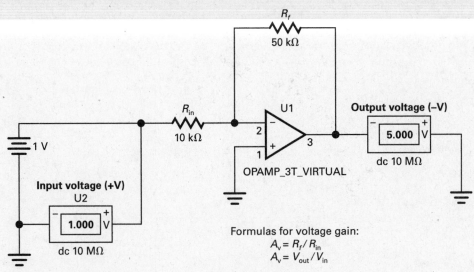

Fig. 14-8 Op-amp test circuit using circuit simulation.

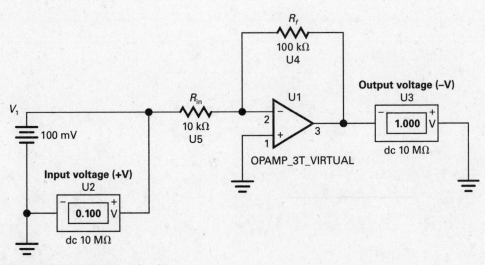

Fig. 14-9 Op-amp test circuit.

14-2 LAB EXPERIMENT: D/A CONVERTER

OBJECTIVE

To construct and test an elementary digital-to-analog (D/A) converter.

MATERIALS

Qty.

1 741 op-amp (μA741) IC
1 10-kΩ, ¼-W resistor
1 37.5-kΩ, ¼-W resistor
   (combined 4.7 kΩ and 33 kΩ)
1 150-kΩ, ¼-W resistor
1 voltmeter (DMM)
4 logic switches

Qty.

1 18.7-kΩ, ¼-W resistor
   (combine 18 kΩ and 680 Ω)
1 75-kΩ, ¼-W resistor
1 dual 9-V, 12-V, or 15-V
   dc power supply (or two
   separate supplies)

SYSTEM DIAGRAM

A wiring diagram for a digital-to-analog converter is shown in Fig. 14-10. The accuracy of the D/A converter depends on the values of the resistors and the V_{in} from the logic switches. The V_{in} is shown as 3 V on the schematic but may vary from 2.8 to 5 V depending on the design of the logic switches on the equipment you are using. This experiment is best done with circuit simulation (see Circuit Simulation 14-3 for this alternative).

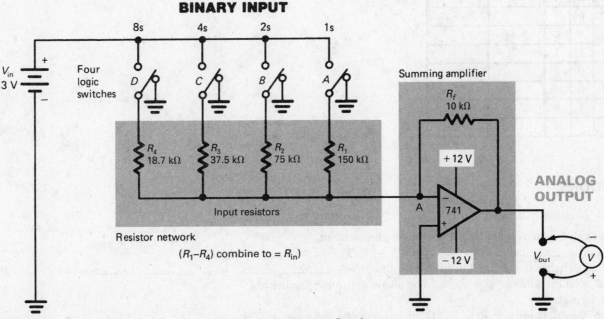

Fig. 14-10 Wiring diagram for a digital-to-analog converter (hardware version).

PROCEDURE

1. Insert the 741 op-amp IC into the mounting board.
2. Power OFF. Connect the dual power supply (+12 and −12) to the IC. Dynalogic DT-1000 Trainer provides +12- and −12-V dc power.
3. Wire the D/A converter shown in Fig. 14-10. Use combinations of resistors to get close to the resistance values. The four switches (A, B, C, and D) will be connected to input logic switches. These switches output about 3 to 5 V for a logical 1 and nearly 0 V for a logical 0.

4. Power ON. Operate the D/A converter by applying the binary inputs on the left in Table 14-2. Remember that the accuracy of your converter depends on the values of the resistors and the V_{in} from the logic switches. Record the results in the center column.
5. Calculate the voltage gain for each row in Table 14-2. Measure the input voltage from the logic switches, and use this for the input voltage V_{in}. Record your calculations in the right column of Table 14-2.
6. Power OFF. Leave this circuit set up for use in a later experiment.

TABLE 14-2 D/A Converter Results

INPUT				OUTPUT	Calculated gain
Binary				Analog	
8s	4s	2s	1s	Measured volts	$A_v = \dfrac{V_{out}}{V_{in}}$
D	C	B	A		
0	0	0	0		
0	0	0	1		
0	0	1	0		
0	0	1	1		
0	1	0	0		
0	1	0	1		
0	1	1	0		
0	1	1	1		
1	0	0	0		
1	0	0	1		
1	0	1	0		
1	0	1	1		
1	1	0	0		
1	1	0	1		
1	1	1	0		
1	1	1	1		

QUESTIONS

Complete questions 1 to 6.

1. With only switch A at a logical 1 in this circuit, the *gain* is _____, based upon the formula $A_v = R_f/R_{in}$.

1. _____

2. With all switches at a logical 1 in this circuit, the *gain* is _____, based upon the formula $A_v = R_f/R_{in}$. You also will use the parallel resistance formula.

2. _____

3. If the feedback resistor R_f is 5 kΩ in this circuit, the *gain* will be _____ with all switches at a logical 1 ($A_v = R_f/R_{in}$).

3. _____

4. If the input voltage were 3.6 V, what would be the output voltage V_{out} in question 3 ($V_{out} = V_{in} \times A_v$)?

4. _____

5. The maximum output voltage of the D/A converter can be increased by _____ (decreasing, increasing) the value of the feedback resistor R_f.

5. _____

6. A D/A converter consists of what *two* sections?
   a. A/D converter
   b. Comparator
   c. Resistor network
   d. Summing amplifier

6. _____ _____

14-3 CIRCUIT SIMULATION: D/A CONVERTER

OBJECTIVE

To construct and test an elementary analog-to-digital (D/A) converter using electronic circuit simulation software.

MATERIALS

• electronic circuit simulation software on computer system.

SYSTEM DIAGRAM

A wiring diagram for an elementary digital-to-analog converter is shown in Fig. 14-11. This is the same simple D/A converter featured in the previous lab but using circuit simulation instead of hardware.

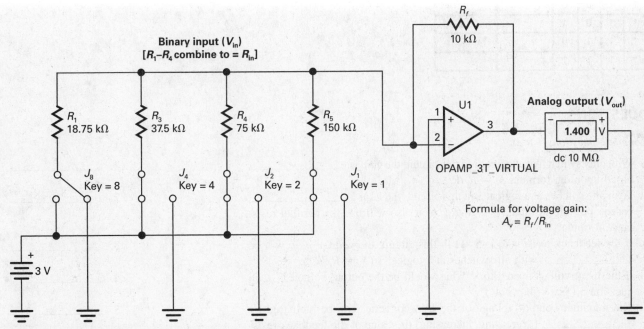

Fig. 14-11 Wiring diagram for digital-to-analog converter.

PROCEDURE

1. Construct the D/A converter circuit shown in Fig. 14-11 using the electronic circuit simulation software.
2. Operate the simulated D/A converter circuit by applying the binary inputs on the left in Table 14-3. Record the results in the center column.
3. Calculate the voltage gain for each row in Table 14-3. Remember that the input voltage is 3 V in this circuit simulation. Record your calculations in the right column of Table 14-3.
4. Try changing a few values in your simulated circuit. Change the input voltage to 3.6 V and the feedback resistor R_f to 5 kΩ. Observe the results.
5. Show your simulated D/A converter to your instructor.

TABLE 14-3 D/A Converter Results

INPUT				OUTPUT	Calculated gain
Binary				Analog	
8s	4s	2s	1s	Measured volts	$A_v = \dfrac{V_{out}}{V_{in}}$
D	C	B	A		
0	0	0	0		
0	0	0	1		
0	0	1	0		
0	0	1	1		
0	1	0	0		
0	1	0	1		
0	1	1	0		
0	1	1	1		
1	0	0	0		
1	0	0	1		
1	0	1	0		
1	0	1	1		
1	1	0	0		
1	1	0	1		
1	1	1	0		
1	1	1	1		

QUESTIONS

Complete questions 1 to 6.

1. With only switch *A* at a logical 1 in this circuit, the *gain* is _____, based upon the formula $A_v = R_f/R_{in}$.

2. With all switches at a logical 1 in this circuit, the *gain* is _____, based upon the formula $A_v = R_f/R_{in}$. You also will use the parallel resistance formula.

3. If the feedback resistor R_f is 5 kΩ in this circuit, the *voltage gain* will be _____ with all switches at a logical 1 ($A_v = R_f/R_{in}$).

4. If the input voltage were 3.6 V, what would be the output voltage V_{out} in question 3 ($V_{out} = V_{in} \times A_v$)?

5. The maximum output voltage of the D/A converter can be increased by _____ (decreasing, increasing) the value of the feedback resistor R_f.

6. The D/A converter featured in Fig. 14-11 consists of what *two* functional sections?
   a. A/D converter
   b. Comparator
   c. Resistor network
   d. Summing amplifier

1. _____

2. _____

3. _____

4. _____

5. _____

6. _____

14-4 LAB EXPERIMENT: AN ELEMENTARY DIGITAL VOLTMETER

OBJECTIVE

To construct and test an analog-to-digital (A/D) converter used as a simple digital voltmeter.

MATERIALS

Qty.

1	7408 two-input AND gate IC
1	7493 counter IC
1	logic switch
1	seven-segment LED display
2	10-kΩ, ¼-W resistors
1	37.5-kΩ, ¼-W resistor (combine 4.7 kΩ and 33 kΩ)
1	75-kΩ, ¼-W resistor
1	4-V zener diode
1	dual 9-V, 12-V, or 15-V dc power supply (or two separate supplies)
1	voltmeter (DMM)

Qty.

1	7447 BCD-to-seven-segment decoder IC
2	741 op-amp ICs (μA741)
1	clock (free-running)
7	150-Ω, ¼-W resistors
1	18.7-kΩ, ¼-W resistor (combine 18 kΩ and 680 Ω)
1	47-kΩ, ¼-W resistor
1	150-kΩ, ¼-W resistor
1	5-V dc regulated power supply
1	variable 0- to 10-V dc power supply

SYSTEM DIAGRAM

A wiring diagram for a very simple one-digit voltmeter is shown in Fig. 14-12. The digital voltmeter is an A/D converter. This circuit contains both digital and linear (analog) ICs. The 7408, 7447, and 7493 ICs are digital units, while the 741 ICs are linear or analog in nature. It is common in larger systems to find both digital and linear ICs in the same unit.

The purpose for including this simplified digital voltmeter is to study the fundamental operation of a digital voltmeter circuit. Also, the digital voltmeter circuit demonstrates how digital and analog devices work together in a hybrid electronic system. Real-world digital voltmeters and DMMs are based on commercial LSI CMOS ICs.

This experiment is best performed using circuit simulation (see Circuit Simulation 14-5 for this alternative).

PROCEDURE

1. Insert the two 741 op-amp ICs; the 7408, 7447, and 7493 ICs; and the seven-segment display into the mounting board.
2. Power OFF. Connect power (V_{CC} and GND) to the 7408, 7447, and 7493 ICs. Connect +5 V (V_{CC}) only to the seven-segment display. Connect the dual supply (+12 and −12 V) to the two 741 ICs. The Dynalogic DT-1000 Trainer provides +12- and −12-V dc power.
3. Wire the digital voltmeter (A/D converter with a digital readout) in Fig. 14-12.
4. Use the free-running clock for the *CLK* input. Use an input switch for the reset. Use the 0- to 10-V dc power supply for the analog input voltage V_{in}.
5. Completely wire the voltmeter.
6. Power ON. Set the input voltage V_{in} at +5 V. Reset the counter (display will read 0). Move the reset input to count position (logical 0). The display should count to 5 and *stop*. Try different voltages at V_{in}. Reset and count up. If the digital readout seems too high or low, adjust the accuracy by changing the value of feedback resistor R_f.

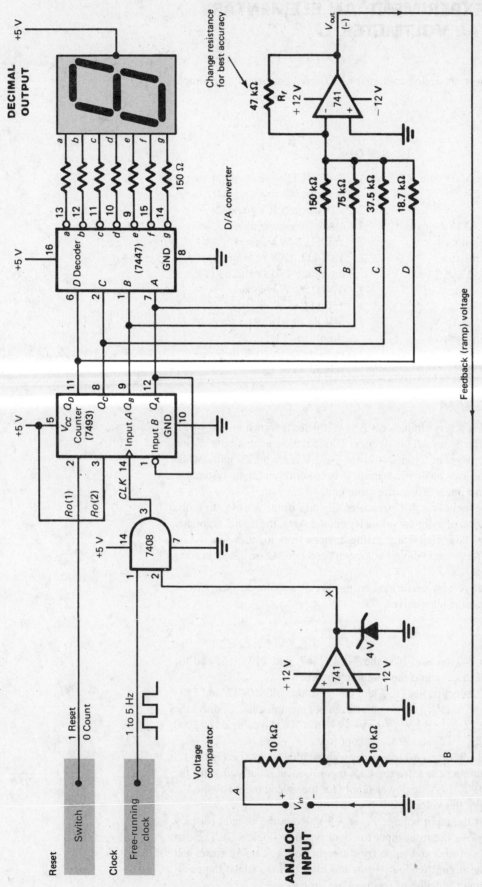

Fig. 14-12 Wiring diagram for a simple digital voltmeter (hardware version).

370

7. If the V_{in} is set higher than the dual power-supply voltages to the op amps, what happens? Try it!

8. Connect a voltmeter at the V_{out} position on the D/A converter. Observe the voltage as you reset and count. Record your observations.

_____

9. Connect the voltmeter at point X of the comparator. Observe the voltages as you reset and count. Record your observations.

_____

10. Have your instructor approve your digital voltmeter.

11. Power OFF. Take down the circuit, and return all equipment to its proper place.

QUESTIONS

Complete questions 1 to 10.

1. Draw a block diagram of a digital voltmeter. Label the input V_{in}. Label the A/D converter, the decoder, and the display.

2. Draw a block diagram of the sections of an A/D converter. Label the input, clock, voltage comparator, AND gate, counter, D/A converter, and 4-bit binary outputs.

3. Output X of the comparator is at _____ (0, −4, +4) V when the voltmeter is counting.

4. Output X of the comparator is at _____ (0, 4) V when the voltmeter finishes counting.

5. The output of the comparator stops the counting by _____ (disabling, enabling) the AND gate.

6. When the voltage at input B of the comparator becomes _____ (greater, less) than at input A, the output goes to logical 0.

7. What is the purpose of the zener diode in the comparator?

8. With one pulse per second coming from the CLK, how long does it take the voltmeter to read 5 V?

9. The digital voltmeter is basically an encoder called a(n) _____.

10. The noninverting input of the op amp is used in the comparator. The _____ input of the op amp was used in the D/A converter in this circuit.

3. _____

4. _____

5. _____

6. _____

7. _____
   _____

8. _____

9. _____

10. _____

14-5 CIRCUIT SIMULATION: DIGITAL VOLTMETER

OBJECTIVE

To construct and test an elementary digital voltmeter using electronic circuit simulation software.

MATERIALS

• electronic circuit simulation software on computer system.

SYSTEM DIAGRAM

A wiring diagram for an elementary digital voltmeter is shown in Fig. 14-13. This is the same simple digital voltmeter featured in the previous lab but using a circuit simulation instead of hardware.

PROCEDURE

1. Construct the digital voltmeter circuit shown in Fig. 14-13 using the electronic circuit simulation software.
2. Set the V_1 battery to 5 V ($+V_{in} = 5$ V).
3. Power ON. Operate the experimental digital voltmeter circuit. Reset the counter to 0 using the R key (H = reset, L = count). Toggle the clock input switch slowly, watching the output voltage on the seven-segment display using the space bar. Also observe the two voltmeters at the left (V_{in} and negative feedback voltage). The voltage display should count upward until the negative feedback voltage is greater than the positive input voltage (V_{in}). Try the reset/count-up sequence several times.
4. Power OFF. Change the input voltage. V_{in} can range from 1 to 15 V. The hexadecimal seven-segment display uses the characters 0–9, A, b, C, d, E and F.
5. Power ON. Operate the experimental digital voltmeter.
6. To increase the accuracy of the experimental digital voltmeter, you can adjust the value of the feedback resistor (R_f) in the D/A converter section.
7. Show your instructor simulation, and be prepared to answer questions about the circuit.

QUESTIONS

Complete questions 1 to 7.

1. Fundamentally, this digital voltmeter is a(n) _____ (A/D converter, ZIP converter).

   1. _____

2. Refer to Fig. 14-13. Clock pulses from the clock _____ (are blocked by, pass through) the AND gate when the output of the comparator (point X) goes HIGH.

   2. _____

3. Refer to Fig. 14-13. If the input voltage (V_{in}) is +5 V while the negative feedback voltage from the D/A converter is at −2 V, the voltage comparator's output will be _____ (HIGH, LOW).

   3. _____

4. Refer to Fig. 14-13. If the input voltage (V_{in}) is +5 V while the negative feedback voltage from the D/A converter is at −5.3 V, the voltage comparator's output will be _____ (HIGH, LOW).

   4. _____

5. Refer to Fig. 14-13. If the input voltage (V_{in}) is +12 V while the negative feedback voltage from the D/A converter is at −10 V, the voltage comparator's output will be _____ (HIGH, LOW).

   5. _____

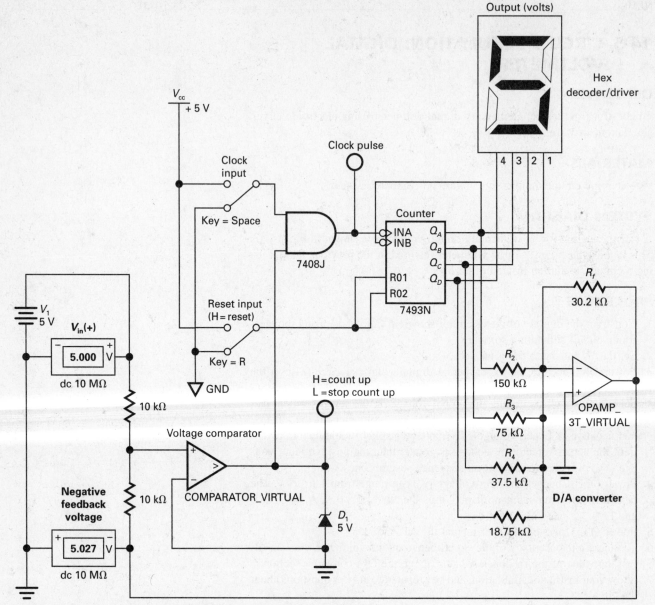

Fig. 14-13 Wiring diagram for an experimental digital voltmeter.

6. Refer to Fig. 14-13. The output of the voltage comparator controls the _____ (AND gate, D/A converter).

7. Refer to Fig. 14-13. When the negative feedback voltage is greater than the input voltage (V_{in}), then the comparator's output goes HIGH and the NAND gate blocks clock pulses from reaching the counter. (T or F)

6. _____

7. _____

14-6 LAB EXPERIMENT: USING THE CMOS ADC0804 A/D CONVERTER IC

OBJECTIVES

1. To construct and test an 8-bit A/D converter using the ADC0804 CMOS IC.
2. To use an oscilloscope to observe the "start conversion" pulse.
3. To calculate the conversion rate of the A/D converter in the test circuit.

MATERIALS

Qty. | **Qty.**

1 ADC0804 8-bit A/D converter CMOS IC
8 LED indicator-light assemblies
1 10-kΩ potentiometer (linear)
1 10-kΩ, ¼-W resistor

1 150-pF capacitor
1 5-V dc regulated power supply
1 DMM (digital multimeter)
1 triggered oscilloscope
1 keypad (N.O. switch)

SYSTEM DIAGRAM

A test circuit for the ADC0804 8-bit A/D converter is drawn in Fig. 14-14. The ADC0804 IC is wired as a "free-running" A/D converter in that the short negative pulse emitted by the $\overline{INTR}$ output at the end of each A/D conversion is fed back to the $\overline{WR}$ input. This "start conversion" pulse starts a new analog-to-digital conversion.

The N.O. push-button switch shown in Fig. 14-14 is used only to start the conversion process after power up. Changing the position of the potentiometer R_2 changes the voltage being fed into the ADC0804 IC. The reference voltage

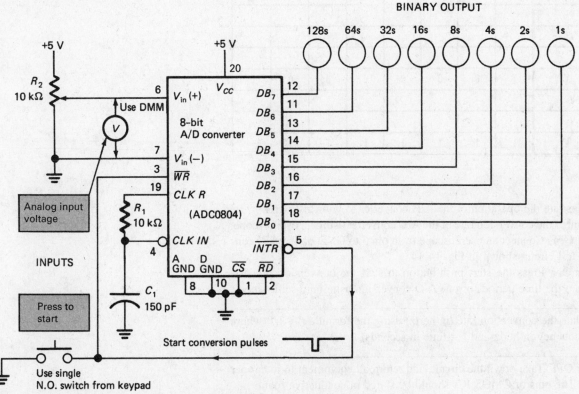

Fig. 14-14 Wiring of test circuit for the ADC0804 A/D converter IC.

being used in this test circuit is the 5-V dc from the regulated power supply. Each increase of about 0.020 V in the analog input voltage increases the binary count by one (5 V/255 = 0.0196 V or about 0.020 V). An accurate DMM will be useful for checking the operation of the A/D converter circuit shown in Fig. 14-14.

PROCEDURE

1. Insert the ADC0804 CMOS IC into the mounting board.

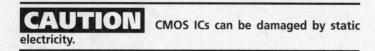 **CAUTION** CMOS ICs can be damaged by static electricity.

2. Power OFF. Connect power (V_{CC}, A GND, and D GND) to the ADC0804 IC.
3. Wire the entire A/D converter circuit shown in Fig. 14-14. Use a single key from the keypad for the N.O. push-button switch. Use eight LED indicator-light assemblies for the output. For best results, use an accurate DMM to measure the analog input voltage to the ADC0804 IC.
4. Power ON. Press the push-button switch once to start the conversion process. Turn the potentiometer, and observe the DMM readings and the binary outputs. Observe and record selected binary outputs and meter measurements in Table 14-4.

TABLE 14-4 A/D Converter Test Results

INPUT	OUTPUT							
Measured Analog Voltage	Binary							
	128s	64s	32s	16s	8s	4s	2s	1s
0.040 V								
0.500 V								
1.000 V								
2.000 V								
V	1	0	0	0	0	0	0	0
3.000 V								
4.000 V								
V	1	1	1	1	1	1	1	1

5. Demonstrate the operation of the A/D converter to your instructor.
6. Determine the conversion rate of the A/D converter using an oscilloscope.
7. Power OFF. Connect an oscilloscope from pin 3 to GND on the A/D converter test circuit shown in Fig. 14-14.
8. Power ON. Press the start push-button to start the conversion process. Measure the time period for one A/D conversion using an oscilloscope.

9. Calculate the conversion rate (in hertz) using the formula $f = 1/t$ (where f = frequency in hertz and t = time in seconds).

10. Power OFF. Take down the circuit, and return all equipment to its proper place. The pins of CMOS ICs should be stored in conductive foam.

QUESTIONS

Complete questions 1 to 10.

1. The ADC0804 is an 8-bit A/D converter with _____ (binary, decimal) outputs.

2. The $\overline{INTR}$ output of the ADC0804 IC emits a _____ (negative, positive) pulse at the end of each A/D conversion. This pulse is fed back to the $\overline{WR}$ input to _____ (end, start) the next analog-to-digital conversion.

3. What is the resolution of the ADC0804 A/D converter IC (number of bits)?

4. Refer to Fig. 14-14. An input voltage of about 0.020 V would cause an output of binary _____.

5. Refer to Fig. 14-14. An input voltage of about _____ V would cause an output of binary 11111111.

6. Refer to Fig. 14-14. What is the reference voltage in this A/D converter circuit?

7. The ADC0804 A/D converter IC has a resolution of _____ (4, 8) bits and a fast conversion time of about _____ microseconds. (see the ADC0804 data sheet—use Internet search if necessary.)

8. The ADC0804 A/D IC can operate as a "stand-alone" A/D converter and is also compatible with microprocessors. (T or F) (See ADC0804 data sheet.)

9. The ADC0804 A/D converter IC uses the successive-approximation technique during analog-to-digital conversion. (T or F) (See ADC0804 data sheet.)

10. The outputs of the ADC0804 A/D converter IC use _____ (open-emitter, three-state) output latches which make them compatible with some computer data buses. (See ADC0804 data sheet.)

1. _____

2. _____ _____

3. _____

4. _____

5. _____
_____

6. _____

7. _____ _____

8. _____

9. _____

10. _____

14-7 LAB EXPERIMENT: DIGITAL LIGHT METER

OBJECTIVES

1. Construct and test a digital light meter circuit with binary readouts using the ADC0804 A/D converter IC.
2. Construct and test a second digital light meter circuit with decimal readouts.

MATERIALS

Qty.   **Qty.**

1   555 timer IC   1   seven-segment LED display
1   7447 BCD-to-seven-segment   7   150-Ω, ¼-W resistors
    decoder/driver TTL IC   1   1-kΩ potentiometer (linear)
1   ADC0804 8-bit A/D   1   1-kΩ, ¼-W resistor
    converter CMOS IC   1   10-kΩ, ¼-W resistor
1   photocell, cadmium sulfide   1   470-kΩ, ¼-W resistor
8   LED indicator-light   1   150-pF capacitor
    assemblies   1   1-μF electrolytic capacitor
                                              1   5-V dc regulated power
                                                  supply

SYSTEM DIAGRAMS

A wiring diagram for a *digital light meter* is drawn in Fig. 14-15. A cadmium sulfide photocell R_3 serves as the input transducer. The photocell converts light energy into a varying resistance. As the light intensity striking the surface of the photocell increases, its resistance decreases. The photocell will have maximum resistance under low-light conditions. The potentiometer R_2 can be used to calibrate the light meter. Note that it is the voltage drop across the potentiometer R_2 that serves as the analog input voltage to the ADC0804 A/D converter IC.

The 555 timer is wired as an astable MV which generates a clock signal of about 1 Hz. Each LOW-to-HIGH transition of the clock pulse at the $\overline{WR}$ input to the ADC0804 chip starts a new analog-to-digital conversion. Because of the slow clock speed, the digital light meter samples the light intensity only about once each second.

The schematic diagram for a second digital light meter is shown in Fig. 14-16. The new circuit is the same as the old except that the updated light meter features an easy-to-read decimal output. Note that only the four most significant bits of the A/D converter's outputs are decoded by the 7447 BCD-to-seven-segment decoder/driver IC. The ADC0804 IC is being used as a low-resolution 4-bit A/D converter in this new light meter circuit. Potentiometer R_2 can be adjusted to scale the output so that very low light will cause the display to read 0 and high-intensity light will cause the display to read a high of 9.

PROCEDURE

1. Insert the ADC0804 CMOS IC and the 555 IC into the mounting board.

 **CMOS ICs can be damaged by static electricity.**

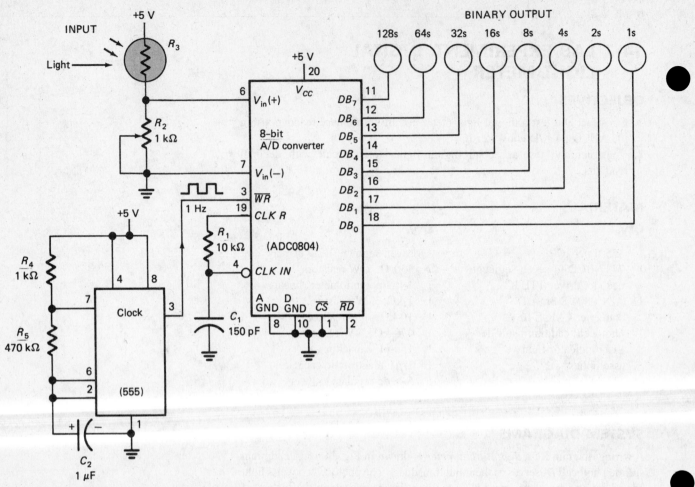

Fig. 14-15 Wiring diagram for a digital light meter with binary outputs.

2. Power OFF. Connect power (V_{CC} and GND) to both ICs.
3. Power OFF. Wire the digital light meter shown in Fig. 14-15. Arrange the eight LED indicator-light assemblies in the order shown in Fig. 14-15 so they read in binary. Set the potentiometer near the middle of its range.
4. Power ON. Vary the light striking R_3 by covering, uncovering, and shining a light at the photocell. Adjust potentiometer R_2 if the scaling of the binary output needs adjusting.
5. Show your instructor your first digital light meter circuit.
6. Power OFF. Insert the 7447 IC and seven-segment LED display into the mounting board.
7. Power OFF. Connect power (V_{CC} and GND) to the added 7447 IC. Connect only +5 V to the common anode connection(s) of the seven-segment LED display.
8. Power OFF. Rewire the digital light meter circuit as shown in Fig. 14-16. Be sure the seven 150-Ω limiting resistors between the 7447 IC and the seven-segment LED display are in place.
9. Power ON. Vary the light striking the photocell. Adjust potentiometer R_2 to properly scale the decimal output.
10. Show your instructor your new digital light meter circuit featuring a decimal output. Be prepared to answer questions about the circuit's operation.
11. Power OFF. Take down the circuit, and return all equipment to its proper place. The pins of the CMOS IC should be stored in conductive foam.

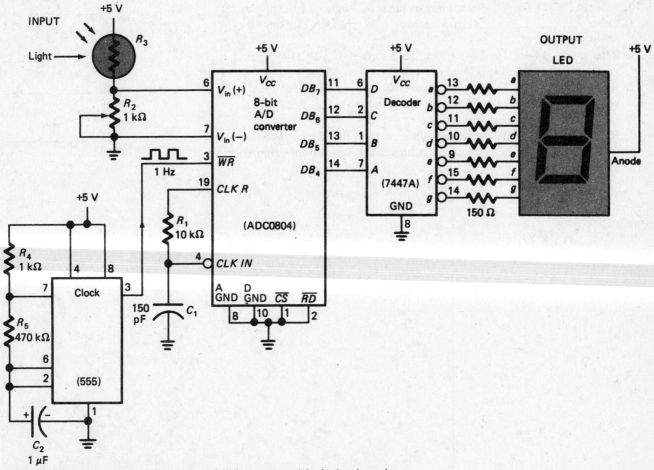

Fig. 14-16 Wiring diagram for a digital light meter with decimal readout.

QUESTIONS

Complete questions 1 to 13.

1. What component serves as the input transducer in the digital light meter circuit in Fig. 14-15?

2. The ADC0804 is being used as an 8-bit A/D converter in Fig. 14-15. However, in Fig. 14-16 it is used as an A/D converter with a resolution of _____-bits.

3. Refer to Fig. 14-15. Increasing the light intensity striking the photocell _____ (decreases, increases) its resistance.

4. Refer to Fig. 14-15. Increasing the light intensity striking the photocell _____ (decreases, increases) the current flowing through series resistances R_2 and R_3.

5. Refer to Fig. 14-15. Increasing the light intensity striking the photocell _____ (decreases, increases) the analog input voltage to the A/D converter IC.

6. Refer to Fig. 14-16. Increasing the light intensity striking the photocell _____ (decreases, increases) the decimal number of the output display.

7. Refer to Fig. 14-16. The light intensity striking the photocell is sampled about _____ (1, 5000) time(s) per second by the A/D converter.

1. _____

2. _____

3. _____

4. _____

5. _____

6. _____

7. _____

8. Refer to Fig. 14-16. Component R_3 converts _____ energy into a varying resistance.

9. What is the reference voltage used by the A/D converter in Fig. 14-16?

10. Refer to Fig. 14-16. A higher reading on the seven-segment LED display means _____ (higher, lower) light intensity striking the photocell.

11. Refer to Fig. 14-16. The 7447 IC can be described as a BCD-to-seven-segment decoder capable of driving an LED display. (T or F)

12. Refer to Fig. 14-16. The seven 150-Ω resistors located between the decoder and seven-segment LED display are referred to as _____ (analog, current-limiting) resistors.

13. Refer to Fig. 14-16. The input *transducer* used in this digital light meter circuit is a _____ (cadmium sulfide, chromel alumel) photocell.

8. _____

9. _____

10. _____

11. _____

12. _____

13. _____

382

14-8 LAB EXPERIMENT: DIGITIZING TEMPERATURE

OBJECTIVES

1. To measure the resistance of a thermistor at various temperatures.
2. To wire and test an elementary circuit that digitizes an analog input from a temperature transducer generating a simple HIGH or LOW output.

MATERIALS

Qty.

1 74LS14 (or 7414) Hex inverter TTL IC
1 thermistor, 10-kΩ at 25°C
1 330-Ω resistor

Qty.

1 10-kΩ potentiometer
1 logic probe
1 DMM (ohmmeter and dc voltmeter)

SYSTEM DIAGRAMS

A *thermistor* is a *temperature-sensitive resistor*. The sketch in Fig. 14-17 shows a typical thermistor with extended leads needed in this experiment. The resistance of the thermistor will be measured at room temperature (about 68°F) and then colder (about 32°F) and hotter (>110°F). The results will indicate that the resistance of the thermistor increases as the temperature decreases. It will also be observed that the change in resistance of the thermistor is not proportional to the change in temperature. It is said that the thermistor has a *nonlinear temperature-versus-resistance characteristic.*

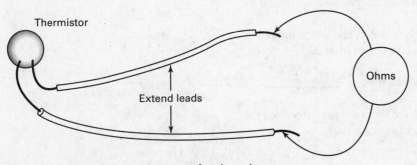

Fig. 14-17 Testing the resistance of a thermistor.

An elementary digitizing circuit is detailed in Fig. 14-18. The thermistor R_3 is the temperature sensor, while the potentiometer R_2 is for calibration. The voltmeter measures the analog input voltage, while the logic probe indicates the digital output. The 74LS14 Schmitt trigger inverter performs the digitizing. The digital output may be either HIGH or LOW.

383

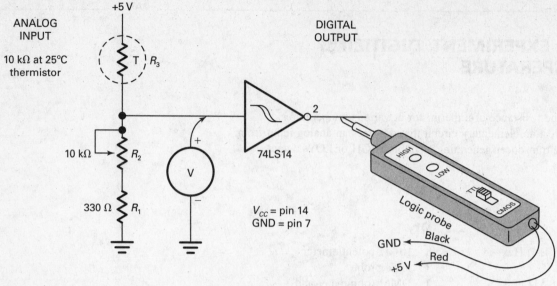

Fig. 14-18 Temperature-sensing input (analog) with digital output.

PROCEDURE

1. Refer to Fig. 14-17. Using an ohmmeter, measure the resistance of the thermistor at room temperature (about 68°F), in ice water (about 32°F), and in hot tap water (>110°F). Record your results.

    Ice water temperature resistance = _____

    Room temperature resistance = _____

    Hot tap water temperature resistance = _____

2. Based on your measured results, the resistance of the thermistor appears to be _____ (linear, nonlinear) when compared to temperature.

3. Power OFF. Wire the digitizing circuit shown in Fig. 14-18. The leads on the thermistor must be extended so they are long enough to reach cups containing ice water and hot tap water. Set the DMM on dc volts. Set the logic probe for TTL.

4. Power ON. Touch the tip of the thermistor in ice water (about 32°F), and calibrate the digitizer circuit so that the dc voltage at the input to the 74LS14 IC is about +0.8 V. The output of the Schmitt trigger inverter should be HIGH.

5. Power ON. Remove the thermistor from ice water, and place in hot tap water (>110°F). Read the input voltage and the output from the logic probe. Record your results.

    Input voltage (thermistor in hot tap water) = _____

    Output (HIGH or LOW) = _____

6. Power ON. Permit the thermistor to adjust the room temperature. Read the input voltage and the logical output. Record your results.

    Input voltage (room temperature) = _____

    Output (HIGH or LOW) = _____

7. Show your instructor your digitizer circuit, and prepare to demonstrate its operation and answer selected questions about its operation.

8. Power OFF. Take down the circuit, and return all equipment to its proper place.

QUESTIONS

Complete questions 1 to 12.

1. Refer to Fig. 14-17. Based on your measurements, as the temperature of the thermistor increases, its resistance _____ (decreases, increases, stays the same).

                                                         1. _____

2. Refer to Fig. 14-18. The temperature transducer in this circuit is the _____ (inverter, thermistor).

3. Refer to Fig. 14-18. The device in this circuit that does the digitizing (analog-to-digital conversion) is the _____ (potentiometer, Schmitt trigger inverter IC).

4. Refer to Fig. 14-18. The device in this circuit that is used for calibration is the _____ (inverter IC, potentiometer).

5. Refer to Fig. 14-18. If the input voltage measured about +0.8 V, the logic probe would indicate a _____ (HIGH, LOW) output.

6. Refer to Fig. 14-18. If the temperature of the thermistor increases to greater than 110°F, the resistance of R_3 _____ (decreases, increases), which causes the input voltage to _____ (decrease, increase). The output of the 74LS14 inverter goes _____ (HIGH, LOW).

7. Refer to Fig. 14-18. If the temperature of the thermistor decreases to about 32°F, the output of the 74LS14 inverter goes _____ (HIGH, LOW).

8. The thermistor used in this lab has a *linear temperature-versus-resistance characteristic.* (T or F)

9. The circuit in Fig. 14-18 seems better suited to the sensing function of a _____ (thermometer, thermostat).

10. Thermistors are very _____ (expensive, inexpensive) sensors. (Use an electronics catalog or Internet search.)

11. The DS1620 temperature sensor IC by Dallas Semiconductor is suitable for implementing either a thermometer or thermostatic controls. (T or F) (Check the DS1620 IC digital thermometer via a Internet search.)

12. A DS1620 temperature sensor IC is commonly
    a. Used as a stand-alone device
    b. Interfaced with a microcontroller and display
    (Look for the DS1620 IC and Parallax with an Internet search.)

2. _____

3. _____

4. _____

5. _____

6. _____
   _____
   _____

7. _____

8. _____

9. _____

10. _____

11. _____

12. _____

APPENDIX A

Pin Diagrams

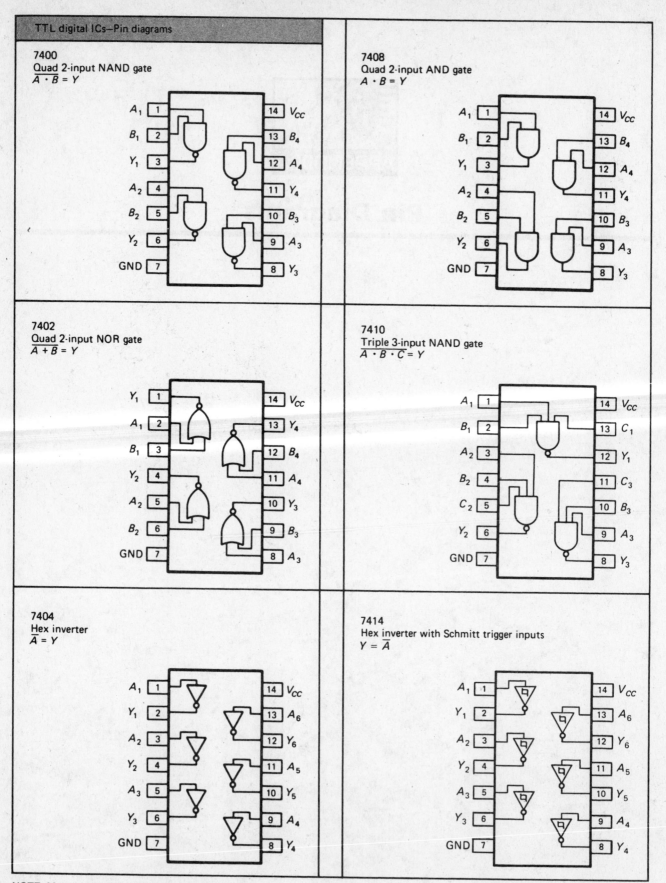

TTL digital ICs—Pin diagrams

7400
Quad 2-input NAND gate
$\overline{A \cdot B} = Y$

7408
Quad 2-input AND gate
$A \cdot B = Y$

7402
Quad 2-input NOR gate
$\overline{A + B} = Y$

7410
Triple 3-input NAND gate
$\overline{A \cdot B \cdot C} = Y$

7404
Hex inverter
$\overline{A} = Y$

7414
Hex inverter with Schmitt trigger inputs
$Y = \overline{A}$

NOTE: May substitute 74LSXX ICs for 74XX.

388

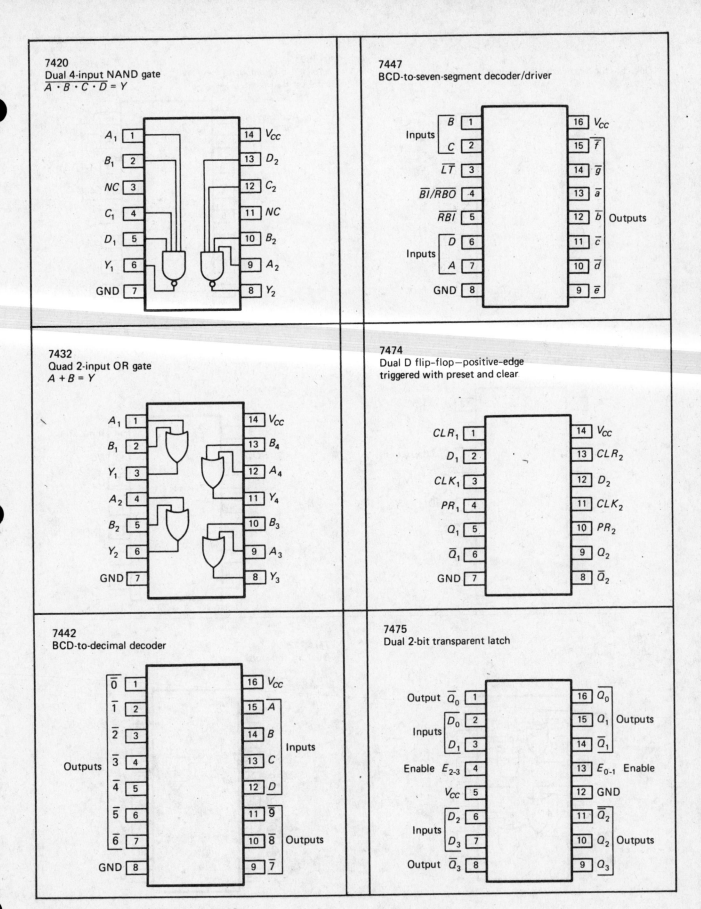

7420
Dual 4-input NAND gate
$\overline{A \cdot B \cdot C \cdot D} = Y$

A_1	1	14	V_{CC}
B_1	2	13	D_2
NC	3	12	C_2
C_1	4	11	NC
D_1	5	10	B_2
Y_1	6	9	A_2
GND	7	8	Y_2

7447
BCD-to-seven-segment decoder/driver

B	1	16	V_{CC}
C	2	15	$\overline{f}$
$\overline{LT}$	3	14	$\overline{g}$
$\overline{BI}/\overline{RBO}$	4	13	$\overline{a}$
$\overline{RBI}$	5	12	$\overline{b}$
D	6	11	$\overline{c}$
A	7	10	$\overline{d}$
GND	8	9	$\overline{e}$

Inputs (1, 2)
Inputs (6, 7)
Outputs (15, 14, 13, 12, 11, 10, 9)

7432
Quad 2-input OR gate
$A + B = Y$

A_1	1	14	V_{CC}
B_1	2	13	B_4
Y_1	3	12	A_4
A_2	4	11	Y_4
B_2	5	10	B_3
Y_2	6	9	A_3
GND	7	8	Y_3

7474
Dual D flip-flop—positive-edge
triggered with preset and clear

CLR_1	1	14	V_{CC}
D_1	2	13	CLR_2
CLK_1	3	12	D_2
PR_1	4	11	CLK_2
Q_1	5	10	PR_2
$\overline{Q}_1$	6	9	Q_2
GND	7	8	$\overline{Q}_2$

7442
BCD-to-decimal decoder

$\overline{0}$	1	16	V_{CC}
$\overline{1}$	2	15	A
$\overline{2}$	3	14	B
$\overline{3}$	4	13	C
$\overline{4}$	5	12	D
$\overline{5}$	6	11	$\overline{9}$
$\overline{6}$	7	10	$\overline{8}$
GND	8	9	$\overline{7}$

Outputs (left), Inputs (15, 14, 13, 12), Outputs (11, 10, 9)

7475
Dual 2-bit transparent latch

Output $\overline{Q}_0$	1	16	Q_0
D_0	2	15	Q_1
D_1	3	14	$\overline{Q}_1$
Enable $E_{2\text{-}3}$	4	13	$E_{0\text{-}1}$ Enable
V_{CC}	5	12	GND
D_2	6	11	$\overline{Q}_2$
D_3	7	10	Q_2
Output $\overline{Q}_3$	8	9	Q_3

Inputs (D_0, D_1), Outputs (Q_0, Q_1, $\overline{Q}_1$), Inputs (D_2, D_3), Outputs (Q_2, Q_3)

NOTE: May substitute 74LSXX for 74XX ICs.

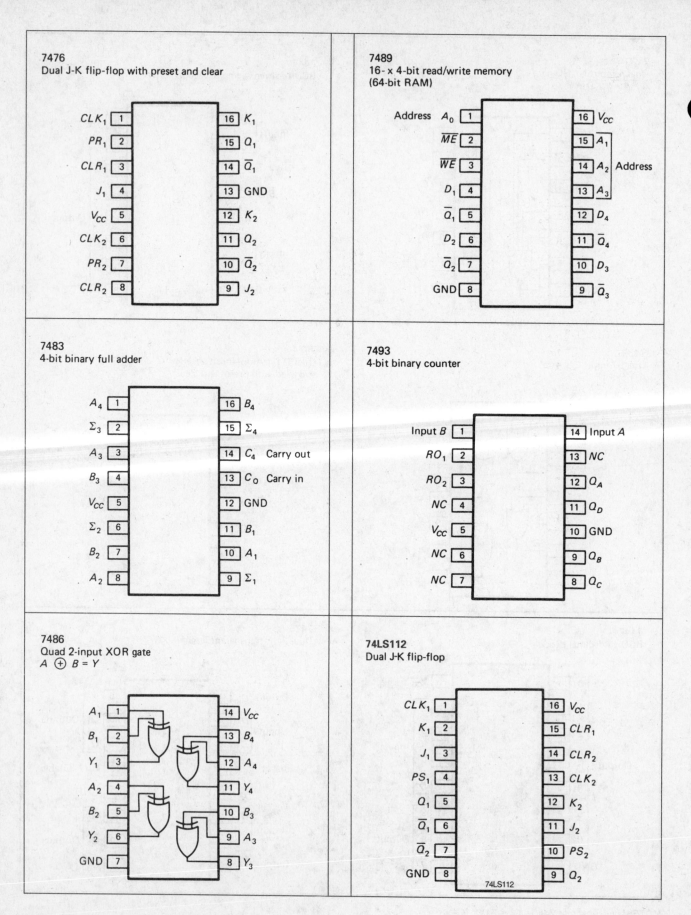

7476
Dual J-K flip-flop with preset and clear

CLK_1	1	16	K_1
PR_1	2	15	Q_1
CLR_1	3	14	$\overline{Q}_1$
J_1	4	13	GND
V_{CC}	5	12	K_2
CLK_2	6	11	Q_2
PR_2	7	10	$\overline{Q}_2$
CLR_2	8	9	J_2

7489
16- x 4-bit read/write memory
(64-bit RAM)

Address A_0 1 — 16 V_{CC}
$\overline{ME}$ 2 — 15 A_1
$\overline{WE}$ 3 — 14 A_2 Address
D_1 4 — 13 A_3
$\overline{Q}_1$ 5 — 12 D_4
D_2 6 — 11 $\overline{Q}_4$
$\overline{Q}_2$ 7 — 10 D_3
GND 8 — 9 $\overline{Q}_3$

7483
4-bit binary full adder

A_4	1	16	B_4
Σ_3	2	15	Σ_4
A_3	3	14	C_4 Carry out
B_3	4	13	C_0 Carry in
V_{CC}	5	12	GND
Σ_2	6	11	B_1
B_2	7	10	A_1
A_2	8	9	Σ_1

7493
4-bit binary counter

Input B	1	14	Input A
RO_1	2	13	NC
RO_2	3	12	Q_A
NC	4	11	Q_D
V_{CC}	5	10	GND
NC	6	9	Q_B
NC	7	8	Q_C

7486
Quad 2-input XOR gate
$A \oplus B = Y$

A_1	1	14	V_{CC}
B_1	2	13	B_4
Y_1	3	12	A_4
A_2	4	11	Y_4
B_2	5	10	B_3
Y_2	6	9	A_3
GND	7	8	Y_3

74LS112
Dual J-K flip-flop

CLK_1	1	16	V_{CC}
K_1	2	15	CLR_1
J_1	3	14	CLR_2
PS_1	4	13	CLK_2
Q_1	5	12	K_2
$\overline{Q}_1$	6	11	J_2
$\overline{Q}_2$	7	10	PS_2
GND	8	9	Q_2

74LS112

NOTE: May substitute 74LSXX for 74XX ICs.

390

74121
Monostable multivibrator (one-shot multivibrator)

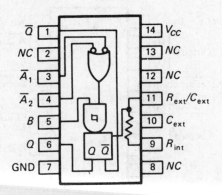

$\overline{Q}$	1		14	V_{CC}
NC	2		13	NC
$\overline{A}_1$	3		12	NC
$\overline{A}_2$	4		11	R_{ext}/C_{ext}
B	5		10	C_{ext}
Q	6		9	R_{int}
GND	7		8	NC

74147
Decimal-to-4-line BCD priority encoder

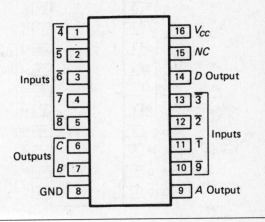

$\overline{4}$	1		16	V_{CC}
$\overline{5}$	2		15	NC
Inputs $\overline{6}$	3		14	D Output
$\overline{7}$	4		13	$\overline{3}$
$\overline{8}$	5		12	$\overline{2}$ Inputs
Outputs C	6		11	$\overline{1}$
B	7		10	$\overline{9}$
GND	8		9	A Output

74125
Quad 3-state buffer
$A = Y$

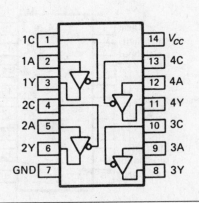

1C	1		14	V_{CC}
1A	2		13	4C
1Y	3		12	4A
2C	4		11	4Y
2A	5		10	3C
2Y	6		9	3A
GND	7		8	3Y

74148
8-line-to-3-line priority encoder

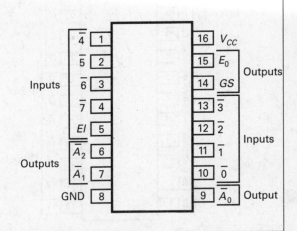

$\overline{4}$	1		16	V_{CC}
$\overline{5}$	2		15	E_0 Outputs
Inputs $\overline{6}$	3		14	GS
$\overline{7}$	4		13	$\overline{3}$
EI	5		12	$\overline{2}$ Inputs
Outputs $\overline{A}_2$	6		11	$\overline{1}$
$\overline{A}_1$	7		10	$\overline{0}$
GND	8		9	$\overline{A}_0$ Output

Truth table 74148 Encoder

	INPUTS								OUTPUTS				
EI	0	1	2	3	4	5	6	7	A2	A1	A0	GS	EO
H	X	X	X	X	X	X	X	X	H	H	H	H	H
L	H	H	H	H	H	H	H	H	H	H	H	H	L
L	X	X	X	X	X	X	X	L	L	L	L	L	H
L	X	X	X	X	X	X	L	H	L	L	H	L	H
L	X	X	X	X	X	L	H	H	L	H	L	L	H
L	X	X	X	X	L	H	H	H	L	H	H	L	H
L	X	X	X	L	H	H	H	H	H	L	L	L	H
L	X	X	L	H	H	H	H	H	H	L	H	L	H
L	X	L	H	H	H	H	H	H	H	H	L	L	H
L	L	H	H	H	H	H	H	H	H	H	H	L	H

H = High L = Low X = Irrelevant

NOTE: May substitute 74LSXX for 74XX ICs.

391

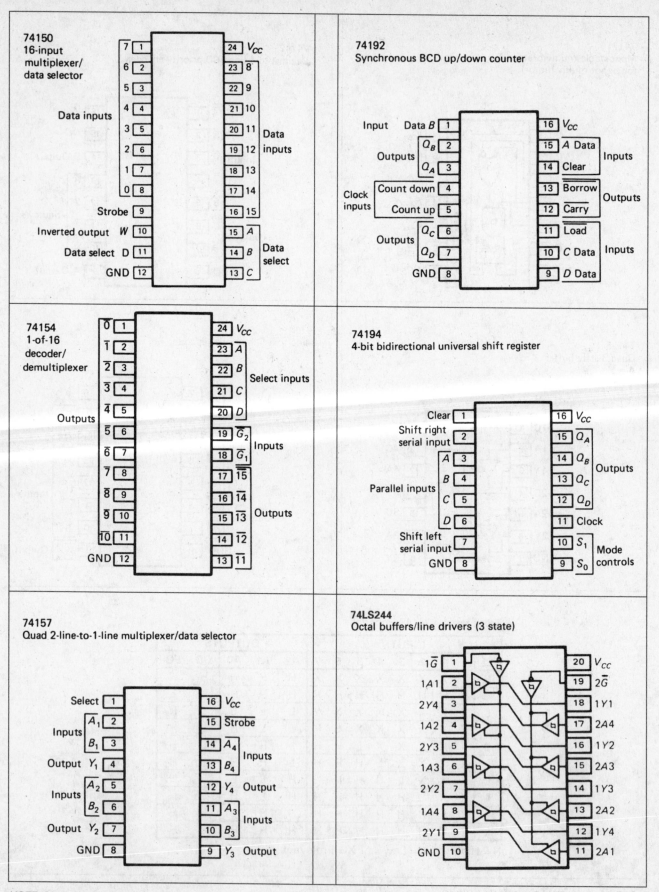

74150
16-input multiplexer/data selector

Data inputs

7	1		24	V_{CC}
6	2		23	8
5	3		22	9
4	4		21	10
3	5		20	11	Data inputs
2	6		19	12
1	7		18	13
0	8		17	14

Strobe — 9 — 16 — 15
Inverted output — W — 10 — 15 — A
Data select D — 11 — 14 — B — Data select
GND — 12 — 13 — C

74192
Synchronous BCD up/down counter

Input Data B — 1 — 16 — V_{CC}
Outputs — Q_B — 2 — 15 — A Data — Inputs
Outputs — Q_A — 3 — 14 — Clear
Clock inputs — Count down — 4 — 13 — Borrow — Outputs
Count up — 5 — 12 — Carry
Outputs — Q_C — 6 — 11 — Load
Q_D — 7 — 10 — C Data — Inputs
GND — 8 — 9 — D Data

74154
1-of-16 decoder/demultiplexer

Outputs

0	1		24	V_{CC}
1	2		23	A
2	3		22	B	Select inputs
3	4		21	C
4	5		20	D
5	6		19	$\overline{G}_2$	Inputs
6	7		18	$\overline{G}_1$
7	8		17	15
8	9		16	14
9	10		15	13	Outputs
10	11		14	12
GND	12		13	11

74194
4-bit bidirectional universal shift register

Clear — 1 — 16 — V_{CC}
Shift right serial input — 2 — 15 — Q_A
A — 3 — 14 — Q_B — Outputs
Parallel inputs — B — 4 — 13 — Q_C
C — 5 — 12 — Q_D
D — 6 — 11 — Clock
Shift left serial input — 7 — 10 — S_1 — Mode controls
GND — 8 — 9 — S_0

74157
Quad 2-line-to-1-line multiplexer/data selector

Select — 1 — 16 — V_{CC}
Inputs — $\overline{A}_1$ — 2 — 15 — Strobe
B_1 — 3 — 14 — $\overline{A}_4$
Output Y_1 — 4 — 13 — B_4 — Inputs
Inputs — $\overline{A}_2$ — 5 — 12 — Y_4 — Output
B_2 — 6 — 11 — $\overline{A}_3$ — Inputs
Output Y_2 — 7 — 10 — B_3
GND — 8 — 9 — Y_3 — Output

74LS244
Octal buffers/line drivers (3 state)

$1\overline{G}$ — 1 — 20 — V_{CC}
$1A1$ — 2 — 19 — $2\overline{G}$
$2Y4$ — 3 — 18 — $1Y1$
$1A2$ — 4 — 17 — $2A4$
$2Y3$ — 5 — 16 — $1Y2$
$1A3$ — 6 — 15 — $2A3$
$2Y2$ — 7 — 14 — $1Y3$
$1A4$ — 8 — 13 — $2A2$
$2Y1$ — 9 — 12 — $1Y4$
GND — 10 — 11 — $2A1$

NOTE: May substitute 74LSXX ICs for 74XX.

392

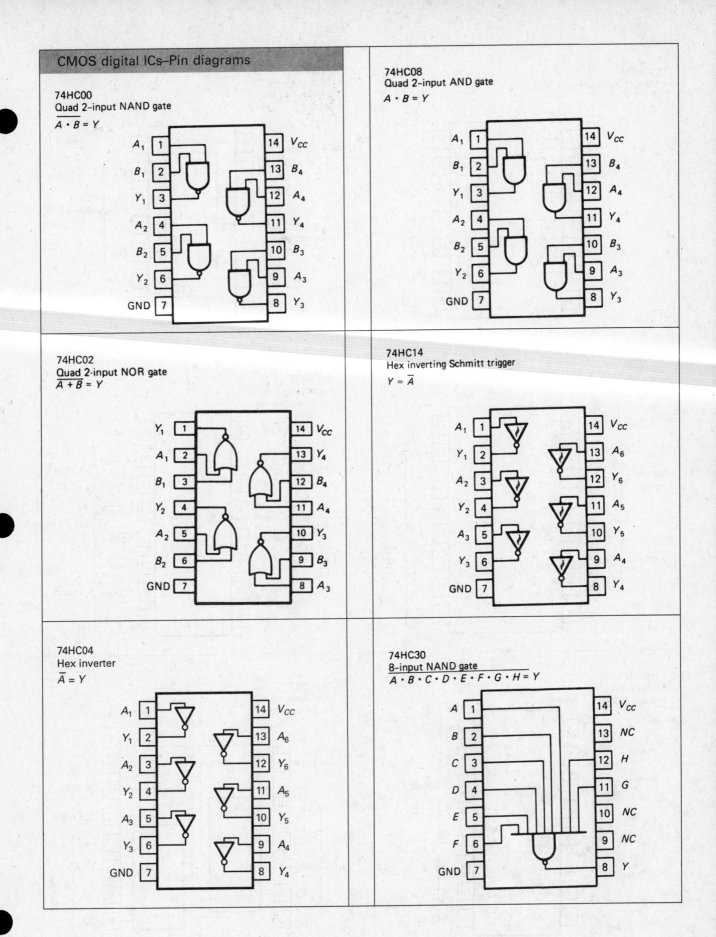

CMOS digital ICs–Pin diagrams

74HC00
Quad 2–input NAND gate
$\overline{A \cdot B} = Y$

Pin			Pin
A_1	1	14	V_{CC}
B_1	2	13	B_4
Y_1	3	12	A_4
A_2	4	11	Y_4
B_2	5	10	B_3
Y_2	6	9	A_3
GND	7	8	Y_3

74HC08
Quad 2–input AND gate
$A \cdot B = Y$

Pin			Pin
A_1	1	14	V_{CC}
B_1	2	13	B_4
Y_1	3	12	A_4
A_2	4	11	Y_4
B_2	5	10	B_3
Y_2	6	9	A_3
GND	7	8	Y_3

74HC02
Quad 2-input NOR gate
$\overline{A + B} = Y$

Pin			Pin
Y_1	1	14	V_{CC}
A_1	2	13	Y_4
B_1	3	12	B_4
Y_2	4	11	A_4
A_2	5	10	Y_3
B_2	6	9	B_3
GND	7	8	A_3

74HC14
Hex inverting Schmitt trigger
$Y = \overline{A}$

Pin			Pin
A_1	1	14	V_{CC}
Y_1	2	13	A_6
A_2	3	12	Y_6
Y_2	4	11	A_5
A_3	5	10	Y_5
Y_3	6	9	A_4
GND	7	8	Y_4

74HC04
Hex inverter
$\overline{A} = Y$

Pin			Pin
A_1	1	14	V_{CC}
Y_1	2	13	A_6
A_2	3	12	Y_6
Y_2	4	11	A_5
A_3	5	10	Y_5
Y_3	6	9	A_4
GND	7	8	Y_4

74HC30
8-input NAND gate
$\overline{A \cdot B \cdot C \cdot D \cdot E \cdot F \cdot G \cdot H} = Y$

Pin			Pin
A	1	14	V_{CC}
B	2	13	NC
C	3	12	H
D	4	11	G
E	5	10	NC
F	6	9	NC
GND	7	8	Y

393

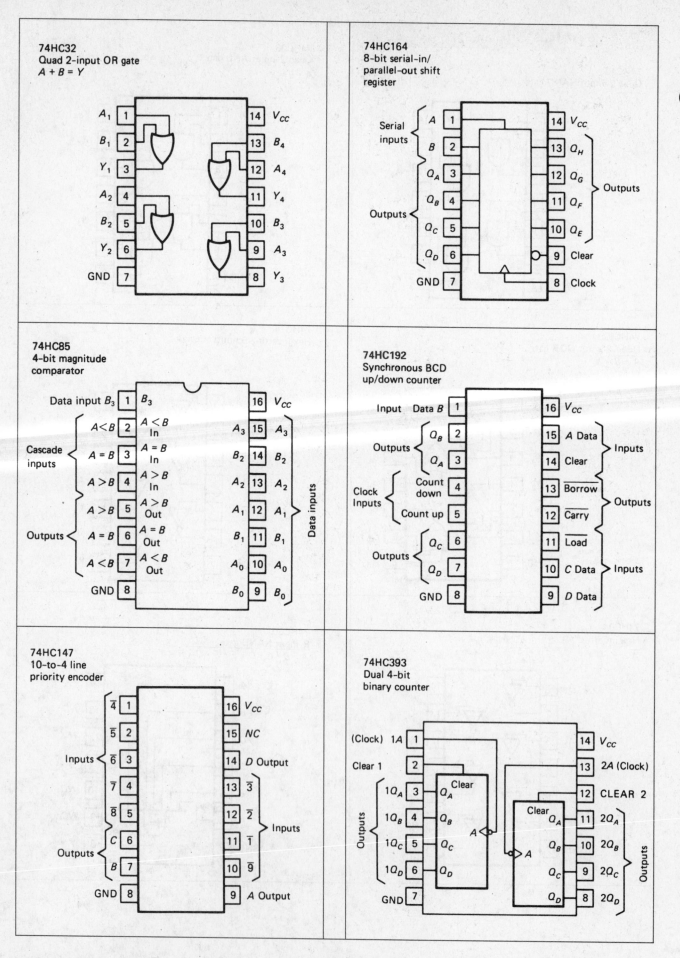

74HC32
Quad 2-input OR gate
$A + B = Y$

74HC164
8-bit serial-in/
parallel-out shift
register

74HC85
4-bit magnitude
comparator

74HC192
Synchronous BCD
up/down counter

74HC147
10-to-4 line
priority encoder

74HC393
Dual 4-bit
binary counter

394

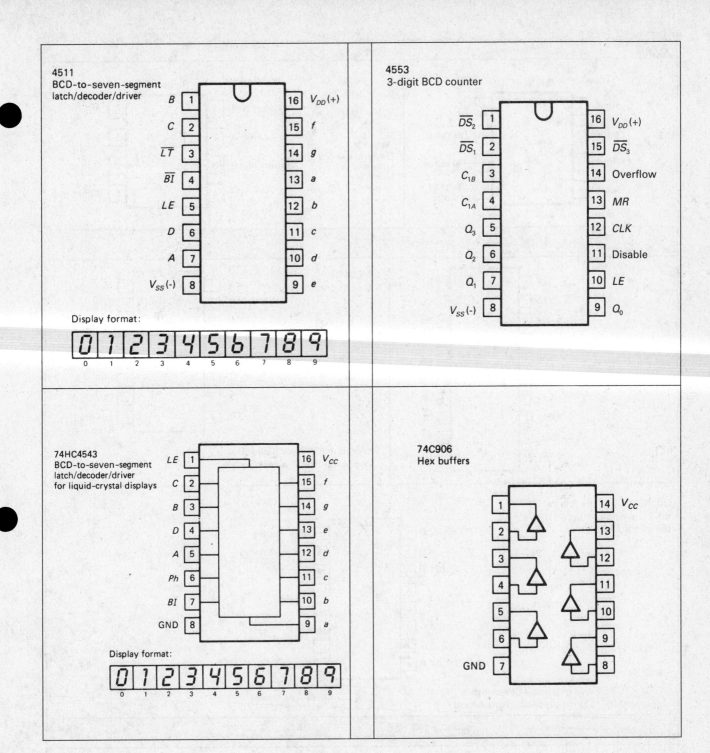

4511
BCD-to-seven-segment
latch/decoder/driver

B	1	16	$V_{DD}(+)$
C	2	15	f
$\overline{LT}$	3	14	g
$\overline{BI}$	4	13	a
LE	5	12	b
D	6	11	c
A	7	10	d
$V_{SS}(-)$	8	9	e

Display format:

0 1 2 3 4 5 6 7 8 9
0 1 2 3 4 5 6 7 8 9

4553
3-digit BCD counter

$\overline{DS}_2$	1	16	$V_{DD}(+)$
$\overline{DS}_1$	2	15	$\overline{DS}_3$
C_{1B}	3	14	Overflow
C_{1A}	4	13	MR
Q_3	5	12	CLK
Q_2	6	11	Disable
Q_1	7	10	LE
$V_{SS}(-)$	8	9	Q_0

74HC4543
BCD-to-seven-segment
latch/decoder/driver
for liquid-crystal displays

LE	1	16	V_{CC}
C	2	15	f
B	3	14	g
D	4	13	e
A	5	12	d
Ph	6	11	c
BI	7	10	b
GND	8	9	a

Display format:

0 1 2 3 4 5 6 7 8 9
0 1 2 3 4 5 6 7 8 9

74C906
Hex buffers

	1	14	V_{CC}
	2	13	
	3	12	
	4	11	
	5	10	
	6	9	
GND	7	8	

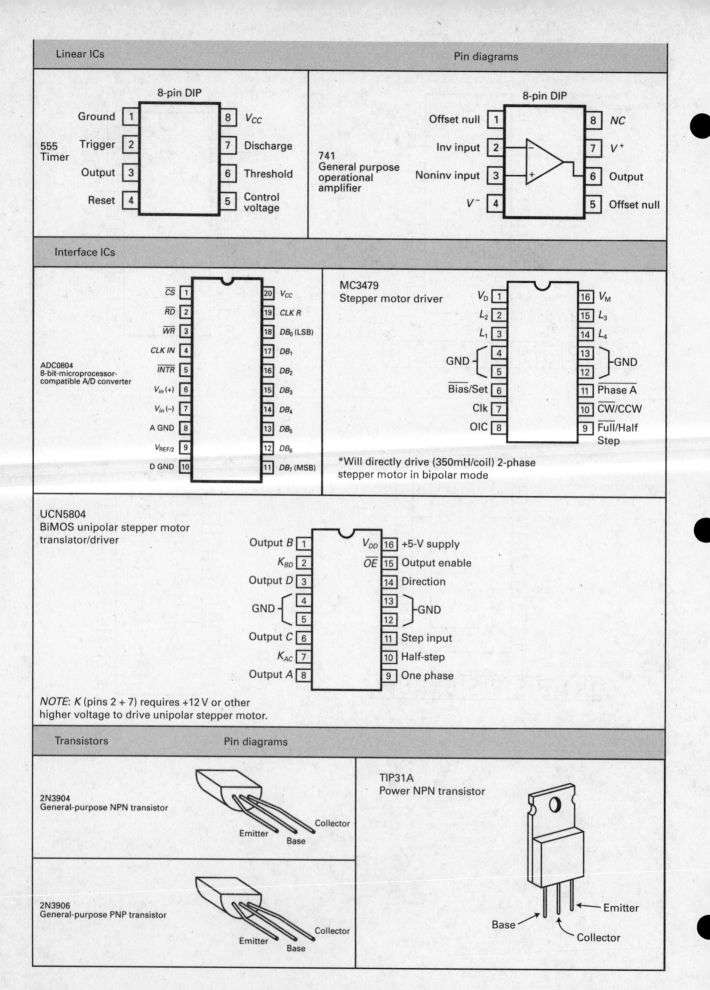

396